# Handbook of 5G Wireless Networks

# Handbook of 5G Wireless Networks

Edited by
Alexander Musk

STATES
ACADEMIC PRESS
www.statesacademicpress.com

Published by States Academic Press,
109 South 5th Street,
Brooklyn, NY 11249, USA



ISBN: 978-1-63989-250-1

**Cataloging-in-Publication Data**

Handbook of 5G wireless networks / edited by Alexander Musk.
p. cm.
Includes bibliographical references and index.
ISBN 978-1-63989-250-1
1. 5G mobile communication systems. 2. Wireless communication systems. I. Musk, Alexander.
TK5103.25 .H36 2022
621.384 56--dc23

For information on all States Academic Press publications
visit our website at www.statesacademicpress.com

STATES
ACADEMIC PRESS

# Contents

# Preface

5G wireless networks are a fifth generation technology standard for broadband cellular networks, deployed by cellular phone companies worldwide. It is the successor of 4G networks. The service area of these networks is divided into small geographical areas called cells. 5G wireless devices in a cell are connected to the internet and telephone network by radio waves through a local antenna in the cell. 5G networks have greater bandwidths compared to the previous networks, and will provide higher download speeds. It is also expected to be used as general internet service providers for laptops and desktops, and aid the advancement of the Internet of things technology. This book brings forth some of the most innovative concepts and elucidates the unexplored aspects of 5G wireless networks. It aims to present researches that have transformed this area of study and aided its advancement. This book will prove to be immensely beneficial to students and researchers in this field.

This book unites the global concepts and researches in an organized manner for a comprehensive understanding of the subject. It is a ripe text for all researchers, students, scientists or anyone else who is interested in acquiring a better knowledge of this dynamic field.

I extend my sincere thanks to the contributors for such eloquent research chapters. Finally, I thank my family for being a source of support and help.

**Editor**

1

# Achievable Energy Efficiency and Spectral Efficiency of Large-Scale Distributed Antenna Systems

Wei Feng, Ning Ge and Jianhua Lu

**Abstract**

In the large-scale distributed antenna system (LS-DAS), a large number of antenna elements are densely deployed in a distributed way over the coverage area, and all the signals are gathered at the cloud processor (CP) via dedicated fiber links for globally joint processing. Intuitively, the LS-DAS can inherit the advantage of both large-scale multiple-input-multiple-output (MIMO) and network densification; thus, it offers enormous gains in terms of both energy efficiency (EE) and spectral efficiency (SE). However, as the number of distributed antenna elements (DAEs) increases, the overhead for acquiring the channel state information (CSI) will increase accordingly. Without perfect CSI at the CP, which is the majority situation in practical applications due to limited overhead, the claimed gain of LS-DAS cannot be achieved. To solve this problem, this chapter considers a more practical case with only the long-term CSI including the path loss and shadowing known at the CP. As the long-term channel fading usually varies much more slowly than the short-term part, the system overhead can be easily controlled under this framework. Then, the EE-oriented and SE-oriented power allocation problems are formulated and solved by fractional programming (FP) and geometric programming (GP) theories, respectively. It is observed that the performance gain with only long-term CSI is still noticeable and, more importantly, it can be achieved with a practical system cost.

**Keywords:** large-scale distributed antenna system (LS-DAS), energy efficiency (EE), spectral efficiency (SE), long-term channel state information (CSI), fractional programming (FP), geometric programming (GP)

## 1. Introduction

The large-scale distributed antenna system (LS-DAS) is a promising candidate technology for the future 5G wireless network. In a LS-DAS, as shown in **Figure 1**, a large number of distributed antenna elements (DAEs) are densely scattered over the coverage area, and the signals from/to all the DAEs are gathered via dedicated fiber links, at the cloud processor (CP), where the globally joint processing is performed [1, 2]. On one hand, the LS-DAS can be regarded as a special large-scale multiple-input-multiple-output (MIMO) system, as shown in **Figure 2**, with *distributed deployment of antenna elements*. On the other hand, it can be regarded as a special heterogeneous small-cell network, as shown in **Figure 3**, with *global inter-cell coordination*. As a consequence, the

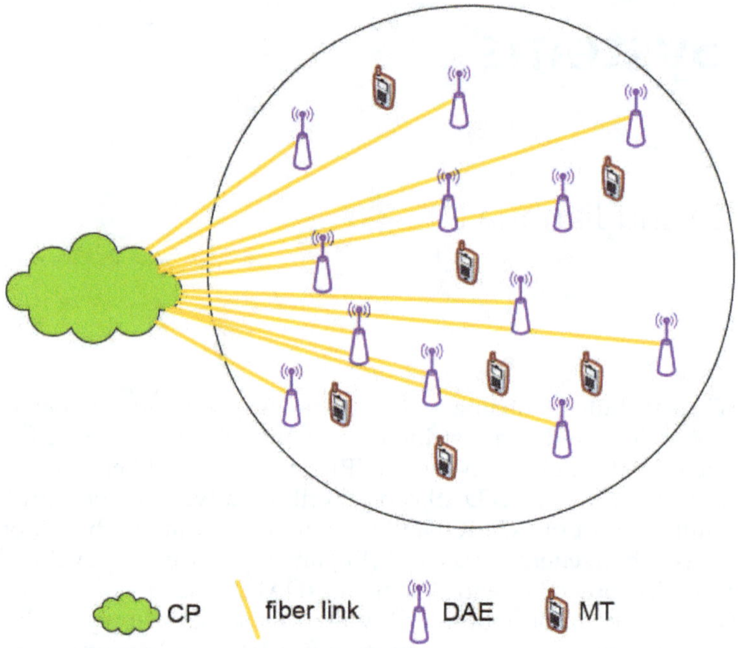


**Figure 1.** Illustration of a large-scale distributed antenna system.

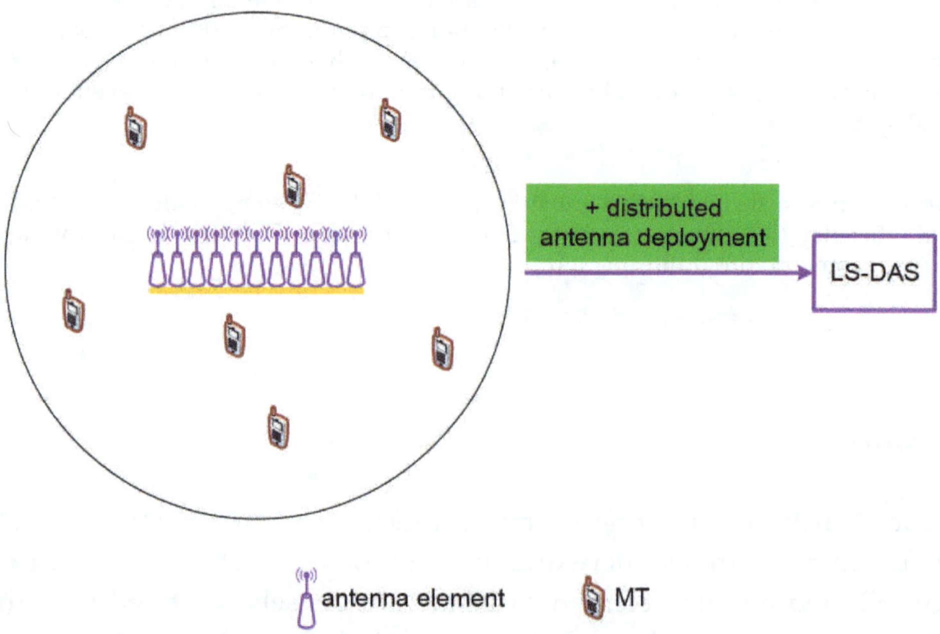


**Figure 2.** Illustration of a traditional large-scale MIMO system.

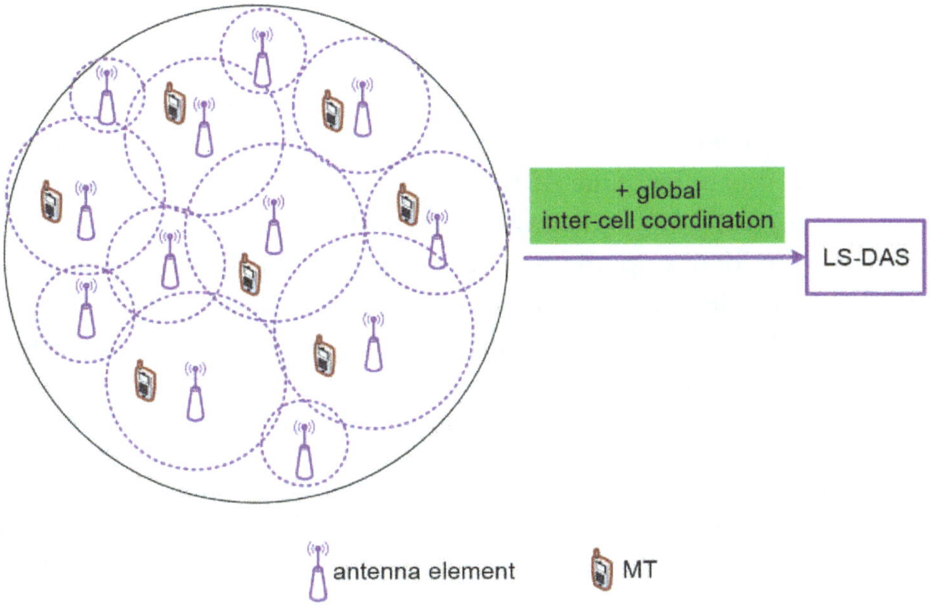


**Figure 3.** Illustration of a traditional heterogeneous small-cell network.

LS-DAS can inherit the advantage of both large-scale MIMO and network densification. Notably, existing studies have already shown that it can offer enormous gains in terms of both energy efficiency (EE) [3, 4] and spectral efficiency (SE) [5, 6].

Due to the distributed deployment of antenna elements, the average access distance of all the mobile terminals (MTs) is reduced. Moreover, due to the global coordination among all the DAEs, the multiplexing gain and diversity gain offered by multiple antenna elements can be obtained [7–9]. These are the main reasons for high EE and SE offered by a LS-DAS. However, to exploit the benefit of LS-DASs, the channel state information (CSI) is crucially required at the CP [10, 11]. Without perfect CSI, the interference among different DAEs will become intractable, and accordingly the system performance will be severely degraded.

In practical applications, the acquisition of full CSI would require an overwhelming amount of system overhead, including the training symbols for channel estimation, the system backhaul for CSI exchanging, and so on. Due to this point, in the literature, some researchers have shown that the system cost of CSI is quite an important issue for evaluating and designing multi-antenna systems. For example, in [12], it has been proved that the optimal number of transmit antennas is equal to the channel coherence interval (CCI). Thus, it will become useless to utilize more antennas than CCI under given channel dynamics. The authors of [13] particularly focused on the cost of CSI for network MIMO systems; they have shown that the optimal number of base stations that can be coordinated exists, which is mainly determined by the CCI in both time and frequency domains. Particularly, for the massive MIMO in frequency division duplex (FDD) mode, it is also very challenging to acquire full CSI at the base station side. In [14], a one-bit feedback scheme was proposed by using a set of predefined precoding vectors. The scheme only performs well in some specific cases, e.g., the multi-antenna channel following one-ring model.

In this chapter, we try to liberate the implementation of LS-DAS from the acquisition of full CSI. We note that the channel of a LS-DAS usually consists of path loss, shadowing, and Rayleigh fading [7–9]. Compared with Rayleigh fading, path loss and shadowing vary much more slowly and can be estimated in a much longer interval than CCI. Thus, it requires a controllable system overhead. In some of the existing studies, path loss and shadowing are classified as large-scale CSI [4, 6]. To distinguish from the *large-scale* in LS-DAS, for clarity, we here use long-term CSI to identify path loss and shadowing. With the knowledge of long-term CSI, the achievable EE and SE will be particularly investigated in the sequel. Different from the reported EE and SE with perfect CSI assumption, which actually cannot be achieved in most practice, our results can be approached with a limited system cost; thus, it is of great significance for the realistic implementation of LS-DASs.

In order to control the computational complexity at the CP, we first divide the whole system into a number of virtual cells (VCs) [5, 15]. As shown in **Figure 4**, the VC is established in a user-centric manner, i.e., each MT chooses only a subset of the surround DAEs for its data transmission. Then, each MT is served by its own VC under the interference from other VCs. To control the interference, the signals of all the VCs are designed in a coordinated fashion at the CP while maximizing the EE or SE of the system. Given VCs, the EE-oriented and the SE-oriented power allocation problems are formulated based on long-term CSI only, both of which are non-convex problems, and thus are difficult to solve. By adopting the fractional programming (FP) theory and the geometric programming (GP) theory, we propose two iterative power allocation algorithms. These algorithms can derive the locally optimal EE and SE of the system, respectively. It is further observed from the simulation results that the

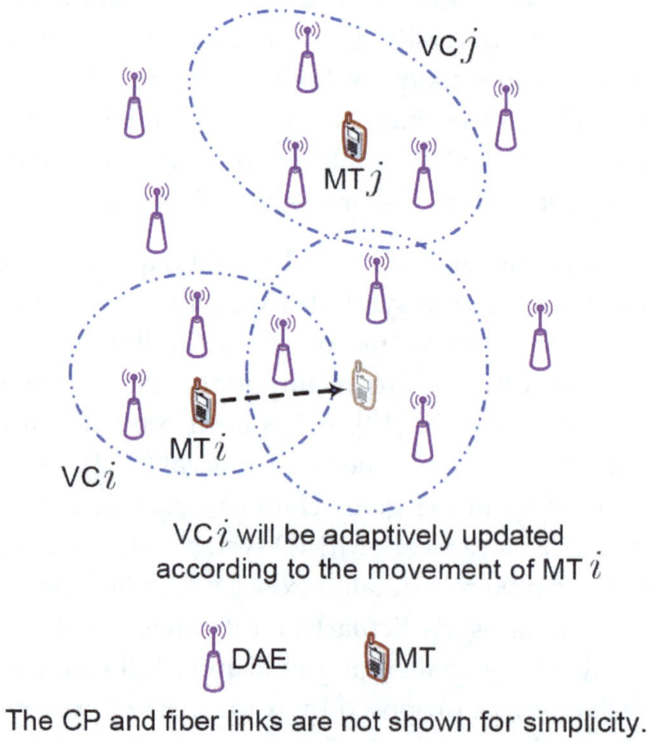


**Figure 4.** Illustration of VCs.

performance gain with only long-term CSI is still remarkable, while it can be achieved with a practical system cost.

The rest of this chapter is organized as follows. The system model of a multiuser LS-DAS is described in Section 2. In the subsequent Sections 3 and 4, the achievable EE and SE are discussed, respectively. Then, we show the simulation results to verify the superiority of the proposed schemes in Section 5. Finally, the conclusion of this chapter is drawn in Section 6.

**Notations**: $\mathbf{I}_n$ denotes an identity matrix with a dimension of $n$, and $\mathbf{O}$ is a zero matrix. $(.)^H$ represents the conjugate transpose operation. $\mathbb{C}^{M\times N}$ denotes the set of complex $M\times N$ matrices, and $\mathcal{CN}$ represents a complex Gaussian distribution. $\mathbf{E}_{(.)}$ represents the expectation operator, and $\mathrm{tr}(\cdot)$ represents the trace operator.

## 2. System Model

We consider a LS-DAS with $K$ MTs. Without loss of generality, all the VCs consist of $N$ DAEs, and the number of antenna elements equipped at each MT is $M$.

For MT $k$, the received signal is

$$\mathbf{y}^{(k)} = \mathbf{H}^{(k)}\mathbf{x}^{(k)} + \sum_{i=1,i\neq k}^{K} \mathbf{H}^{(k,i)}\mathbf{x}^{(i)} + \mathbf{n}^{(k)}, \tag{1}$$

where $\mathbf{H}^{(k)}\in\mathbb{C}^{M\times N}$, $k = 1, 2, ..., K$, represents the channel between the DAEs in VC $k$ and MT $k$, $\mathbf{H}^{(k,i)}\in\mathbb{C}^{M\times N}$, $k = 1, 2, ..., K$, $i = 1, 2, ..., K$, denotes the channel between the DAEs in VC $i$ and MT $k$, $\mathbf{x}^{(i)}\in\mathbb{C}^{N\times 1}$, $i = 1, 2, ..., K$, is the transmitted signal vector for MT $i$, and $\mathbf{n}^{(k)}\in\mathbb{C}^{M\times 1}$, $k = 1, 2, ..., K$, denotes the additive white Gaussian noise with distribution $\mathcal{CN}(0,\sigma^2 I_M)$.

$$\mathbf{E}\left[\mathbf{x}^{(k)}\mathbf{x}^{(k)^H}\right] = \mathbf{P}^{(k)} = \begin{bmatrix} p_1^{(k)} & & \\ & \ddots & \\ & & p_N^{(k)} \end{bmatrix}, \quad k = 1, ..., K. \tag{2}$$

Assuming a total transmit power constraint $P_{\max}^{(k)}$ for MT $k$, we set

$$\sum_{n=1}^{N} p_n^{(k)} \leq P_{\max}^{(k)}. \tag{3}$$

The channel matrix can be modeled as

$$\mathbf{H}^{(k,i)} = \mathbf{S}^{(k,i)}\mathbf{L}^{(k,i)}, \tag{4}$$

where $\mathbf{S}^{(k,i)}$ and $\mathbf{L}^{(k,i)}$ reflect the short-term fading and the long-term fading, respectively. Particularly, the entries of $\mathbf{S}^{(k,i)}$ are all independent and identically distributed (i.i.d.) circular symmetric complex Gaussian variables following $\mathcal{CN}(0, 1)$ distribution.

$$\mathbf{L}^{(k,i)} = \begin{bmatrix} l_1^{(k,i)} & & \\ & \ddots & \\ & & l_N^{(k,i)} \end{bmatrix}, \tag{5}$$

with

$$l_n^{(k,i)} = \sqrt{\left(D_n^{(k,i)}\right)^{-\gamma} S_n^{(k,i)}}, \quad n = 1, 2, \ldots, N, \tag{6}$$

where $D_n^{(k,i)}$ is the transmission distance between the DAE $n$ in VC $i$ and MT $k$, and $\gamma$ is the path loss exponent, and $S_n^{(k,i)}$ represents the shadow fading caused by large objects such as tall buildings or walls.

## 3. Achievable Ee

Given perfect CSI, the authors of [16] have proposed an energy-efficient power allocation scheme for traditional DASs. In [17], further taking the inter-VC interference into consideration, an iterative power allocation scheme was presented to improve the EE of a LS-DAS, via applying the successive Taylor expansion method. In contrast, we investigate the achievable EE with the long-term CSI only in this section.

First of all, the sum rate of the system can be derived according to Eq. (1) as

$$R = \sum_{k=1}^{K} \log_2 \det\left(\mathbf{I}_M + \frac{\mathbf{H}^{(k)}\mathbf{P}^{(k)}\mathbf{H}^{(k)^H}}{\sigma_k^2}\right), \tag{7}$$

where

$$\sigma_k^2 = \sum_{i=1,\, i\neq k}^{K} \sum_{n=1}^{N} [l_n^{(k,i)}]^2 p_n^{(i)} + \sigma^2, \tag{8}$$

is the total interference-plus-noise power at MT $k$.

When only the long-term CSI is known, the average sum rate can be calculated via taking expectation over the short-term channel fading $\Omega = \{\mathbf{S}^{(k)} | k = 1, \ldots, K\}$ as

$$\overline{R} = \sum_{k=1}^{K} \mathbf{E}_{\Omega}\left[\log_2 \det\left(\mathbf{I}_M + \frac{\mathbf{H}^{(k)}\mathbf{P}^{(k)}\mathbf{H}^{(k)^H}}{\sigma_k^2}\right)\right]. \tag{9}$$

Then, the EE of the system, denoted as $\eta$, can be derived as

$$\eta = \frac{\overline{R}}{\rho \sum_{k=1}^{K} \sum_{n=1}^{N} p_n^{(k)} + P_c}, \tag{10}$$

where

$$\rho = \frac{\varepsilon}{\gamma}, \tag{11}$$

with $\varepsilon$ and $\gamma$ denoting the peak-to-average power ratio and the power amplifier efficiency, respectively, and $P_c$ denotes the circuit power consumption [4].

In order to investigate the achievable EE under this framework, we formulate the following optimization problem:

$$\max \eta \tag{12a}$$

$$\text{s.t.} \sum_{n=1}^{N} p_n^{(k)} \leq P_{\max}^{(k)}, \tag{12b}$$

$$p_n^{(k)} \geq 0, k = 1, \ldots, K, n = 1, \ldots, N. \tag{12c}$$

Because of the non-convexity of $\overline{R}$, the problem shown in Eq. (12) is a complicated non-convex problem [18]. To simplify it, we introduce an upper bound to the objective function as

$$\hat{\eta} = \frac{\sum_{k=1}^{K} \log_2 \det \left( \mathbf{I}_N + \frac{M\mathbf{P}^{(k)} \left( \mathbf{L}^{(k)} \right)^2}{\sigma_k^2} \right)}{\rho \sum_{k=1}^{K} \sum_{n=1}^{N} p_n^{(k)} + P_c}, \tag{13}$$

the numerator of which is an upper bound to $\overline{R}$ [10]. Accordingly, the problem in Eq. (12) can be reformulated as

$$\max \hat{\eta} \tag{14a}$$

$$\text{s.t.} \sum_{n=1}^{N} p_n^{(k)} \leq P_{\max}^{(k)}, \tag{14b}$$

$$p_n^{(k)} \geq 0, \ k = 1, \ldots, K, n = 1, \ldots, N. \tag{14c}$$

which is simpler than Eq. (12). However, it is still non-convex [18]. To further solve the problem in Eq. (14), we express

$$\hat{\eta} = \frac{f_1 - f_2}{\rho \sum_{k=1}^{K} \sum_{n=1}^{N} p_n^{(k)} + P_c}, \tag{15}$$

where

$$f_1 = \sum_{k=1}^{K} \log_2 \det\left(\sigma_k^2 \mathbf{I}_N + M\mathbf{P}^{(k)}\mathbf{L}^{(k)2}\right), \tag{16a}$$

$$f_2 = \sum_{k=1}^{K} N \log_2(\sigma_k^2), \tag{16b}$$

both of which are clearly concave functions.

We find that if the numerator of $\hat{\eta}$, i.e., $f_1 - f_2$, can be transformed into a concave form, the problem in Eq. (14) can be recast as a quasi-concave fractional programming problem, further considering the linearity of its denominator [19]. Toward this end, we linearize $f_2$ by applying the first-order Taylor expansion at a given point $\overline{\mathbf{P}}$ as

$$\tilde{f}_2(\mathbf{P}|\overline{\mathbf{P}}) = \sum_{k=1}^{K} N \log_2\left(\sigma_k^2(\overline{\mathbf{P}})\right) + \log_2(e) \sum_{k=1}^{K} \frac{N}{\sigma_k^2(\overline{\mathbf{P}})} tr(G_k[\mathbf{P} - \overline{\mathbf{P}}]), \tag{17}$$

where $\mathbf{P} = \{\mathbf{P}^{(1)}, \ldots, \mathbf{P}^{(k)}\}$ and

$$\mathbf{G}_k = \mathrm{diag}\{\mathbf{G}^{(k,1)}, \ldots, \mathbf{G}^{(k,K)}\}, \tag{18a}$$

$$\mathbf{G}^{(k,i)} = \left(\mathbf{L}^{(k,i)}\right)^2, \ k \neq i,\ k,\ i = 1, \ldots, K, \tag{18b}$$

$$\mathbf{G}^{(k,k)} = \mathbf{O}. \tag{18c}$$

By substituting $\tilde{f}_2(\mathbf{P}|\overline{\mathbf{P}})$ for $f_2(\mathbf{P})$, the problem in Eq. (14) can be approximated as

$$\max \overline{\eta} = \frac{f_1(\mathbf{P}) - \tilde{f}_2(\mathbf{P}|\mathbf{P})}{\rho \sum_{k=1}^{K} \sum_{n=1}^{N} p_n^{(k)} + P_c} \tag{19a}$$

$$s.t. \sum_{n=1}^{N} p_n^{(k)} \leq P_{\max}^{(k)}, \tag{19b}$$

$$p_n^{(k)} \geq 0,\ k = 1, \ldots, K,\ n = 1, \ldots, N, \tag{19c}$$

whose objective function is fortunately fractional with concave numerator and convex denominator [18]. Adopting the FP theory, the problem in Eq. (19) can be optimally solved in an iterative way. In our previous paper [4], we have shown in detail how to solve the problem in Eq. (19). In the following, for brevity, we just present the basic idea and procedure of the iterative algorithm.

We use $t \geq 1$ and $s \geq 1$ to denote the successive Taylor expansion iteration step and the FP iteration step, respectively. After introducing a positive variable $\omega$, the following concave optimization problem can be formulated

$$\max v(\mathbf{P}|\mathbf{P}_{t-1,s-1},\omega) \tag{20a}$$

$$s.t. \sum_{n=1}^{N} p_n^{(k)} \leq P_{\max}^{(k)}, \tag{20b}$$

$$p_n^{(k)} \geq 0,\ k=1,\ \ldots,\ K,\ n=1,\ldots,\ N, \tag{20c}$$

where

$$v(\mathbf{P}|\mathbf{P}_{t-1,s-1},\omega) = f_1(\mathbf{P}) - \tilde{f}_2(\mathbf{P}|\mathbf{P}_{t-1,s-1}) - \omega\rho \sum_{k=1}^{K}\sum_{n=1}^{N} p_n^{(k)} - \omega P_c. \tag{21}$$

Further define

$$V(\omega) = \max v(\mathbf{P}|\mathbf{P}_{t-1,s-1},\omega), \tag{22}$$

---

**Algorithm 1 Iterative power allocation for maximizing EE.**

1. Initialization: $\mathbf{P}_0 = \mathbf{diag}\{\mathbf{P}_0^{(1)},\cdots,\mathbf{P}_0^{(K)}\}$ with $\mathbf{P}_0^{(k)} = \frac{\mathbf{P}_{\max}^{(k)}}{N}\mathbf{I}_N$, $k=1,\ldots,K$, $\mathbf{P}_{0,0}^{(k)} = \mathbf{P}_0^{(k)}$, $k=1,\ldots,K$, $\omega = 0$, and $\xi = 1\times10^{-3}$, $\delta = 1\times10^{-3}$, $t=1, s=1$;
2. Solve Eq. (20), and denote the obtained power matrix by $\mathbf{P}_{0,1}^{(k)}$, $k=1,\ldots,K$, set $\mathbf{P}_1^{(k)} = \mathbf{P}_{0,1}^{(k)}$, $k=1,\ldots,K$, and $\mathbf{P}_1 = \mathbf{diag}\{\mathbf{P}_1^{(1)},\cdots,\mathbf{P}_1^{(K)}\}$;
3. **while** $|\hat{\eta}(\mathbf{P}_t) - \hat{\eta}(\mathbf{P}_{t-1})|/\hat{\eta}(\mathbf{P}_{t-1}) > \xi$ **do**
4. $t = t+1$, $s = 1$, and $\omega = 0$;
5. $\mathbf{P}_{t-1,0}^{(k)} = \mathbf{P}_{t-1}^{(k)}$, $k=1,\ldots,K$;
6. Solve Eq. (20), derived $V(\omega)$ and denote the obtained power matrix by $\mathbf{P}_{t-1,1}^{(k)}$, $k=1,\ldots,K$;
7. **while** $V(\omega) > \delta$ **do**
8. $\omega = \overline{\eta}\left(\mathbf{P}_{t-1,s}^{(k)}|\mathbf{P}_{t-1}^{(k)}\right)$;
9. $s = s+1$;
10. Solve Eq. (20), derived $V(\omega)$ and denote the obtained power matrix by $\mathbf{P}_{t-1,s}^{(k)}$, $k=1,\ldots,K$;
11. **end while**

**12.** $\mathbf{P}_t^{(k)} = \mathbf{P}_{t-1,s}^{(k)}$, $k = 1, ..., K$, and $\mathbf{P}_t = \mathrm{diag}\{\mathbf{P}_t^{(1)}, \cdots, \mathbf{P}_t^{(K)}\}$;

**13. end while**

**14.** Output: $\mathbf{P}_t$.

---

we can propose an iterative power allocation algorithm for maximizing EE, as described in **Algorithm 1**. By adopting **Algorithm 1**, the achievable EE with long-term CSI only can be derived with low computational complexity [4].

## 4. Achievable Se

For traditional single-cell DASs, the achievable SE was studied in [20, 21], which by considering the general DAS with random antenna layout has identified that DAS outperforms colocated multi-antenna systems. In [22], the authors have taken the inter-cell interference into consideration, and they have presented a close-form expression for the achievable EE in a multi-cell environment. However, this work has not considered interference coordination. The authors of [23] took a step further; they have put forward a coordinated power allocation scheme for dealing with the inter-cell interference. Nevertheless, the result was derived by approximately treating the inter-cell interference as Gaussian noise, and thus it is only applicable to the low signal-to-noise-ratio (SNR) situation. In a recent work, the SE of single-cell multiuser LS-DAS was studied [24]. It however also has not considered interference coordination, which is in general inevitable in most practical applications. Different from all the above existing studies, in this section, we investigate the achievable SE of a LS-DAS with long-term CSI only.

With the target of average system sum rate maximization, the problem of SE-oriented power allocation can be formulated as

$$\max \ \overline{R} \tag{23a}$$

$$s.t. \ \sum_{n=1}^{N} p_n^{(k)} \leq P_{\max}^{(k)}, \tag{23b}$$

$$p_n^{(k)} \geq 0, \ k = 1, \ldots, K, \ n = 1, \ldots, N. \tag{23c}$$

As $\overline{R}$ is non-convex, this problem is complicatedly non-convex [18]. Besides, the objective function is actually in an integral form as a result of the expectation operator in $\overline{R}$, and it cannot be directly expressed in a compact closed form, which renders it even more challenging to obtain the optimal solution of Eq. (23).

We try to simplify the formulated problem. To this end, a closed-form approximation for the average system sum rate $\overline{R}$ is leveraged as

$$\overline{R}_{ap} = \sum_{k=1}^{K}\sum_{n=1}^{N}\log_2\left(1+\frac{[l_n^{(k)}]^2 p_n^{(k)} \Upsilon_k^{-1} M}{\sigma_k^2}\right) + M\sum_{k=1}^{K}\log_2(\Upsilon_k) - M\sum_{k=1}^{K}\log_2 e(1-\Upsilon_k^{-1}), \tag{24}$$

where $\Upsilon_k$ satisfies

$$\Upsilon_k = 1 + \sum_{n=1}^{N}\frac{[l_n^{(k)}]^2 p_n^{(k)}}{\sigma_k^2 + [l_n^{(k)}]^2 p_n^{(k)} \Upsilon_k^{-1} M}, \; k = 1, .., K. \tag{25}$$

This approximation can be derived through using the random matrix theory [10], and the introduced parameter $\Upsilon_k$ can be calculated in an iterative way as shown in the following **Algorithm 2**.

According to the existing studies [10], $\overline{R}_{ap}$ is quite a precise approximation for $\overline{R}$. Therefore, we directly use it as the objective function, and the joint power allocation problem can be recast as

$$\max \; \overline{R}_{ap} \tag{26a}$$

$$s.t. \; \sum_{n=1}^{N} p_n^{(k)} \leq P_{\max}^{(k)}, \tag{26b}$$

$$p_n^{(k)} \geq 0, \; k = 1, ..., K, \; n = 1, ..., N, \tag{26c}$$

which is much simplified. However, due to the non-convexity of $\overline{R}_{ap}$ [18], the new problem in Eq. (26) is still non-convex. In the following, we explore the achievable SE of the system by contriving an iterative algorithm, which can find a locally optimal solution of Eq. (26) efficiently.

To eliminate the effect of the introduced parameters $\Upsilon_1, \Upsilon_2, ..., \Upsilon_K$, we first fix $\Upsilon_1, \Upsilon_2, ..., \Upsilon_K$ as constants. Then we can equivalently simplify the objective function in Eq. (26) as

$$\overline{R}_{ap}' = \sum_{k=1}^{K}\sum_{n=1}^{N}\log_2\left(1+\frac{[l_n^{(k)}]^2 p_n^{(k)} \Upsilon_k^{-1} M}{\sigma_k^2}\right). \tag{27}$$

As $\log_2(\cdot)$ is monotonically increasing, the problem shown in Eq. (26) can be equivalently transformed into

$$\min \prod_{k=1}^{K}\prod_{n=1}^{N}\frac{\sigma_k^2}{\sigma_k^2 + [l_n^{(k)}]^2 p_n^{(k)} \Upsilon_k^{-1} M} \tag{28a}$$

$$s.t. \; \sum_{n=1}^{N} p_n^{(k)} \leq P_{\max}^{(k)}, \tag{28b}$$

$$p_n^{(k)} \geq 0,\ k = 1, \ldots, K,\ n = 1, \ldots, N. \tag{28c}$$

Define

$$f_{n,k}(\mathbf{P}) = \sigma_k^2(\mathbf{P}) + [l_n^{(k)}]^2 p_n^{(k)} \Upsilon_k^{-1} M = \sum_{i=1, i \neq k}^{K} \sum_{j=1}^{N} g_j^{(k,i)}(\mathbf{P}) + g_n^{(k)}(\mathbf{P}) + \sigma^2, \quad n = 1, \ldots, N, k = 1, \ldots, K, \tag{29}$$

where

$$g_j^{(k,i)}(\mathbf{P}) = [l_j^{(k,i)}]^2 p_j^{(i)},\ k \neq i, \tag{30}$$

$$g_n^{(k)}(\mathbf{P}) = [l_n^{(k)}]^2 p_n^{(k)} \Upsilon_k^{-1} M, \tag{31}$$

and then, given a feasible point $\overline{\mathbf{P}}$, an approximation of $f_{n,k}(\mathbf{P})$ can be obtained as

$$\tilde{f}_{n,k}(\mathbf{P}|\overline{\mathbf{P}}) = \left( \prod_{i=1, i \neq k}^{K} \prod_{j=1}^{N} \left( \frac{g_j^{(k,i)}(\mathbf{P})}{\alpha_{n,j}^{(k,i)}} \right)^{\alpha_{n,j}^{(k,i)}} \right) \times \left( \frac{g_n^{(k)}(\mathbf{P})}{\alpha_{n,n}^{(k)}} \right)^{\alpha_{n,n}^{(k)}} \times \left( \frac{\sigma^2}{\alpha_{n,k}^0} \right)^{\alpha_{n,k}^0}, \tag{32}$$

where

$$\alpha_{n,j}^{(k,i)} = g_j^{(k,i)}(\overline{\mathbf{P}}) / f_{n,k}(\overline{\mathbf{P}}), \tag{33}$$

$$\alpha_{n,n}^{(k)} = g_n^{(k)}(\mathbf{P}) / f_{n,k}(\overline{\mathbf{P}}), \tag{34}$$

$$\alpha_{n,k}^0 = \sigma^2 / f_{n,k}(\overline{\mathbf{P}}). \tag{35}$$

By using the inequality of arithmetic and geometric means, it is easy to obtain that

$$f_{n,k}(\mathbf{P}) \geq \tilde{f}_{n,k}(\mathbf{P}|\overline{\mathbf{P}}). \tag{36}$$

The equality holds if and only if

$$\mathbf{P} = \overline{\mathbf{P}}. \tag{37}$$

By replacing $f_{n,k}(\mathbf{P})$ with $\tilde{f}_{n,k}(\mathbf{P}|\overline{\mathbf{P}})$, the problem in Eq. (28) can be recast as

$$\min \prod_{k=1}^{K} \prod_{n=1}^{N} \frac{\sigma_k^2}{\tilde{f}_{n,k}}(\mathbf{P}) \tag{38a}$$

$$s.t. \sum_{n=1}^{N} p_n^{(k)} \leq P_{\max}^{(k)}, \tag{38b}$$

$$p_n^{(k)} \geq 0,\ k = 1, \ldots, K,\ n = 1, \ldots, N, \tag{38c}$$

which is a good approximation for the original problem in the neighborhood of $\overline{\mathbf{P}}$. More importantly, it is a standard GP problem [25]; thus, it can be efficiently solved via convex optimization tools, e.g., the interior point algorithm [18].

We use $t \geq 1$ and $s \geq 1$ to denote the updating iteration step of $\Upsilon_k$ and the arithmetic-to-geometric approximation iteration step, respectively. Then the following convex optimization problem is derived

$$\min \prod_{k=1}^{K} \prod_{n=1}^{N} \frac{\sigma_k^2}{\tilde{f}_{n,k}} (\mathbf{P}|\mathbf{P}^{s-1}, \Upsilon_k^t) \tag{39a}$$

$$s.t. \ \sum_{n=1}^{N} p_n^{(k)} \leq P_{\max}^{(k)}, \tag{39b}$$

$$p_n^{(k)} \geq 0, k = 1, ..., K, n = 1, ..., N. \tag{39c}$$

Accordingly, we propose an iterative power allocation algorithm for maximizing SE as described in **Algorithm 2**. In the algorithm, $\Upsilon_k, k = 1, ..., K$ and $\mathbf{P}$ are updated in an alternate way. By adopting the algorithm, the achievable SE with long-term CSI only can be derived with low computational complexity [6].

## 5. Simulation Results

In this section, we illustrate the EE and SE performance of the proposed schemes by simulations. To be general, we consider a circular coverage area with a radius of 500 m. There are 20 DAEs randomly deployed in the coverage area with a two-dimension uniform distribution. The number of MTs is set as $K = 3$. The number of antenna elements equipped at each MT is set as $M = 3$. In order to fully exploit the spatial degree of freedom of each MT and, in the meantime, well control the system complexity, we set the size of each VC as $N = M = 3$. As for the channel parameters, we set $\gamma = 4$ (path loss exponent), $\sigma^2 = -107$ dBm (noise power), and the shadowing standard deviation is 8 dB. Without loss of generality, we consider the same transmit power constraint for all MTs, i.e., $P_{\max}^{(1)} = P_{\max}^{(2)} = P_{\max}^{(3)}$. Particularly, 100 randomly selected system topologies are considered in the simulation, and the averaged results are shown in the following.

First, the achievable EE of different schemes is compared in **Figure 5**. Both the scheme presented in reference [16] and the simplest equal power allocation scheme are considered. It can be seen from **Figure 5** that the proposed scheme outperforms the other ones, especially when the transmit power constraint goes larger. The scheme proposed in [16] has not considered interference coordination; thus, in a multi-VC setting, its performance is worse than the proposed scheme, although it has assumed the perfect CSI as the CP. In contrast, although using the long-term CSI only, the proposed scheme can still offer the highest EE performance. We can also observe from **Figure 5** that the key point for high EE is to set proper transmit power, i.e., when the transmit power has reached a corresponding point, it should no longer be

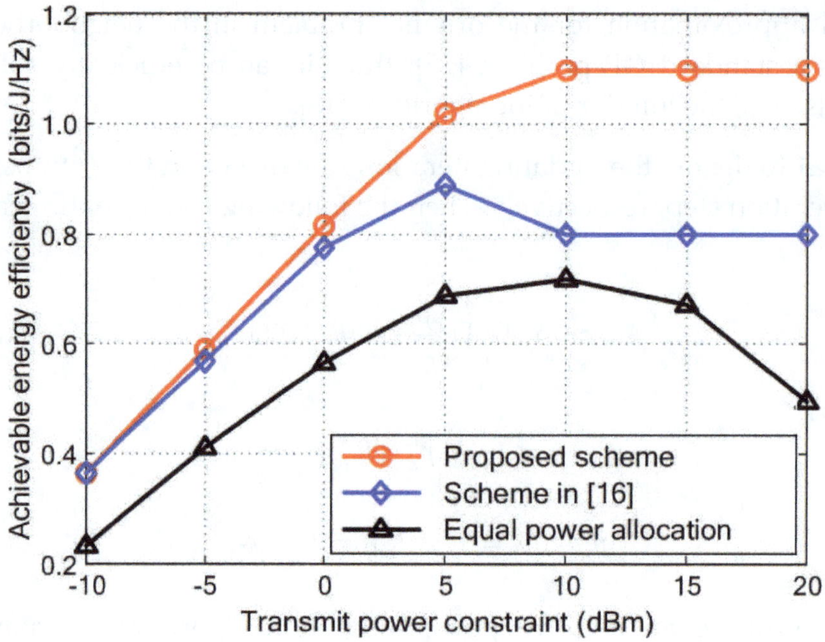


**Figure 5.** Comparison of achievable EE by different schemes.

increased even though the power consumption constraint goes larger. Intuitively, this observation can be explained by the fact that when the transmit power goes larger, the sum rate gain will become smaller and smaller due to the impact of interference; thus, the EE of the scheme will fall instead of rising.

---

**Algorithm 2 Iterative power allocation for maximizing SE.**

---

Initialization: Set $\mathbf{P}^0 = \{[p_1^{(1)}]^0, [p_2^{(1)}]^0, \ldots, [p_N^{(K)}]^0\}$, where $[p_n^{(k)}]^0 = \frac{P_{\max}^{(k)}}{N}$, $k = 1, \ldots, K$, $n = 1, \ldots, N$, and $\varepsilon = 1\times10^{-4}$, $\delta = 1\times10^{-3}$, $s = 1$;

**for** $k = 1$ to $K$ **do**

$t = 1$;

$\Upsilon_k^0 = 1$;

$\Upsilon_k^1 = 1 + \sum_{n=1}^{N} \frac{[l_n^{(k)}]^2 [p_n^{(k)}]^0}{\sigma_k^2(\mathbf{P}^0) + [l_n^{(k)}]^2 [p_n^{(k)}]^0 [\Upsilon_k^0]^{-1} M}$;

**while** $|\Upsilon_k^t - \Upsilon_k^{t-1}| > \varepsilon$ **do**

$t = t + 1$;

$\Upsilon_k^t = 1 + \sum_{n=1}^{N} \frac{[l_n^{(k)}]^2 [p_n^{(k)}]^0}{\sigma_k^2(\mathbf{P}^0) + [l_n^{(k)}]^2 [p_n^{(k)}]^0 [\Upsilon_k^{t-1}]^{-1} M}$;

**end while**

Output $\Upsilon_k' = \Upsilon_k^t, k = 1, \ldots, K$.

**end for**

**12.** Solve Eq. (39) with $\Upsilon_k = \Upsilon_k^{'}, k = 1, ..., K$, and denote the obtained power matrix by $\mathbf{P}^1$;

**13.** **while** $|\overline{R}_{ap}(\mathbf{P}^s) - \overline{R}_{ap}(\mathbf{P}^{s-1})| / \overline{R}_{ap}(\mathbf{P}^{s-1}) > \delta$ **do**

**14.** **for** $k = 1$ to $K$ **do**

**15.** $t = 1$;

**16.** $\Upsilon_k^0 = 1$;

**17.** $\Upsilon_k^1 = 1 + \sum_{n=1}^{N} \frac{[l_n^{(k)}]^2 [p_n^{(k)}]^s}{\sigma_k^2(\mathbf{P}^s) + [l_n^{(k)}]^2 [p_n^{(k)}]^s [\Upsilon_k^0]^{-1} M}$;

**18.** **while** $|\Upsilon_k^t - \Upsilon_k^{t-1}| >$ **do**

**19.** $t = t + 1$;

**20.** $\Upsilon_k^t = 1 + \sum_{n=1}^{N} \frac{[l_n^{(k)}]^2 [p_n^{(k)}]^s}{\sigma_k^2(\mathbf{P}^s) + [l_n^{(k)}]^2 [p_n^{(k)}]^s [\Upsilon_k^{t-1}]^{-1} M}$;

**21.** **end while**

**22.** Output $\Upsilon_k^{'} = \Upsilon_k^t, k = 1, ..., K$.

**23.** **end for**

**24.** $s = s + 1$;

**25.** Solve Eq. (39) with $\Upsilon_k = \Upsilon_k^{'}, k = 1, ..., K$, and denote the obtained power matrix by26: $\mathbf{P}^s$;

**26.** **end while**

**27.** Output: $\mathbf{P}^s$.

---

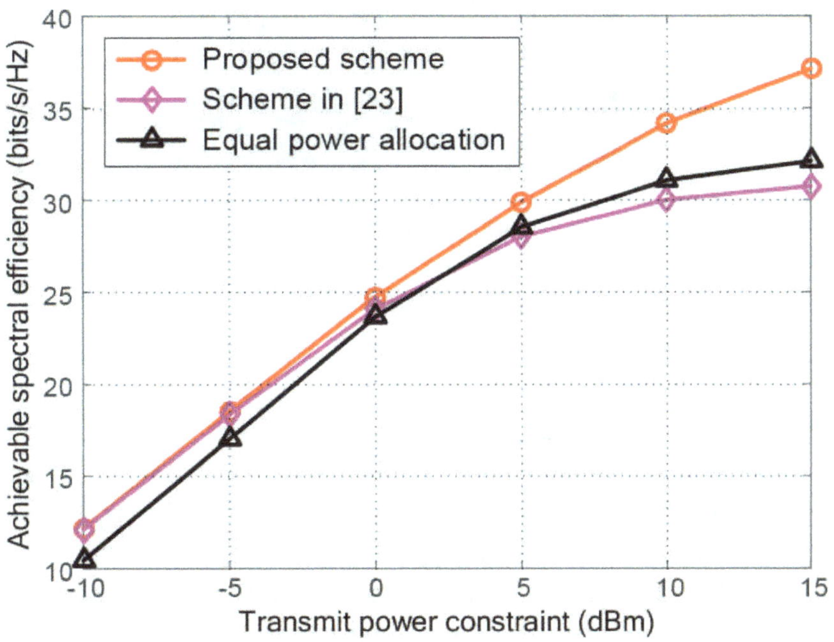


**Figure 6.** Comparison of achievable SE by different schemes.

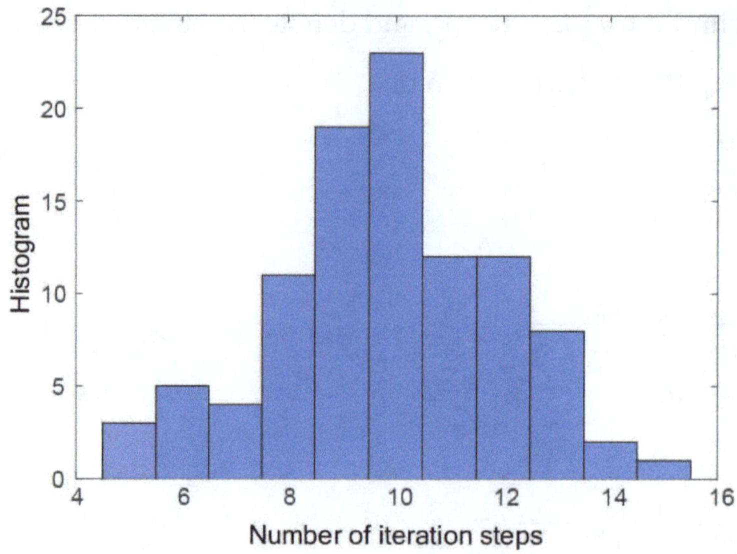


**Figure 7.** Histogram of the number of iteration steps for **Algorithm 1**.

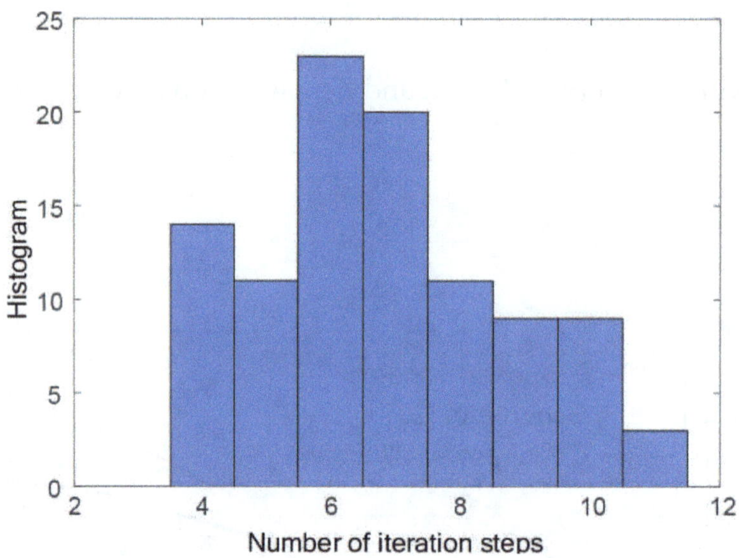


**Figure 8.** Histogram of the number of iteration steps for **Algorithm 2**.

Then, we evaluate the performance of the proposed scheme in terms of achievable SE. The scheme presented in reference [23] and equal power allocation scheme are taken into comparison. The results are shown in **Figure 6**. We can find that the proposed scheme performs the best among the three schemes. The scheme presented in [23] is only applicable to the low SNR condition; thus, the performance gas between it and the proposed scheme goes larger when the transmit power constraint increases, which implies that the impact of inter-VC interference becomes bigger. The results identify that it is still effective for enhancing the SE of the system when only the long-term CSI is available.

According to the discussion in [4, 6], the proposed **Algorithms 1** and **2** are assured to converge to a local optimum. The histogram of the number of iteration steps is illustrated in **Figures 7** and **8**, for **Algorithms 1** and **2**, respectively. We can observe from the figures that 15 iteration steps are enough for the convergence of **Algorithm 1** and that for **Algorithm 2** is 11.

## 6. Conclusions

The LS-DAS is a promising candidate technology for the future 5G wireless network, due to its remarkable gains in terms of both EE and SE. In this chapter, we try to liberate the implementation of LS-DAS from the acquisition of full CSI. With the knowledge of long-term CSI, including the path loss and shadow fading, the achievable EE and SE have been investigated. Different from the reported EE and SE with perfect CSI condition, which actually cannot be achieved in most practice, our results can be achieved with a limited system cost; thus, it is of great significance for the realistic implementation of LS-DASs. We also use the concept of VC to control the computational complexity at the CP. Accordingly, we design the transmit power of all the VCs in a coordinated fashion, to control the interference and finally maximize EE or SE of the system. Particularly, the EE-oriented and the SE-oriented power allocation problems are formulated based on long-term CSI only, both of which are non-convex problems, and thus are difficult to solve. By adopting the FP theory and the GP theory, we propose two iterative power allocation algorithms. These algorithms can derive the locally optimal EE and SE of the system, respectively. It is further observed from the simulation results that the performance gain with only long-term CSI is still remarkable, while it can be achieved with a practical system overhead.

## Acknowledgements

This work was supported in part by the National Science Foundation of China for Young Scholars under grant no. 61201186 and the National Basic Research Program of China under grant no. 2013CB329001 and the National Science Foundation of China under grant no. 61132002.

## Author details

Wei Feng*, Ning Ge and Jianhua Lu

*Address all correspondence to: fengw@mails.tsinghua.edu.cn

Department of Electronic Engineering, Tsinghua National Laboratory for Information Science Technology, Tsinghua University, Beijing, PR China

## References

[1] Dai L. A comparative study on uplink sum capacity with co-located and distributed antennas. IEEE Journal on Selected Areas in Communications. 2011;29:1200–1213. DOI:10.1109/JSAC.2011.110608

[2] Feng W, Chen Y, Shi R, Ge N, and Lu J. Exploiting macro-diversity in massively distributed antenna systems: a controllable coordination perspective. IEEE Transactions on Vehicular Technology. DOI:10.1109/TVT.2015.2506720

[3] Feng W, Chen Y, Ge N, and Lu J. Optimal energy-efficient power allocation for distributed antenna systems with imperfect CSI. IEEE Transactions on Vehicular Technology. 2016;65:7759–7763. DOI:10.1109/TVT.2015.2497140

[4] Wang J, Feng W, Chen Y, and Zhou S. Energy efficient power allocation for multicell distributed antenna systems. IEEE Communications Letters. 2016;20:177–180. DOI:10.1109/LCOMM.2015.2498608

[5] Wang J and Dai L. Downlink rate analysis for virtual-cell based large-scale distributed antenna systems. IEEE Transactions on Wireless Communications. 2016;15:1998–2011. DOI:10.1109/TWC.2015.2497678

[6] Wang Y, Feng W, Zhao Y, Zhou S, and Wang J. Joint power allocation for multi-cell distributed antenna systems with large-scale CSIT. In: Proceedings of the IEEE International Conference on Communications (ICC'12); 10–15 June 2012; Ottawa. IEEE; 2012. pp. 6786–6791.

[7] Roh W, Paulraj A. Outage performance of the distributed antenna systems in a composite fading channel. In: Proceedings of the IEEE Vehicular Technology Conference (VTC 2002-Fall); September 2002; Vancouver. IEEE; 2002. pp. 1520–1524.

[8] Roh W, Paulraj A. MIMO channel capacity for the distributed antenna systems. In: Proceedings of the IEEE Vehicular Technology Conference (VTC 2002-Fall); September 2002; Vancouver. IEEE; 2002. pp. 706–709.

[9] Feng W, Li Y, Gan J, Zhou S, Wang J, and Xia M. On the deployment of antenna elements in generalized multi-user distributed antenna systems. Mobile Networks and Applications. 2011;16:35–45. DOI:10.1007/s11036-009-0214-1

[10] Feng W, Wang Y, Ge N, Lu J, and Zhang J. Virtual MIMO in multi-cell distributed antenna systems: coordinated transmissions with large-scale CSIT. IEEE Journal on Selected Areas in Communications. 2013;31:2067–2081. DOI:10.1109/JSAC.2013.131009

[11] Wang Y, Feng W, Xiao L, Zhao Y, and Zhou S. Coordinated multicell transmission for distributed antenna systems with partial CSIT. IEEE Communications Letters. 2012;16:1044–1047. DOI:10.1109/LCOMM.2012.050912.120383

[12] Hassibi B, Hochwald BM. How much training is needed in multiple-antenna wireless links? IEEE Transactions on Information Theory. 2003;49:951–963. DOI: 10.1109/TIT.2003.809594

[13] Caire G, Ramprashad SA, and Papadopoulos HC. Rethinking network MIMO: Cost of CSIT, performance analysis, and architecture comparisons. In: Proceedings of the Information Theory and Applications Workshop (ITA'10); 31 January–5 February 2010; San Diego. IEEE; 2010. pp. 1–10.

[14] Zhang Y, Feng W, and Ge N. Dealing with large overhead in FDD massive MIMO systems: a one-bit feedback scheme. In: Proceedings of International Conference on Telecommunications (ICT'16); 16–18 May 2016; Thessaloniki. IEEE; 2016. pp. 1–5.

[15] Feng W, Ge N, and Lu J. Hierarchical transmission optimization for massively dense distributed antenna systems. IEEE Communications Letters. 2015;19:673–676. DOI:10.1109/LCOMM.2015.2401584

[16] Chen X, Xu X, and Tao X. Energy efficient power allocation in generalized distributed antenna system. IEEE Communications Letters. 2012;16:1022–1025. DOI:10.1109/LCOMM.2012.051512.120241

[17] Wang J, Wang Y, Feng W, Su X, and Zhou S. An iterative power allocation scheme for improving energy efficiency in massively dense distributed antenna systems. In: Proceedings of the IEEE Vehicular Technology Conference (VTC Spring'16); 15–18 May 2016; Nanjing. IEEE; 2016. pp. 1–5.

[18] Boyd S, Vandenberghe L, Convex Optimization. New York: Cambridge University; 2004. 730 p. ISBN:0 521 83378 7

[19] Dinkelbach W. On nonlinear fractional programming. Management Science. 1967;13:492–498. DOI:10.1287/mnsc.13.7.492

[20] Zhuang H, Dai L, Xiao L, and Yao Y. Spectral efficiency of distributed antenna systems with random antenna layout. Electronics Letters. 2003;39:495–496. DOI:10.1049/el:20030327

[21] Feng W, Zhang X, Zhou S, Wang J, and Xia M. Downlink power allocation for distributed antenna systems with random antenna layout. In: Proceedings of the IEEE Vehicular Technology Conference (VTC Fall'09); 20–23 September 2009; Anchorage. IEEE; 2009. pp. 1–5.

[22] Choi W, Andrews JG. Downlink performance and capacity of distributed antenna systems in a multicell environment. IEEE Transactions on Wireless Communications 2007;6:69–73. DOI:10.1109/TWC.2007.05207

[23] Feng W, Wang Y, Zhao M, Zhou S, and Wang J. Practical power allocation for clustered distributed antenna systems in the low SNR regime. AEU - International Journal of Electronics and Communications 2010;65:595–598. DOI:10.1016/j.aeue.2010.07.005

[24] Li J, Wang D, Zhu P, and You X. Spectral efficiency analysis of single-cell multi-user large-scale distributed antenna system. IET Communications 2014;8:2213–2221. DOI:10.1049/iet-com.2013.0855

[25] Boyd SP, Kim SJ, Hassibi A, and Vandenbarghe L. A tutorial on geometric programming. Optimization and Engineering. 2007;8:67–128. DOI: 10.1007/s11081-007-9001-7

2

# Energy Efficiency for 5G Multi-Tier Cellular Networks

Md. Hashem Ali Khan and Moon Ho Lee

**Abstract**

The heterogeneous cellular network (HCN) is most significant as a key technology for future fifth-generation (5G) wireless networks. The heterogeneous network consists of randomly macrocell base stations (MBSs) overlaid with femtocell base stations (FBSs). Stochastic geometry has been shown to be a very powerful tool to model, analyze, and design networks with random topologies such as wireless ad hoc, sensor networks, and multi-tier cellular networks. HCNs can be energy-efficiently designed by deploying various BSs belonging to different networks, which has drawn significant attention to one of the technologies for future 5G wireless networks. In this chapter, we propose switching off/on systems enabling the BSs in the cellular networks to efficiently consume the power by introducing active/sleep modes, which is able to reduce the interference and power consumption in the MBSs and FBSs on an individual basis as well as improve the energy efficiency of the cellular networks. We formulate the minimization of the power consumption for the MBSs and FBSs as well as an optimization problem to maximize the energy efficiency subject to throughput outage constraints, which can be solved by the Karush-Kuhn-Tucker (KKT) conditions according to the femto tier BS density. We also formulate and compare the coverage probability and the energy efficiency in HCN scenarios with and without coordinated multi-point (CoMP) to avoid coverage holes.

**Keywords:** heterogeneous cellular networks, stochastic geometry, poisson point process (PPP), different sleeping policy, CoMP, energy efficiency, power consumption

## 1. Introduction

Looking ahead to the year 2020 and beyond, there will be explosive growth in mobile data traffic. The existing cellular networks are experiencing some basic challenges such as higher data rates, excellent end-to-end performance, user coverage in hot-spots and crowded areas with lower latency energy consumption and amount of expenditure per information transfer.

The fifth-generation (5G) cellular networks are envisioned to overcome these challenges. It is expected that 5G systems will have the ability to adopt a multi-tier architecture consisting of macrocells, different types of licensed small cells, relays, and device-to-device (D2D) networks to serve users with different quality-to-service (QoS) requirements in an energy efficient manner [1]. It is expected that 5G wireless communication technologies will attain 1000 times higher mobile data volume per unit area, 10–100 times number of connecting devices and longevity of battery 10 times, user data rate, and 5 times reduced latency [2]. A key attribute of 5G networks is that the expected cell data rate will be of the order of 10 Gb/s, whereas average data rate for single 4G networks is 1 Gb/s. Therefore, such a heterogeneous cellular network (HCN) architecture has drawn significant research attention and been recognized as a key technology for future 5G wireless networks. An HCN consisting of $K$ tiers [3] is considered, in which each tier models base stations (BSs) of a particular class such as femtocells, picocells, microcells, or macrocells as shown in **Figure 1a**. The energy efficiency (EE) of small cell networks is of great concern as the BS density will be significantly increased. We study that

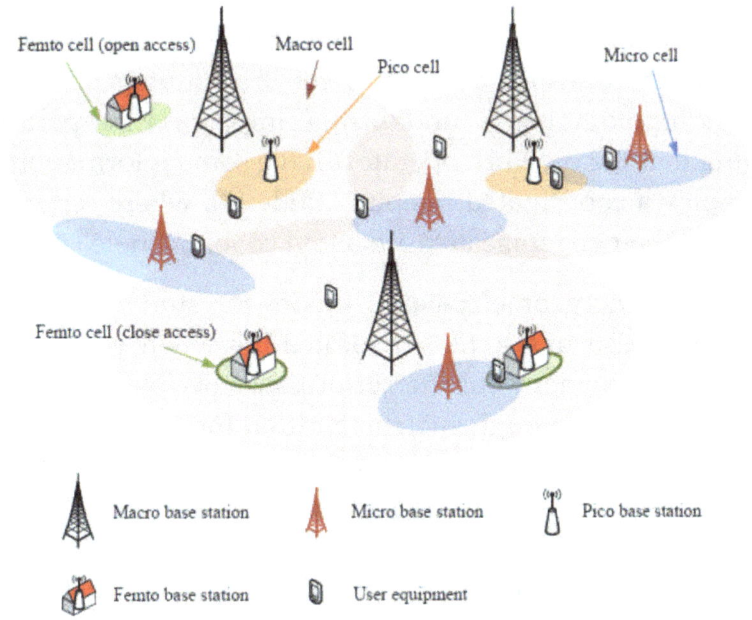


(a)

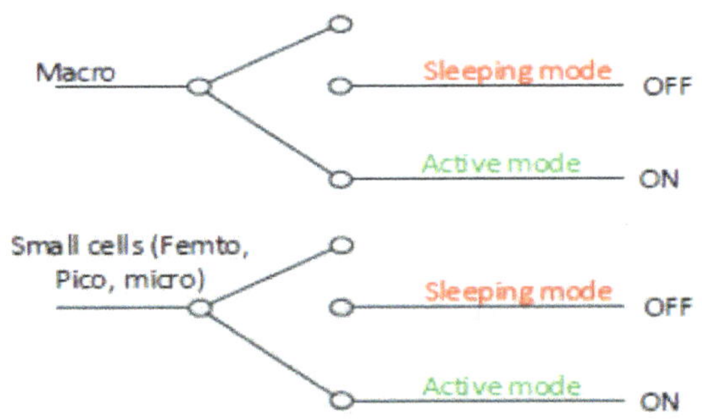


(b)

**Figure 1.** (a) Heterogeneous cellular networks [11] and (b) switching system for BSs power consumption.

the optimal energy efficiency of a two-tier heterogeneous network consists of a macrocell and many small cells under coverage performance constraints for different deployments. The other more important challenge is the greater energy consumption in HCNs because of the dense and randomly deployment of femto BSs (FBSs). In order to realize the aspect of green wireless networks, energy efficiency is an important tool. Because of the increasing share of wireless systems, the total energy expended in communications and networking systems are deemed important. Report shows that total amount of global carbon dioxide emission is originated from information and communication technologies (ICT), more than 9% of emits from wireless and mobile communication [4]. However, within the sleep mode, some key issues must be considered. When BSs are switched off, radio coverage and QoS must be still guaranteed. As BSs are densely deployed, users in sleeping BS coverage can be served by neighboring active BSs by slightly increasing BS transmit power [5]. For sleep mode operation, small cells can always be managed by operators. Nowadays, efforts have been made related to power saving in cellular networks with the introduction of sleep modes [6–8] for BSs. Power consumption is reduced by using sleep mode in low traffic [9] as a case study for saving the energy of macro BSs (MBSs). In a wireless network where multiple links share the same radio spectrum, the signal-to-interference-plus-noise ratio (SINR) at any receiver is a function of the locations of the transmitting nodes and the transmit powers of the transmitters using the same channel. Therefore, the network topology has a fundamental impact on the performance of wireless networks. By assuming that the network operators have some information of the traffic usage patterns, they can employ a coordinated sleeping mode [9], where certain MBSs will be shut off, while others increase their coverage areas to avoid coverage hole [10].

Thus, we provide a stochastic geometry-based model for studying the BSs cooperation in downlink HCNs, which consists of two tiers of located BSs where each tier is characterized by different density and power and develops the performance of coverage probability. We investigate the energy saving problem through switching off/on for MBS and FBS in HCNs. We also derive two-tier HCNs under different sleeping policies and formulate the power consumption minimization for MBS and FBS. An optimization problem is formulated to maximize the energy efficiency subject to throughput outage constraints and solved by the Karush-Kuhn-Tucker (KKT) conditions in terms of femto tier BS density. BSs in sleeping mode might cause coverage holes, which have a negative impact on the connectivity of the network, combined coordinated multi-point (CoMP) and BS sleeping scheme in HCNs for energy efficiency. We introduce the energy efficiency performance based on two-state Markovian wireless channel model.

## 2. System model

We consider a HCN composed by $K$ independent network tiers of BSs with different deployment densities and transmit powers in **Figure 1a**. We assume that the BSs in the $i$th tier are spatially distributed as a Poisson point process (PPP) $\varphi$ of density $\lambda$, transmit at a power $P_i$, and have a SINR target of threshold $T$. The locations of the BSs in the two tiers are distributed as two spatial PPPs in the $\mathbb{R}^2$ Euclidian space denoted by $\phi_M$ and $\phi_F$, with densities $\lambda_M$ and $\lambda_F$, respectively. The probability density function (pdf) is given by $f(r) = 2\pi\lambda r \exp(-\lambda\pi r^2)$.

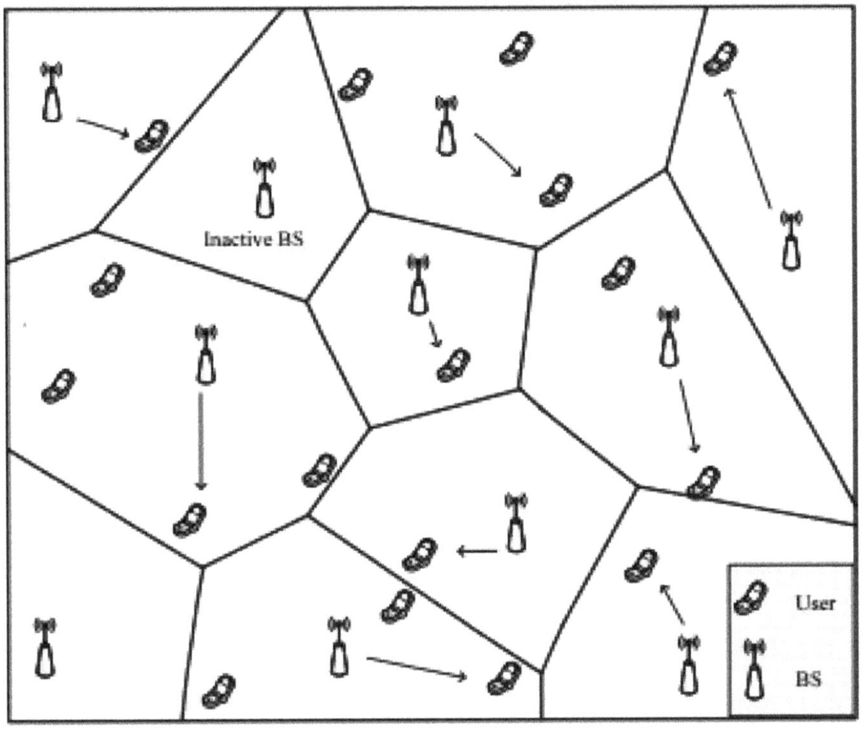


**Figure 2.** Poisson distributed BSs and mobiles, with each mobile associated with the nearest BS. The cell boundaries are shown and form a Voronoi tessellation [12].

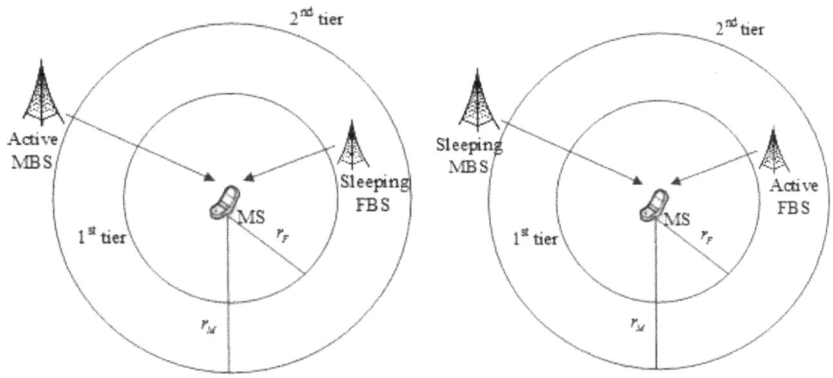


**Figure 3.** The activity level of BSs and location of users.

We focus on a typical user located and assume that a subset of the total ensemble of BSs cooperates by jointly transmitting a message to this tagged receiver, if we consider a nearest BS connectivity model, where a mobile tried to connect with its closest BS. This results in a Voronoi tessellation of the plane corresponding to the BS locations. In this case, the service area of a BS is the Voronoi cell associated with it (in **Figure 2**). When femtocells operate in closed access mode, only registered femtocells user can be allowed to contact to FBSs. On other hand, in open access mode, both macrocell user and unregistered femtocells user can be allowed to contact to FBSs, and then, the coverage region of FBS includes femtocells user and macrocell user connecting to femtocell as shown in **Figure 3**. We can see that $r_M$ and $r_F$ are the distances of MBS and FBS from user. From our proposed scheme, when the FBS is in sleeping mode, the user communicates with the active MBS. On the contrary, the user communicates with the active FBS as shown in **Figure 3**.

### 2.1. Signal-to-interference-plus-noise ratio

We denote a BS by its location, while the user is at the origin 0. For downlink transmission of a MBS to the typical user 0, the SINR experienced by a macrocell user is given by:

$$SINR = \frac{P_i h_i r^{-\alpha}}{\sum\limits_{i=1,\ i\neq j} P_j h_j |r_i|^{-\alpha} + \sigma^2}, \tag{1}$$

where $h$ is channel, the background noise is assumed to be additive white Gaussian with variance $\sigma^2$ and $\alpha$ being the path loss exponent.

### 2.2. Power consumption

Without employing any sleeping mode at each base station in the $i$th tier, the average power consumption of the $i$th tier heterogeneous networks is given by

$$P_{\text{Het},i} = \lambda_i (P_{io} + \Delta_i \beta P_i). \tag{2}$$

In a two-tier cellular network, the total power consumption comes from macrocell tier and femtocell tier, which are expressed as:

$$P_{\text{total}} = \underbrace{\lambda_M (P_{M0} + \Delta M \beta P_{MBS})}_{\text{macro}-tier} + \pi r_M^2 \underbrace{\lambda_F (P_{F0} + \Delta F \beta P_{FBS})}_{\text{femto}-tier}, \tag{3}$$

where $P_{M0}$ and $P_{F0}$ are the static power expenditure of the MBS and FBS, and $\Delta M$, and $\Delta F$ are the slope of the load-dependent power consumption in MBS and FBS, respectively. $\beta$ is the power control coefficient of MBS and FBS. $P_{MBS}$ and $P_{FBS}$ are the transmit powers of MBSs and femto BSs, respectively.

### 2.3. Network energy efficiency

The throughput outage probability defined as the probability that a user in the macro (femto) tier is unable to achieve a certain minimum target throughput as follows:

$$\begin{aligned} \varepsilon_M(\lambda_F) &= 1\text{-}\mathbb{P}\Big(B_M \ln(1 + SINR_M) > T_M\Big) \\ \varepsilon_F(\lambda_F) &= 1\text{-}\mathbb{P}\Big(B_F \ln(1 + SINR_F) > T_F\Big). \end{aligned} \tag{4}$$

Network energy efficiency can be defined as the ratio of the total amount of throughput and total power consumption in the network. The energy efficiency (EE) function can be written as:

$$\begin{aligned} EE &= \frac{\lambda_M C_M + \lambda_F \pi r_M^2 C_F}{P_M + \pi r_M^2 P_F} \\ &= \frac{\lambda_M(1-\varepsilon_M)\log_2(1+SINR_M) + \lambda_F \pi r_M^2 (1-\varepsilon_F)\log_2(1+SINR_F)}{\lambda_M(P_{M0}+\Delta MP_M) + \lambda_F \pi r_M^2 (P_{F0}+\Delta FP_F)}, \end{aligned} \tag{5}$$

where $C$ is the throughput and $\varepsilon$ is coverage probability of macro and femto users, respectively.

## 3. Coverage probability

In this section, we use stochastic geometry theory to analyze the coverage performance of MBS and FBS system under different allocation strategies. Under orthogonal deployment, the spectrum allocation for MBS and FBS is orthogonal, which avoids the cross-tier interference [4]. The received SINR of macro-mobile station (MS) located at the cell boundary is given by:

$$SINR_M = \frac{P_{M,tr} h_M r_M^{-\alpha}}{\sigma^2}. \tag{6}$$

To guarantee the coverage performance of macrocell, the received SINR of the MS at the macrocell edge should satisfy the following equation:

$$\mathbb{P}[SINR_M \geq T_M] = \mathbb{P}\left[\frac{P_{M,tr} h_M r_M^{-\alpha}}{\sigma^2} \geq T_M\right]. \tag{7}$$

There is no interference coordination in femtocell. So, inter-tier interference will provide in femtocell. The received SINR of MS at femtocell edge is written as:

$$SINR_F = \frac{P_{F,tr} h_F r_F^{-\alpha}}{I_F + \sigma^2}. \tag{8}$$

Similar way, the received SINR of the MS at the femtocell edge should satisfy the following equation:

$$\mathbb{P}[SINR_F \geq T_F] = \mathbb{P}\left[\frac{P_{F,tr} h_F r_F^{-\alpha}}{I_F + \sigma^2} \geq T_F\right] = \mathbb{P}\left[h_F \geq \frac{T_F r_F^{\alpha}}{P_{F,tr}}(I_F + \sigma^2)\right]. \tag{9}$$

Conditioning on the nearest BS being at a distance $r$ from the typical user, the probability of coverage averaged over the plane is written as:

$$\begin{aligned} p_c(T,\lambda,\alpha) &= \mathbb{E}_r[\mathbb{P}[SINR > T|r]] = \int_{r>0} \mathbb{P}[SINR > T|r]f_r(r)dr \\ &= \int_{r>0} \mathbb{P}\left[\frac{h_F r^{-\alpha}}{\sigma^2 + I_F + I_M} > T|r\right] e^{-\lambda\pi r^2} 2\pi\lambda r dr \\ &= \int_{r>0} e^{-\lambda\pi r^2} \mathbb{P}[hr^{-\alpha} > T_F(\sigma^2 + I_F + I_M)|r] 2\pi\lambda r dr. \\ &= \int_{r>0} e^{-\lambda\pi r^2} \mathbb{P}[h > Tr^{\alpha}(\sigma^2 + I_F + I_M)|r] 2\pi\lambda r dr \end{aligned} \tag{10}$$

Using the fact that $h \approx \exp(\mu)$, the coverage probability can be expressed as:

$$\begin{aligned} \mathbb{P}[h > Tr^{\alpha}(\sigma^2 + I_F + I_M)|r] &= \mathbb{E}_{I_\psi}[\mathbb{P}[h > Tr^{\alpha}(\sigma^2 + I_F + I_M)|r, I_r]] \\ &= \mathbb{E}_{I_r}\left[\exp\left(-\mu Tr^{\alpha}(\sigma^2 + I_F + I_M)\right)|r\right] = e^{-\mu Tr^{\alpha}\sigma^2} \mathcal{L}_{I_F}(\mu Tr^{\alpha}) \mathcal{L}_{I_M}(\mu Tr^{\alpha}), \end{aligned} \tag{11}$$

where $\mathcal{L}_{I_F}(s)$ and $\mathcal{L}_{I_M}(s)$ are the Laplace transform of random variable $I_\varphi$ evaluated at the distance to the closest BS from the origin. This gives a coverage expression:

$$p_c(T,\lambda,\alpha) = \int_{r>0} e^{-\lambda\pi r^2} e^{-\mu Tr^{\alpha}\sigma^2} \mathcal{L}_{I_F}(\mu Tr^{\alpha}) \mathcal{L}_{I_M}(\mu Tr^{\alpha}) 2\pi\lambda r dr. \tag{12}$$

The definition of Laplace transform yields [13]

$$\begin{aligned} \mathcal{L}_{I_r}(s) &= \mathbb{E}_{I_\varphi}[e^{-sI_\varphi}] = \mathbb{E}_{I_\varphi}[\exp(-s\sum_i g_i R_i^{-\alpha})] \\ &= \mathbb{E}_{I_\varphi}[\prod_i \exp(-sg_i R_i^{-\alpha})] = \mathbb{E}_{I_r}[\prod_i \mathbb{E}_g[\exp(-sg_i R_i^{-\alpha})]] \\ &= \exp\left(-2\pi\lambda\int_r^{\infty}\left(1 - \mathbb{E}_g[\exp(-sg_i R_i^{-\alpha})]\right)vdv\right). \end{aligned} \tag{13}$$

Now, we have

$$\mathcal{L}_{I_\varphi}(s) = \mathbb{E}_{\phi,\{g_i\}}[\prod_{i\in\phi} E_{g_i}[\exp(-sg_i R_i^{-\alpha})]] = \mathbb{E}_\Phi\left[\prod_{i\in\phi}\frac{\mu}{\mu + sR_i^{-\alpha}}\right] = \exp\left(-2\pi\lambda\int_r^{\infty}\left(1 - \frac{\mu}{\mu + sv^{-\alpha}}\right)vdv\right). \tag{14}$$

Let $g_i \approx \exp(\mu)$ and $s = \mu Tr^{\alpha}$.

$$\mathcal{L}_{I_\varphi}(\mu Tr^{\alpha}) = \exp\left(-2\pi\lambda\int_r^{\infty}\frac{T}{T + (r/v)^{\alpha}}vdv\right), \tag{15}$$

Again, $u = (v/rT^{1/\alpha})^2$, then we get

$$\mathcal{L}_{I_\varphi}(\mu Tr^{\alpha}) = \exp\left(-2\pi\lambda T^{2/\alpha}\int_{T^{-2/\alpha}}^{\infty}\frac{1}{1 + u^{\alpha/2}}du\right) = \exp\left(-2\pi\lambda\rho(T,\alpha)\right), \tag{16}$$

where $\rho(T,\alpha) = T^{2/\alpha}\int_{T^{-2/\alpha}}^{\infty}\frac{1}{1 + u^{\alpha/2}}du.$

Putting (16) into (12) with gives the desired result.

## 4. Propose base stations sleep mode strategies

We know that the coverage probability is independent of the sleeping mode. However, we need to maintain the coverage of the cellular networks when we implement sleeping mode in MBSs through power control small cells as shown in **Figures 1b** and **3**. In Ref. [9], authors introduced active/sleep (on/off) modes in MBSs and improved the energy efficiency in cellular networks. In this chapter, we consider the HCNs comprised of macrocell and femtocell tiers. We propose switching off/on systems for the efficient power consumption at the BSs in the cellular networks, which introduce active/sleep modes in the MBSs and FBSs. The active/sleep modes reduce the interference and power consumption as well as improve the energy efficiency of the cellular networks. We derive the two-tier HCNs under different sleeping policies as well as formulate power consumption minimization for the MBSs and FBSs. An optimization problem is formulated to maximize the energy efficiency subject to throughput outage constraints as well as solved by the KKT conditions in terms of the femto tier BS density. Thus, the total power consumed by each BS in the macro and femto tiers is modeled as follows:

$$P_M = \begin{cases} P_{M0} + \Delta M\beta P_{MBS}, & \text{for } \textit{active mode} \\ 0_M, & \text{for } \textit{sleeping mode} \end{cases}. \\ P_F = \begin{cases} P_{F0} + \Delta F\beta P_{FBS}, & \text{for } \textit{active mode} \\ 0_F, & \text{for } \textit{sleeping mode} \end{cases} \tag{17}$$

From Eq. (17), we can see that the MBS and FBS are active modes, and the maximum power is consumed by BSs. Otherwise, power consumption is zero when it is in sleeping mode.

### 4.1. Random sleeping

In random sleeping strategy, we take it as a Bernoulli trial, that is, each BS actives with probability $q$ and sleeps with probability $1 - q$ independently for macro and femto BSs [9, 14]. Then, the sleep modes of other BSs are determined according to the distances between a BS and user. Power consumption of random sleeping problem is formulated as follows:

$$P_{RS}(MBS) = \lambda_M q_M (P_{MO} + \Delta M\beta P_{MBS}) + \lambda_M (1-q_M) P_{sleep}, \tag{18}$$

and

$$P_{RS}(FBS) = \lambda_F q_F (P_{FO} + \Delta F\beta P_{FBS}) + \lambda_F (1-q_F) P_{sleep}. \tag{19}$$

The power is consumed in the macro tier and femto tier BS when operating in the active and sleep mode, and then the total average power is given by:

$$P_{total} = \underbrace{\lambda_M q_M(P_{M0} + \Delta M\beta P_M) + \lambda_M(1-q_M)P_{sleep}}_{\text{macro}-tier} + \pi r_M^2 \cdot \underbrace{\lambda_F(P_{F0} + \Delta F\beta P_F) + \lambda_F(1-q_F)P_{sleep}}_{\text{femto}-tier} \quad (20)$$

Thus, the energy efficiency of the network for random sleeping is given by:

$$EE = \frac{\lambda_M(1-\varepsilon_M)\log_2(1 + SINR_M) + \pi r_M^2 \lambda_F(1-\varepsilon_F)\log_2(1 + SINR_F)}{\lambda_M q_M(P_{M0} + \Delta M\beta P_M) + \lambda_M(1-q_M)P_{sleep} + \pi r_M^2 \lambda_F(P_{F0} + \Delta F\beta P_F) + \lambda_F(1-q_F)P_{sleep}}. \quad (21)$$

The network energy efficiency is expressed in the units of nats/Joule. The numerator in Eq. (21) is the total average throughput achieved by all the users in the two-tier network, and the denominator is the total power consumption use of Eqs. (18), (19) and (20).

### 4.2. Strategic sleeping

The sleep mode strategy can be considered as a load-aware policy and can incorporate traffic profile in the optimization problem. By applying strategic sleeping, the average power consumption can be expressed as:

$$P_{SS}(MBS) = \lambda_M\Big(E\{s\}(P_{MO} + \Delta M\beta_M P_{MBS}) + \lambda_M(1-E\{s\})P_{sleep}\Big), \quad (22)$$

and

$$P_{SS}(FBS) = \lambda_F\Big(E\{s\}(P_{FO} + \Delta F P_{MBS}) + \lambda_F(1-E\{s\})P_{sleep}\Big). \quad (23)$$

In case of random sleeping mode, a network is developed that is adaptive to the fluctuating activity levels during the day. The strategic sleeping mode can go one step further. It can model a network that is adaptive to fluctuating activity levels within the location [9]. In addition, the strategic sleeping model can measure the impact of cooperation among MBSs. The energy efficiency of the network for strategic sleeping is given by:

$$EE = \frac{\lambda_M(1-\varepsilon_M)\log_2(1 + SINR_M) + \lambda_F \pi r_M^2(1-\varepsilon_F)\log_2(1 + SINR_F)}{\underbrace{\lambda_M\Big(E\{s\}(P_{MO} + \Delta M\beta P_{MBS}) + \lambda_M(1-E\{s\})P_{sleep}\Big)}_{\text{macro}-tier} + \underbrace{\pi r_M^2\Big(\lambda_F\Big(E\{s\}(P_{FO} + \Delta F\beta P_{MBS}) + \lambda_F(1-E\{s\})P_{sleep}\Big)\Big)}_{\text{femto}-tier}}. \quad (24)$$

Similar way, the network energy efficiency is expressed as the numerator in Eq. (24) of the total average throughput achieved by all the users in the two-tier network and the denominator of the total power consumption use of Eqs. (22) and (23).

### 4.3. Optimization problem

To solve the following multi-objective optimization problem [14]:

$$
\begin{aligned}
&\max_{\lambda_F}\ EE(\lambda_F) \\
s.t.\quad & 1-\mathbb{P}\Big(B_M \ln(1+SINR_M) > T_M\Big) \le \varepsilon_M, \\
& 1-\mathbb{P}\Big(B_F \ln(1+SINR_F) > T_F\Big) \le \varepsilon_F
\end{aligned}
\tag{25}
$$

where $\varepsilon_M$ and $\varepsilon_F$ denote the outage objectives guaranteeing a minimum target throughput for each user in the macro and femto tier, respectively. The optimal femto tier BS density $\lambda_F^*$ that maximizes the energy efficiency of network subject to the downlink outage constraints is given by $\lambda_F^*$

$$
\lambda_F^* = \begin{cases}
[\lambda_{EE,F}] & \text{for } \mu_M^* = 0,\ \ \mu_F^* = 0 \ \text{(both inactive)} \\
\lambda_M(1-q)\zeta^{-1} & \text{for } \mu_M^* > 0,\ \ \mu_F^* = 0 \ \text{(macro active \& femto inactive)} \\
\lambda_F - \lambda_M q \zeta^{-1} & \text{for } \mu_M^* = 0,\ \ \mu_F^* > 0 \ \text{(macro inactive \& femto active)} \\
\lambda_F(1-q) & \text{for } \mu_M^* > 0,\ \ \mu_F^* > 0 \ \text{(both active)}
\end{cases},
\tag{26}
$$

where $\mu_M^*$ and $\mu_F^*$ are the Lagrange multipliers and $\zeta = (P_F/P_M)^{2/\alpha}$ is power ratio of BSs. The optimization problem in Eq. (25) is determined by satisfying the KKT conditions as follows:

$$
\begin{aligned}
\mathcal{L}(\lambda_{EE}, \mu_M, \mu_F, \lambda_F) = EE(\lambda_F) - \mu_M[1-\mathbb{P}\Big(B_M \ln(1+SINR_M) > T_M\Big) - \varepsilon_M]. \\
-\mu_F[1-\mathbb{P}\Big(B_F \ln(1+SINR_F) > T_F\Big) - \varepsilon_F]
\end{aligned}
\tag{27}
$$

The KKT conditions are then listed as follows:

$$
\begin{aligned}
&\frac{\partial \mathcal{L}(\lambda_F^*)}{\partial \lambda_F} = 0, \\
&1-\mathbb{P}\Big(B_M \ln(1+SINR_M) > T_M\Big) \le \varepsilon_M \\
&1-\mathbb{P}\Big(B_F \ln(1+SINR_F) > T_F\Big) \le \varepsilon_F
\end{aligned}
\tag{28}
$$

$$
\begin{aligned}
&\mu_M^*[1-\mathbb{P}\Big(B_M \ln(1+SINR_M) > T_M\Big) - \varepsilon_M] = 0. \\
&\mu_F^*[1-\mathbb{P}\Big(B_F \ln(1+SINR_F) > T_F\Big) - \varepsilon_F] = 0 \\
&\mu_M^* > 0,\ \ \mu_F^* > 0
\end{aligned}
\tag{29}
$$

Based on the listed KKT conditions, evaluating each possible scenario for which $\mu_M^*$ and $\mu_F^*$ are either active or inactive gives the optimal femto tier BS density $\lambda_F^*$.

# 5. Combined coordinated multi-point (CoMP) transmission and BS sleeping scheme

In this section, we also evaluate the performance of the combined CoMP and BS sleeping scheme in a two-tier HCNs. The first tier is deployed as MBSs with a density of $\lambda_M$, and the second tier is deployed as FBSs with a density of $\lambda_F$.

## 5.1. BS cooperation

BS sleeping has been proved to be an effective technique for saving energy consumption in cellular networks. However, BSs in sleeping mode might cause coverage holes, which have a negative impact on the connectivity of the network. We conduct a stochastic geometry analysis to evaluate the performance of the proposed combined CoMP and BS sleeping scheme in HCNs for energy efficiency [10]. We apply CoMP to avoid coverage holes when the target SINR cannot be reached. Applying stochastic geometry tools, we formulate and compare the coverage probability and the energy efficiency in HCN scenarios with and without CoMP.

The cooperative set is composed of the closest BSs in each network tier to the user. The density of CoMP is the same as the tier contains BSs with the lowest density. The probability of CoMP happens is equal to the probability of awake MBSs $q$, and its density is $q\lambda_M$. We assume that the awake MBSs can always cooperate with FBSs to transmit, so that $n = K = 2$. Here, $n$ is the number of cell cooperatives. The following lemma gives the coverage probability of the combined CoMP and BSs sleeping control.

**Theorem** [10]: In two-tier HCNs with CoMP and BSs sleeping, the coverage probability of a randomly located user is given by:

$$
\begin{aligned}
p_{c\_CoMP} &= 4\pi^2 q^2 \lambda_M \lambda_F \int \exp\left(-2\pi q \lambda_M s_1^{2/\alpha} F(r_1 s_1^{-1/\alpha})\right) \times \\
&\exp\left(-2\pi q \lambda_F s_2^{2/\alpha} F(r_2 s_2^{-1/\alpha})\right) \times \exp\left(-\pi q(\lambda_M r_1^2 + \lambda_F r_2^2)\right) r_1 r_2 dr_1 r_2,
\end{aligned} \tag{30}
$$

where $s_i = \frac{TP_i}{P_1 r_1^{-\alpha} + P_2 r_2^{-\alpha}}$ for $r_i \geq 0$, $i = \{1, 2\}$ and $F(x) = \int_x^\infty \frac{r}{1 + r^\alpha} dr$.

The energy efficiency of the networks for BS cooperation

$$
EE = \frac{\lambda_M p_{c\_CoMP} \log_2(1 + SINR_M) + \pi r_M^2 \lambda_F p_{c\_CoMP} \log_2(1 + SINR_F)}{\underbrace{\lambda_M q_M (P_{M0} + \Delta M \beta P_M) + \lambda_M (1 - q_M) P_{sleep}}_{\text{macro}-tier} + \pi r_M^2 \underbrace{\left(\lambda_F (P_{F0} + \Delta F \beta P_F) + \lambda_F (1 - q_F) P_{sleep}\right)}_{\text{femto}-tier}}. \tag{31}
$$

From Eq. (31), we can see that the energy efficiency is related to the coverage probability and the power consumption of whole networks.

### 5.2. BS non-cooperation

The typical user only connects to the nearest BS, which belongs to first tier in a non-CoMP scenario [10]. Then, the coverage probability in the case of BS non-cooperation is given by:

$$p_{c\_\mathrm{Non-CoMP}} = \frac{1}{1 + T^{2/\alpha} 2F(T^{-1/\alpha}) + \frac{T^{2/\alpha}}{\mathrm{sinc}(2/\alpha)} \frac{q\lambda_F}{q\lambda_M} \frac{P_2^{2/\alpha}}{P_1^{2/\alpha}}}. \tag{32}$$

Thus, the energy efficiency of the networks for BS non-cooperation is given by:

$$EE = \frac{\lambda_M p_{c\_Non-CoMP} \log_2(1 + SINR_M) + \pi r_M^2 \lambda_F p_{c\_Non-CoMP} \log_2(1 + SINR_F)}{\underbrace{\lambda_M q_M (P_{M0} + \Delta M \beta P_M) + \lambda_M (1 - q_M) P_{sleep}}_{\mathrm{macro}-tier} + \pi r_M^2 \underbrace{\left(\lambda_F (P_{F0} + \Delta F \beta P_F) + \lambda_F (1 - q_F) P_{sleep}\right)}_{\mathrm{femto}-tier}}. \tag{33}$$

From Eqs. (30) and (32), we can see that the coverage probability depends on both the sleep strategy and BSs density ratio.

## 6. Markovian wireless networks

The BS can be in either of the two operational states: ON or OFF. If BS is ON, the energy increases with the energy harvesting rate and decreases according to the number of users served by that BS. However, if the BS is OFF, it does not serve any users.

### 6.1. Uncoordinated

In this class of strategies, the decision to toggle the operational state, that is, turn a BS ON or OFF, is taken by the BS independently of the operational states of the other BSs.

### 6.2. Coordinated

In this class of strategies, the decision to toggle the state of a particular BS is dependent upon the states of the other BSs.

### 6.3. Energy efficiency of two-cell cellular networks

To investigate the basic energy efficiency performance of two-cell cellular network, in this case, a user's channel of two-cell cellular network is modeled into good and bad states due to channel conditions [15]. Moreover, a transition from one state to the next state only depends on the current state with the state space $\{0, 1\}$, where '0' corresponds to a good state and '1' corresponds to a bad state in **Figure 4**. Based on properties of Markovian processes, a channel transition probability matrix is given by:

$$q^{(n)} = \begin{bmatrix} q_{00}^{(n)} & q_{01}^{(n)} \\ q_{10}^{(n)} & q_{11}^{(n)} \end{bmatrix} = \begin{bmatrix} q_{00} & q_{01} \\ q_{10} & q_{11} \end{bmatrix}^{(n)}, \tag{34}$$

where $q_{i,j}$, $i$ and $j \in \{0,\ 1\}$, is a one-step transition probability from the state $i$ into the state $j$, and $q_{i,j}^{(n)}$, $i$ and $j \in \{0,\ 1\}$, is a probability from the initial state $i$ into the state $j$ after $n$ steps transition. The energy efficiency for multicell cellular networks is given by:

$$EE_{multicell} = \sum_{i=1}^{K} \log_2\left(1 + \frac{P_i \| h_i \|_F^2}{\sigma_i^2 + \sum\limits_{j=1,\ i\neq j} P_j \| h_{i,j} \|_F^2}\right) / \sum_{i=1}^{K} P_i. \tag{35}$$

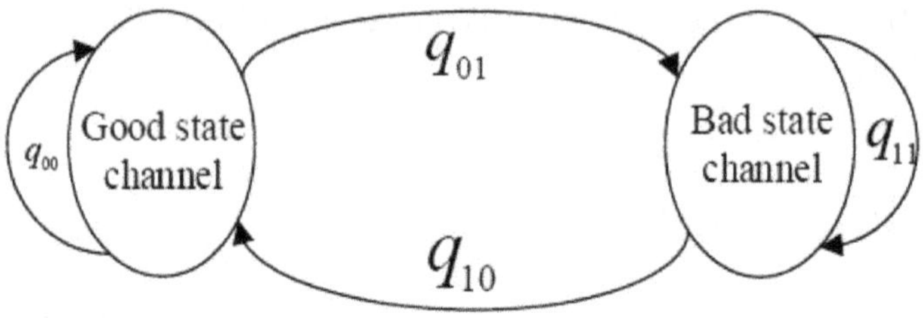


**Figure 4.** State transition diagram of two-state Markovian wireless channel.

The wireless channels of multicell cellular network are assumed as two-state Markovian wireless channels, due to the memory-less property of two-state Markovian wireless channel model [15]. Furthermore, after an $n$ steps state transition in two-state Markovian wireless channels, a model of energy efficiency of multicell cellular network is given by:

$$EE_{multicell} = \frac{\sum\limits_{i=1}^{K} \left\{ \log_2\left(1 + \frac{P_i \| h_i^{\text{good}} \|_F^2}{\sigma_i^2 + \sum\limits_{j=1,\ i\neq j} P_j \| h_{i,j}^{\text{good}} \|_F^2}\right) q_{00}^{(n)} + \log_2\left(1 + \frac{P_i \| h_i^{\text{bad}} \|_F^2}{\sigma_i^2 + \sum\limits_{j=1,\ i\neq j} P_j \| h_{i,j}^{\text{bad}} \|_F^2}\right) q_{01}^{(n)} \right\}}{\sum\limits_{i=1}^{K} P_i}. \tag{36}$$

To analyse the impact of cell number on the energy efficiency of multicell cellular networks; for a good state channel, $h_i^{\text{good}} = 0.9$ and $h_{i,j}^{\text{good}} = 0.1$; for a bad state channel, $h_i^{bad} = 0.6$ and $h_{i,j}^{bad} = 0.4$; $n$ steps transition probabilities of two-state Markovian channels are fixed as $P_{00}^{(n)} = 0.8$ and $P_{01}^{(n)} = 0.2$; and the noise is $\sigma_i^2 = 0.1$. Moreover, an initial state transition probability matrix of two-state Markovian chain channels is shown as:

$$q = \begin{bmatrix} q_{00} & q_{01} \\ q_{10} & q_{11} \end{bmatrix} = \begin{bmatrix} 0.8 & 0.2 \\ 0.6 & 0.4 \end{bmatrix} = \begin{bmatrix} 4/5 & 1/5 \\ 3/5 & 2/5 \end{bmatrix}. \tag{37}$$

## 7. Numerical results

In this section, we present numerical evaluations of the integral expressions for the coverage probability and energy efficiency performance. We focus on the two network tiers consisting of a macro tier overlaid with a femto tier. The assumed parameter values for two-tier HCNs are based on the values used in **Table 1**. We assume that $\alpha = 4$ and that the first tier has spatial intensity $\lambda_1 = (500^2\pi)^{-1}$ and available power $P_1 = 25$, while the second tier has spatial intensity $\lambda_2 = 5\lambda_1$ and available power $P_2 = P_1/25$.

**Figure 5** illustrates the effect of the SINR threshold $T$ on the coverage probability. By comparing the performance of the cooperative scheme to the baseline of no cooperation scheme, we observe that around 0 dB cooperation yields relative gains in coverage probability of up to about 30% compared to non-cooperative. The coverage probability can be directly related to the ergodic rate of communication from the cooperating BSs to the typical receiver.

**Figure 6** plots the coverage probability versus noise $\sigma^2$ for different sleeping strategies. The sleeping strategy is modeled as 0 and 1, respectively. As shown in **Figure 6**, in strategic

| Symbol | Description | Value |
|---|---|---|
| $B$ | Bandwidth | 180 kHz |
| $\alpha$ | Path loss exponent | 4 |
| $T_M$ | SINR threshold for macro | 8 dB |
| $T_F$ | SINR threshold for femto | 5 dB |
| $P_{MBS}$ | Macro BS transmit power | 20 W |
| $P_{FBS}$ | Femto BS transmit power | 2 W |
| $r_M$ | Macro range | 300 m |
| $r_F$ | Femto range | 15 m |
| $P_{MO}$ | Static power MBS | 130 W |
| $P_{FO}$ | Static power FBS | 4.8 W |
| $\Delta M$ | Slope of MBS | 4.7 |
| $\Delta F$ | Slope of FBS | 8 |
| $P_{M-sleep}$ | Sleeping power MBS | 75 W |
| $P_{F-sleep}$ | Sleeping power FBS | 5 W |
| $\lambda_M$ | Density of MBS | $1\times10^{-4}\ m^{-2}$ |
| $\lambda_F$ | Density of FBS | $1\times10^{-2}\ m^{-2}$ |

**Table 1.** Network parameter values.

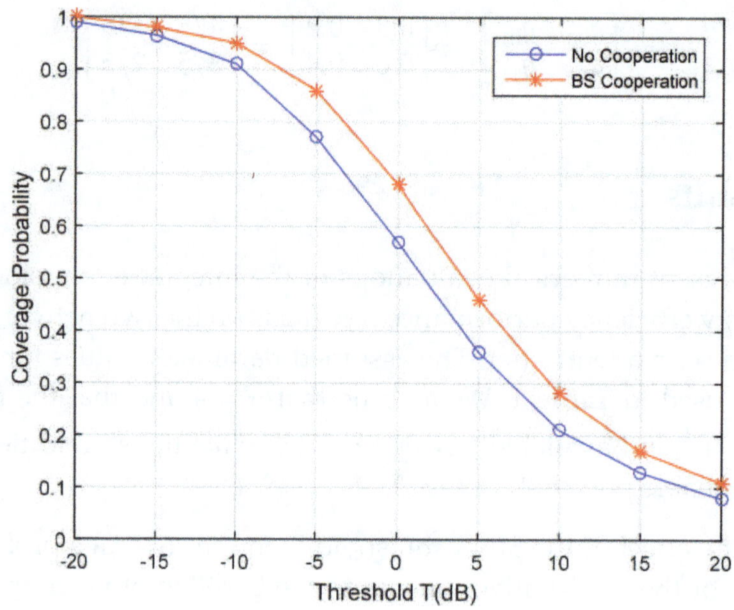


**Figure 5.** Comparison of the coverage probabilities for BS cooperation and no cooperation against the threshold in dB.

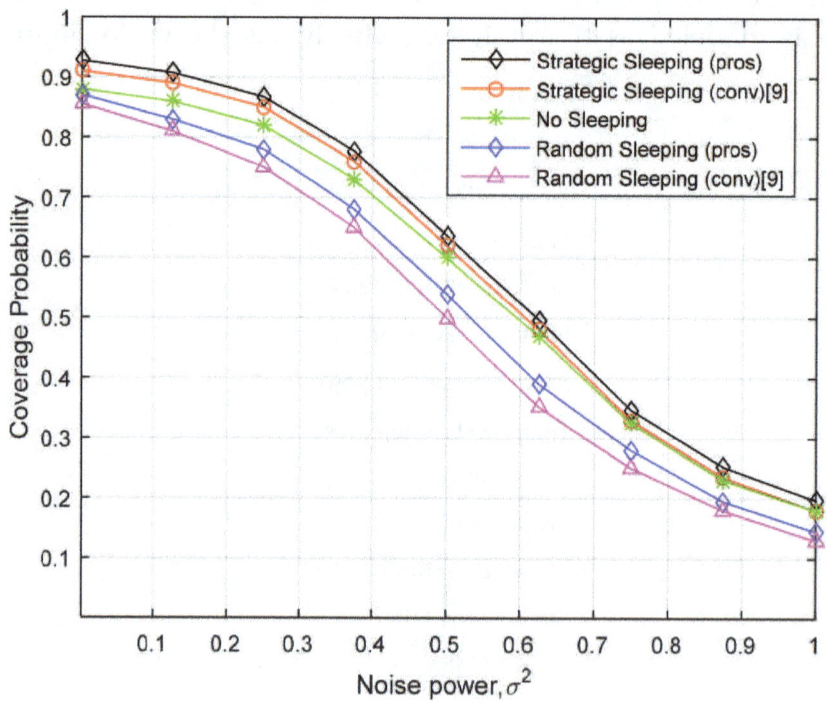


**Figure 6.** Coverage probabilities for different sleeping strategies.

sleeping mode, the coverage probability is marginally better than no sleeping mode. It can also be said that strategic sleeping has a bigger margin of improvement over no sleeping when $\sigma^2 \to 0$. Finally, it can be seen that strategic sleeping is always better than random sleeping for the same fraction of sleeping MBSs and FBSs.

**Figure 7** shows the maximum two-tier achieved energy efficiency versus density. The assumed parameter values for the two-tier HCNs are based on the values used in **Table 1**. In general, the maximum two-tier energy efficiency decreases with increasing density. Note that, we show the

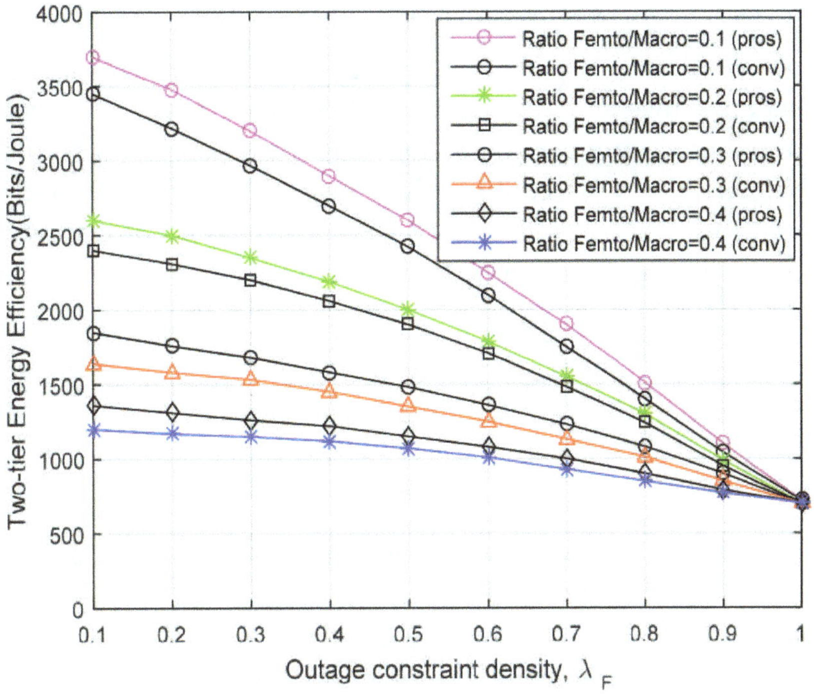


**Figure 7.** Two-tier network energy efficiency versus density.

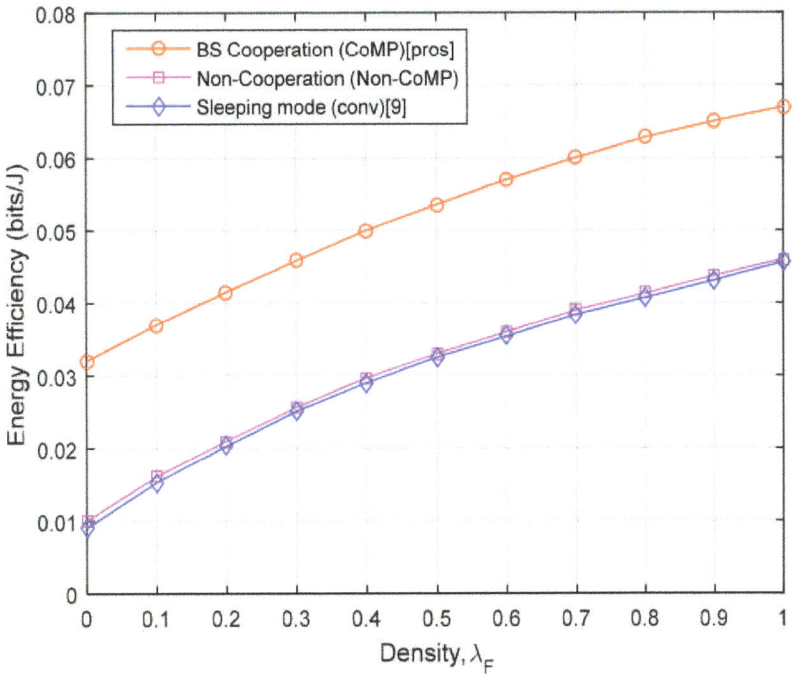


**Figure 8.** Energy efficiency versus density for the CoMP and non-CoMP.

energy efficiency curves close to the points for $P_{FBS}/P_{MBS} = 0.1,\ 0.2,\ 0.3$ and 0.4. The observations made from **Figure 7** underscore the impact of the femto-to-macro BS power consumption factor on the ability to maximize the two-tier energy efficiency while satisfying the outage objectives.

**Figure 8** shows the energy efficiency of the CoMP and non-CoMP schemes versus density. It is observed that the energy efficiency improves according to the density. The proposed scheme of combined CoMP and BSs sleeping mode is increased by 2% of energy efficiency from non-CoMP schemes. Numerical results confirm that the combined CoMP and BS sleeping can improve the energy efficiency as well as increase the coverage probability compared with implementing BS sleeping only. Moreover, the performance of non-CoMP is almost same as the macro BS sleeping only [9].

## 8. Conclusion

In this chapter, we provide energy efficiency of two-tier network through deploying sleeping strategy in MBSs and FBSs. The MBS and FBS are switching off/on systems, that is, it reduces power consumption and interference and improves the energy efficiency of HCNs. Power consumption is formed into optimization problems, which is determined by the optimal density of femto tier BS. BSs in sleeping mode might cause coverage holes, which have a negative impact on the connectivity of the network. Thus, we proposed combined CoMP and BS sleeping scheme in HCNs for energy efficiency to avoid coverage holes. Numerical results show that the proposed sleeping strategy can effectively increase energy efficiency. We also analyze the energy efficiency performance of cellular network based on two-state Markovian wireless channels.

## Acknowledgements

This work was supported by the MEST 2015R1A2A1A05000977, NRF, Korea.

## Author details

Md. Hashem Ali Khan and Moon Ho Lee*

*Address all correspondence to: moonho@jbnu.ac.kr

Division of Electronics and Information Engineering, Chonbuk National University, Jeonju, Republic of Korea

## References

[1] Hossain E., Rasti M., Tabassum H., Abdelnasser A. Evolution toward 5G multi-tier cellular wireless networks: An interference management perspective. IEEE Wireless Communications. 2014; **21**(3): pp. 118–127. doi: 10.1109/MWC.2014.6845056

[2] Metis. Scenarios, Requirements and KPIs for 5G Mobile and Wireless System. ICT-317669, Metis Project. May 2013.

[3] Dhillon D. S., Ganti R. K., Andrews J. G. Modelling and analysis of K-tier downlink heterogeneous cellular networks. IEEE Journal of Selected Areas in Communications. 2012; **30**(3): pp. 550–560. doi:10.1109/JSAC.2012.120405

[4] Zhang X., Su Z., Yan Z., Wang W. Energy efficiency study for two-tier heterogeneous networks under coverage performance constraints. Mobile Networks and Applications. 2013; 18(4): pp. 567–577. doi:10.1007/s11036-013-0435-1

[5] Liu C., Natarajan B., Xia H. Small cell base station sleep strategies for energy efficiency. IEEE Transaction on Vehicular Technology. 2016; **65**(3): pp. 1652–1661. doi:10.1109/TVT.2015.2413382

[6] Ashraf I., Boccardi F., Ho L. Sleep mode techniques for small cell deployments. IEEE Communications Magazine. 2011; **49**(8): pp. 72–79.

[7] Ashraf I., Boccardi F., Ho L. Power savings in small cell deployments via sleep mode techniques. Proc. IEEE 21st Int. Symp. PIMRC Workshops. 2010; pp. 307–311.

[8] Li Y., Celebi H., Daneshmand M., Wang C., Zhao W. Energy efficient femtocell networks: Challenges and opportunities. IEEE Wireless Communication. 2013; **20**(6): pp. 99–105.

[9] Soh Y. S., Quek T. Q. S., Kountouris M., Shin H. Energy efficient heterogeneous cellular networks. IEEE Journal on Selected Areas in Communications. 2013; **31**(5): pp. 840–850. doi:10.1109/JSAC.2013.130503

[10] He A., Liu D., Chen Y., Zhang T. Stochastic geometry analysis of energy efficiency in HetNets with combined CoMP and BS sleeping. In: IEEE 25th Annual International Symposium on Personal, Indoor and Mobile Radio Communications (PIMRC); 2–5 Sept. 2014; Washington DC. IEEE; pp. 1798–1802. doi:10.1109/PIMRC.2014.7136461

[11] Chenlong J. Energy Efficient Design of Heterogeneous Cellular Networks Using Stochastic Geometry. PhD thesis. National University of Singapore, 2015.

[12] Li C., Zhang J., Letaief K. B. Throughput and energy efficiency analysis of small cell networks with multi-antenna base stations. IEEE Transaction on Wireless Communications. 2014; 13(5): pp. 2505–2517. doi:10.1109/TWC.2014.031714.131020

[13] Andrews J. G., Baccelli F., Ganti R. K. A tractable approach to coverage and rate in cellular networks. IEEE Transaction on Communications. 2011; 59(11): pp. 3122–3134.

[14] Rao J. B., Fapojuwo A. O. An analytical framework for evaluating spectrum/energy efficiency of heterogeneous cellular networks. IEEE Transactions on Vehicular Technology. 2016; **65**(5): pp. 3568–3584. doi:10.1109/TVT.2015.2448593

[15] Ge X., Cao C., Jo M., Chen M., Hu J., Humar I. Energy efficiency modelling and analysing based on multicell and multi-antenna cellular networks. KSII Transactions on Internet and Information Systems. 2010; **4**(4): pp. 560–574. doi:10.3837/tiis.2010.08.007

# 3

# Beamforming in Wireless Networks

Mohammad-Hossein Golbon-Haghighi

**Abstract**

This chapter is about the beamforming approach in wireless 5G networks, which involves communication between multiple source-destination pairs. The relays can be multiple-input multiple-output (MIMO) and/or distributed single-input single-output (SISO), and full channel state information of source-relays and relay-destinations are assumed to be available. Our design consists of a two-step amplify-and-forward (AF) protocol. The first step includes signal transmission from the sources to the relays, and the second step contains transmitting a version of the linear precoded signal to the destinations. Beamforming is investigated only in relay nodes to reduce end user's hardware complexity. Accordingly, the optimization problem is defined to find the relay beamforming coefficients that minimize the total relay transmit power by keeping the signal-to-interference-plus-noise ratio (SINR) of all destinations above a certain threshold value. It is shown that this optimization problem is a non-convex, and can be solved efficiently.

**Keywords:** beamforming, 5G wireless networks, MIMO, optimization

## 1. Introduction

Recently, cooperative communication has become one of the appealing techniques that can be used in 5G wireless relay networks to achieve spatial diversity and multiplexing, which overcomes the channel impairments caused by several fading effects and destructive interference. Though various cooperative communication schemes exist [1, 2], the AF scheme is more attractive due to its simplicity since the relays simply forward the amplitude phase-adjusted version of received signals to destinations. In Ref. [2], a distributed beamforming relay system with a single transmitter-receiver pair, and several relaying nodes have been proposed. The authors assumed that perfect channel state information (CSI) is available at all relay nodes. Although the same scenario is investigated in Ref. [3], the second-order statistics of all channel coefficients are assumed to be available at the relays. Furthermore, the beamforming weights are obtained in order to maximize the signal-to-noise ratio (SNR) at destination subject to holding the relay power above a certain threshold value.

In the past three decades, code-division-multiple-access (CDMA) systems have been extensively investigated as the one of the important candidates for transmitting data over single channels while sharing a fixed bandwidth among a large number of users [4]. The design of receivers to increase the number of supported users, in these systems, has been explored in Ref. [5, 6]. In Ref. [6], joint channel estimation and data detection based on an expectation-maximization (EM) algorithm [7] is proposed. The authors have shown that the proposed receiver achieves a near-optimum performance with modest complexity. Furthermore, the authors in Ref. [5] designed a double stage linear-detection receiver to increase the number of supported users on the system. This design requires complex processing at the receiver's side instead of using a precoding scheme at the transmitter where more hardware complexity is tolerable. Therefore, the authors in Ref. [8] studied a MIMO CDMA system implementing zero-forcing beamforming (ZFBF) as an efficient precoding technique.

Though various complex multiuser detection techniques that can be used in CDMA systems [9], the unconventional matched filter receiver is chosen at destination nodes due to the intractability of the precoding design when other forms of detectors are used. In this article, we have focused on the optimization of the beamforming weights applied to the outputs of matched filter banks to minimize the total relay transmit power subject to a target SINR of all destinations. Our proposed distributed CDMA-relay network can easily overcome the other multiplexing schemes such as space division-multiple access (SDMA), time division-multiple access (TDMA) or frequency division-multiple access (FDMA). The SDMA schemes [10] in which sources, destinations and relays are distributed in the space, have two disadvantages. First, these schemes should have a significant number of relays in proportion to their users to be able to overcome channel impairments at destinations. Although the SDMA scheme with the limited number of relays cannot compensate the interference power, our CDMA schemes can easily satisfy the network QoS due to their ability to decrease the interference effect at destinations. So, the second disadvantage of SDMA is the inefficient use of hardware communication resources. In the SDMA scheme, if the number of users increases, the network data rate can significantly decrease. Therefore, the number of relays should be considerably increased to be able to satisfy the QoS constraints, which is costly for the network operator.

**Notation**: We denote the complex conjugate, transpose, Hermitian (conjugate transpose) and inner product operators by $(\cdot)^*$, $(\cdot)^T$, $(\cdot)^H$ and $\langle \cdot \rangle$, respectively. We use $E\{\cdot\}$ to denote statistical expectation. trace$\{\cdot\}$ and Rank$\{\cdot\}$ represent the trace and rank of the matrix, respectively. Vec$(\cdot)$ is the vectorization operator stacking all columns of a matrix on top of each other; $\otimes$ represent the Kronecker product of two matrices and $A \geq 0$ stands for semi-definite conic inequality that means $A$ is a non-negative semi-definite matrix.

## 2. 5G wireless system and equations

Consider a wireless relay network with $d$ pairs of source-destination (peers) communicating without a direct link through $R$ MIMO or SISO relay antennas. In this chapter, a two-step AF protocol is used. In the first step, each source user broadcasts its spread symbol toward the relays. A matched filter is applied in each relay in order to retrieve the source's signals. In the second step, the adjusted and spread signals by the relays are transmitted to destinations.

## 3. MIMO relay networks

In this section, a peer-to-peer MIMO-relay network with $d$ pairs of source-destination nodes is considered, as shown in **Figure 1**. It is assumed that all source and destination nodes are equipped with one SISO antenna and each source attempts to maintain communication with its corresponding destination. It is assumed that there is no direct link between source and destination pairs due to path loss and deep shadowing and all nodes are working in a half-duplex mode. We use a two-step AF protocol. During the first step, each source broadcasts its signals to MIMO-relay. Then, after applying the beamforming weights at MIMO-relay, the adjusted signals transmit to all destinations.

Let $s_k$ stands for the $k^{\text{th}}$ source symbol that is assumed to be independent of the other sources, that is, $E\{s_k s_l^*\} = P_k \delta_{kl}$. Denote the channel coefficient from the $k^{\text{th}}$ source to the $r^{\text{th}}$ relay as $f_{rk}$ and the channel coefficient from $r^{\text{th}}$ relay to $k^{\text{th}}$ destination as $g_{rk}$. Then, the received signal at the $r^{\text{th}}$ relay is given by:

$$\chi_r = \sum_{l=1}^{d} f_{rl} s_l + \omega_r, \; r \in \{1, \ldots, R\}, \tag{1}$$

where $\omega_r$ is the noise at the $r^{\text{th}}$ relay. For simplicity, Eq. (1) can be rewritten as:

$$\boldsymbol{\chi} = \sum_{l=1}^{d} \mathbf{f}_l s_l + \boldsymbol{\omega}, \tag{2}$$

where $\boldsymbol{\chi} \triangleq [\chi_1, \chi_2, \ldots, \chi_R]^T$, $\boldsymbol{\omega} \triangleq [\omega_1, \omega_2, \ldots, \omega_R]^T$, $\mathbf{f}_l \triangleq [f_{1l}, f_{2l}, \ldots, f_{Rl}]^T$.

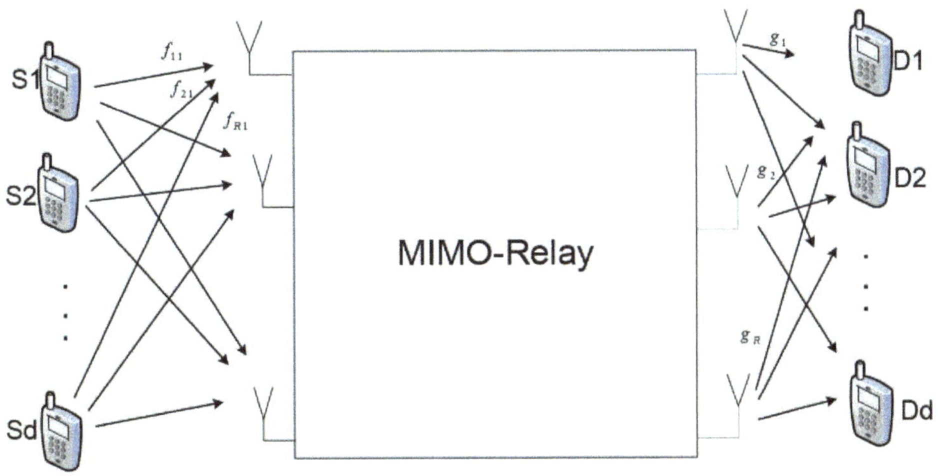


**Figure 1.** A MIMO-relay network (from M.H. Golbon et al. [11]).

The received signal in MIMO relay has been processed by the beamforming weights, that is, $\mathbf{W} \in \mathbb{C}^{R \times R}$, which should be designed appropriately. Finally, each MIMO-relay antenna transmits the following signal to destinations:

$$\gamma = \mathbf{W}\chi \in \mathbb{C}^{R\times 1} \tag{3}$$

The $r^{\text{th}}$ entry of $\gamma$ is the signal transmitted by $r^{\text{th}}$ MIMO antennas. Finally, the received signal at the $k^{th}$ destination is given by

$$y_k = \mathbf{g}_k{}^T\gamma + \zeta_k \tag{4}$$

where $\zeta_k(t)$ is the noise at the $k^{\text{th}}$ receiver. We can easily rewrite Eq. (4) as:

$$\begin{aligned} y_k &= \mathbf{g}_k{}^T\mathbf{W}\chi + \zeta_k = \mathbf{g}_k{}^T\mathbf{W}\left(\sum_{l=1}^{d}\mathbf{f}_l s_l + \omega\right) + \zeta_k \\ &= \mathbf{g}_k{}^T\mathbf{W}\sum_{l=1}^{d}\mathbf{f}_l s_l + \mathbf{g}_k{}^T\mathbf{W}\omega + \zeta_k \\ &= \underbrace{\mathbf{g}_k{}^T\mathbf{W}\mathbf{f}_k s_k}_{\text{desired received signal}} + \underbrace{\mathbf{g}_k{}^T\mathbf{W}\sum_{l=1,\, l\neq k}^{d}\mathbf{f}_l s_l}_{\text{interference part}} + \underbrace{\mathbf{g}_k{}^T\mathbf{W}\omega + \zeta_k}_{\text{noise part}} \end{aligned} \tag{5}$$

The three last terms of Eq. (5) are the desired received signal, interference and noise at the $k^{th}$ destination, respectively. The object of the network beamforming is to minimize the total relay transmit power subject to maintaining every destination SINR above a pre-defined threshold value $\gamma_{th}$ (as a QoS parameter of the network). In this case, the instantaneous SINR for $k^{th}$ destination simply becomes the desired signal power of the desired signal to the power of interference plus noise. So, the optimization problem can now be written as

$$\begin{aligned} &\underset{\mathbf{w}}{\text{Minimize}}\ P_R \\ &\text{Subject to}\ \text{SINR}_k \geq \gamma_{th}^k,\ \ k \in \{1, 2, \ldots, d\} \end{aligned} \tag{6}$$

where $P_R$ is the total relay transmit power, $w$ stands for beamforming weights, $\text{SINR}_k$ and $\gamma_k$ denote the received SINR and the target SINR (threshold value) at the $k^{\text{th}}$ destination node, respectively.

First, using Eq. (3), the total relay transmit power can be calculated as

$$\begin{aligned} P_R &= E\{\gamma^H\gamma\} \\ &= E\{\chi^H\mathbf{W}\mathbf{W}^H\chi\} = \text{trace}\{\mathbf{W}^H\mathbf{R}_x\mathbf{W}\} \end{aligned} \tag{7}$$

where $\mathbf{R}_x \triangleq E\{\chi\chi^H\}$ and it can be calculated as:

$$\mathbf{R}_x = \sum_{l=1}^{d} P_l E\left\{\mathbf{f}_l\mathbf{f}_l{}^H\right\} + \sigma_\omega^2\mathbf{I}_{R\times R} \tag{8}$$

For any conforming matrices **M**, **N** and **Z**, the following equation holds

$$\text{trace}\left(\mathbf{M}\mathbf{Z}^{H}\mathbf{N}\mathbf{Z}\right) = \text{vec}(\mathbf{Z})^{H}\left(\mathbf{M}^{T}\otimes\mathbf{N}\right)\text{vec}(\mathbf{Z}) \tag{9}$$

Therefore, Eq. (7) can be rewritten as the following quadratic form:

$$\begin{aligned} P_R &= \text{vec}(\mathbf{W})^{H}\underbrace{(\mathbf{I}_{R\times R}\otimes\mathbf{R}_x)}_{\mathbf{T}}\text{vec}(\mathbf{W}) \\ &= \mathbf{w}^{H}\mathbf{T}\mathbf{w} \end{aligned} \tag{10}$$

where $\mathbf{w} \triangleq \text{vec}(\mathbf{W})$ and $\mathbf{T} \triangleq (\mathbf{I}_{R\ \times\ R} \otimes \mathbf{R}_x)$.

Using Eq. (5), the desired signal power at the $k^{\text{th}}$ destination can be obtained as:

$$\begin{aligned} P_{S_k} &= P_k E\left(\mathbf{f}_k^{H}\mathbf{W}^{H}\ \mathbf{g}_k^{*}\mathbf{g}_k{}^{T}\mathbf{W}\mathbf{f}_k\right) \\ &= P_k\,\text{vec}(\mathbf{W})^{H}\underbrace{\left(\mathbf{R}_{\mathbf{f}_k}{}^{T}\otimes\mathbf{R}_{\mathbf{g}_k}\right)}_{\mathbf{R}_k}\text{vec}(\mathbf{W}) \\ &= \mathbf{w}^{H}\mathbf{R}_k\mathbf{w} \end{aligned} \tag{11}$$

where $\mathbf{R}_{\mathbf{f}_k}\triangleq E\left(\mathbf{f}_k\mathbf{f}_k^{H}\right)$, $\mathbf{R}_{\mathbf{g}_k}\triangleq E\left(\mathbf{g}_k^{*}\mathbf{g}_k^{T}\right)$ and $\mathbf{R}_k\triangleq P_k\left(\mathbf{R}_{\mathbf{f}_k}{}^{T}\otimes\mathbf{R}_{\mathbf{g}_k}\right)$.

Also, using Eq. (5), the received noise power at $k^{\text{th}}$ destination can be calculated as:

$$\begin{aligned} P_{N_k} &= E\left(\boldsymbol{\omega}^{H}\mathbf{W}^{H}\mathbf{g}_k{}^{*}\mathbf{g}_k{}^{T}\mathbf{W}\boldsymbol{\omega}\right) + \sigma^2_{\varsigma_k} \\ &= \sigma^2_{\omega}\,\text{trace}\left\{\mathbf{W}^{H}\mathbf{R}_{\mathbf{g}_k}\mathbf{W}\right\} + \sigma^2_{\varsigma_k} \\ &= \text{vec}(\mathbf{W})^{H}\underbrace{\left(\mathbf{I}\otimes\mathbf{R}_{\mathbf{g}_k}\right)}_{\mathbf{N}_k}\text{vec}(\mathbf{W}) \\ &= \mathbf{w}^{H}\mathbf{N}_k\mathbf{w} + \sigma^2_{\varsigma_k} \end{aligned} \tag{12}$$

where $\mathbf{N}_k\triangleq\left(\mathbf{I}\otimes\mathbf{R}_{\mathbf{g}_k}\right)$.

Finally, the power of the received interference at the $k^{\text{th}}$ destination can be computed as

$$\begin{aligned} P_{I_k} &= E\left(\left(\sum_{l=1,\,l\neq k}^{d}\mathbf{f}_l s_l\right)^{H}\mathbf{W}^{H}\mathbf{g}_k{}^{*}\mathbf{g}_k{}^{T}\mathbf{W}\left(\sum_{l=1,\,l\neq k}^{d}\mathbf{f}_l s_l\right)\right) \\ &= \text{trace}\left\{\underbrace{\left(P_k E\left\{\sum_{\substack{l,m=1\\ l,m\neq k}}^{d}\mathbf{f}_l\mathbf{f}_m{}^{H}\right\}\right)}_{\mathbf{F}_k}\mathbf{W}^{H}\mathbf{R}_{\mathbf{g}_k}\mathbf{W}\right\} \\ &= \text{vec}(\mathbf{W})^{H}\left(\mathbf{F}_k{}^{T}\otimes\mathbf{R}_{\mathbf{g}_k}\right)\text{vec}(\mathbf{W}) \\ &= \mathbf{w}^{H}\mathbf{I}_k\mathbf{w} \end{aligned} \tag{13}$$

where $\mathbf{F}_k\triangleq P_k E\left\{\sum_{\substack{l,m=1\\ l,m\neq k}}^{d}\mathbf{f}_l\mathbf{f}_m{}^{H}\right\}$ and $\mathbf{I}_k\triangleq\left(\mathbf{F}_k{}^{T}\otimes\mathbf{R}_{\mathbf{g}_k}\right)$.

In this case, the instantaneous SINR for $k^{\text{th}}$ destination simply becomes the desired signal power of the desired signal to the power of interference plus noise. So, the optimization problem can now be written as

$$\begin{aligned}&\underset{\mathbf{w}}{\text{Minimize}}\ \mathbf{w}^H\mathbf{T}\mathbf{w}\\&\text{Subject to } \text{SINR}_k = \frac{\mathbf{w}^H\mathbf{R}_k\mathbf{w}}{\mathbf{w}^H(\mathbf{N}_k+\mathbf{I}_k)\mathbf{w}+\sigma^2_{\varsigma_k}} \geq \gamma^k_{th}\\&k\in\{1,2,\ldots,d\}\end{aligned} \tag{14}$$

Since $\mathbf{w}^H(\mathbf{N}_k+\mathbf{I}_k)\mathbf{w}+\sigma^2_{\varsigma_k}\geq 0$, the constraints of the optimization problem can be formulated as

$$\mathbf{w}^H\left(\mathbf{R}_k-\gamma^k_{th}(\mathbf{N}_k+\mathbf{I}_k)\right)\mathbf{w}\geq\gamma^k_{th}\sigma^2_{\varsigma_k} \tag{15}$$

In this problem, if all the matrices $\mathbf{R}_k-\gamma^k_{th}(\mathbf{N}_k+\mathbf{I}_k)$ are negative semi-definite for all $k$, the problem is convex and can be solved uniquely. However, the feasible set of our optimization problem is empty since $\mathbf{w}^H\left(\mathbf{R}_k-\gamma^k_{th}(\mathbf{N}_k+\mathbf{I}_k)\right)\mathbf{w}\leq 0$ for all $\boldsymbol{k}$ and $\mathbf{w}$. Therefore, $\mathbf{R}_k-\gamma^k_{th}(\mathbf{N}_k+\mathbf{I}_k)$ is non-negative definite matrix which results in non-convex inequality constraints, hence the quadratically constrained quadratic programming (QCQP) problem is non-convex and NP-hard in general. However, we will show that a simple near optimal solution can be found in our problem. First, we replaced our QCQP problem with a semi-definite programming (SDP) problem. Let us define $\mathbf{D}_k\triangleq\mathbf{R}_k-\gamma^k_{th}(\mathbf{N}_k+\mathbf{I}_k)$ , $\mathbf{X}\triangleq\mathbf{w}\mathbf{w}^H$ and using the fact that trace($AB$) = trace($BA$) (when A is an $m \times n$ and B is an $n \times m$ matrix), the optimization problem Eq. (14), can recast to

$$\begin{aligned}&\underset{\mathbf{X}}{\text{Minimize}}\ \text{trace}(\mathbf{T}\mathbf{X})\\&\text{Subject to } \text{trace}(\mathbf{D}_k\mathbf{X})\geq\gamma^k_{th}\sigma^2_{\varsigma_k}, k\in\{1,\ldots,d\}\\&\text{Rank}(\mathbf{X})=1\ , \mathbf{X}\geq 0\end{aligned} \tag{16}$$

This optimization problem is non-convex, because the Rank(**X**) = 1 constraint is non-convex. We relax the problem by ignoring this non-convex constraint and convert it to a convex SDP problem. The following semi definite representation (SDR) form is the relaxed version of the problem Eq. (16).

$$\begin{aligned}&\underset{\mathbf{X}}{\text{Minimize}}\ \text{trace}(\mathbf{T}\mathbf{X})\\&\text{Subject to } \text{trace}(\mathbf{D}_k\mathbf{X})\geq\gamma^k_{th}\sigma^2_{\varsigma_k},\ \ k\in\{1,\ldots,d\}\\&\mathbf{X}\geq 0\end{aligned} \tag{17}$$

The optimal value of the relaxed problem is a lower bound of the optimal value of SDP problem (Eq. 16).Well-known semi-definite problem solvers such as SeDuMi or CVX can solve the above problem in polynomial time using interior point methods. If the optimal value of Eq. (17), that is, $\mathbf{X_{opt}}$, is rank one, then its principal eigenvector is exactly the optimal solution of the original optimization problem. Since the solution of Eq. (17) is not always rank one, one can use randomization techniques [10] to obtain an approximate solution of the original problem from the solution of the relaxed problem. The randomization technique is finding the best solution from the candidate sets of beamforming vectors generated from $\mathbf{X_{opt}}$ [12]. Luo et al. [13] and Chang et al. [14] analyzed the accuracy of these techniques for different semidefinite problems, and it has been found that the randomization technique has acceptable performance in practical scenarios [15]. Therefore, the eigenvalue decomposition of $\mathbf{X_{opt}}$ can be calculated as $\mathbf{X_{opt}} = VDV^H$. Then the candidate sets of beamforming vectors is generating as $x_c = VD^{(1/2)}p_c$, where $p_c$ is a circularly symmetric complex, and zero mean, unit variance white Gaussian vector, that is, $p_c \in \mathbb{C}^{R \times 1} \sim \mathbb{C}N(0, 1)$. Hence, it can be easily recognized that the vector $x_c$ satisfies $E\{x_c x_c^H\} = \mathbf{X_{opt}}$. This candidate vector generation should perform several times and in each iteration, any vector (or scaled version) that satisfies SINR constraints of problem Eq. (17) is saved as a candidate vector ($x_c^{'}$) along with corresponding objective values. The vector generation should be repeated for a predefined number of times. The final minimum solution can be achieved by a simple minimization over the obtained objective values as an approximate solution of the problem.

Then, solving problem Eq. (16) from $x_c$ becomes finding a proper scaling factor of $\sqrt{\beta} \geq 0$. Applying $\beta$ to Eq. (17), the following problem will be attained

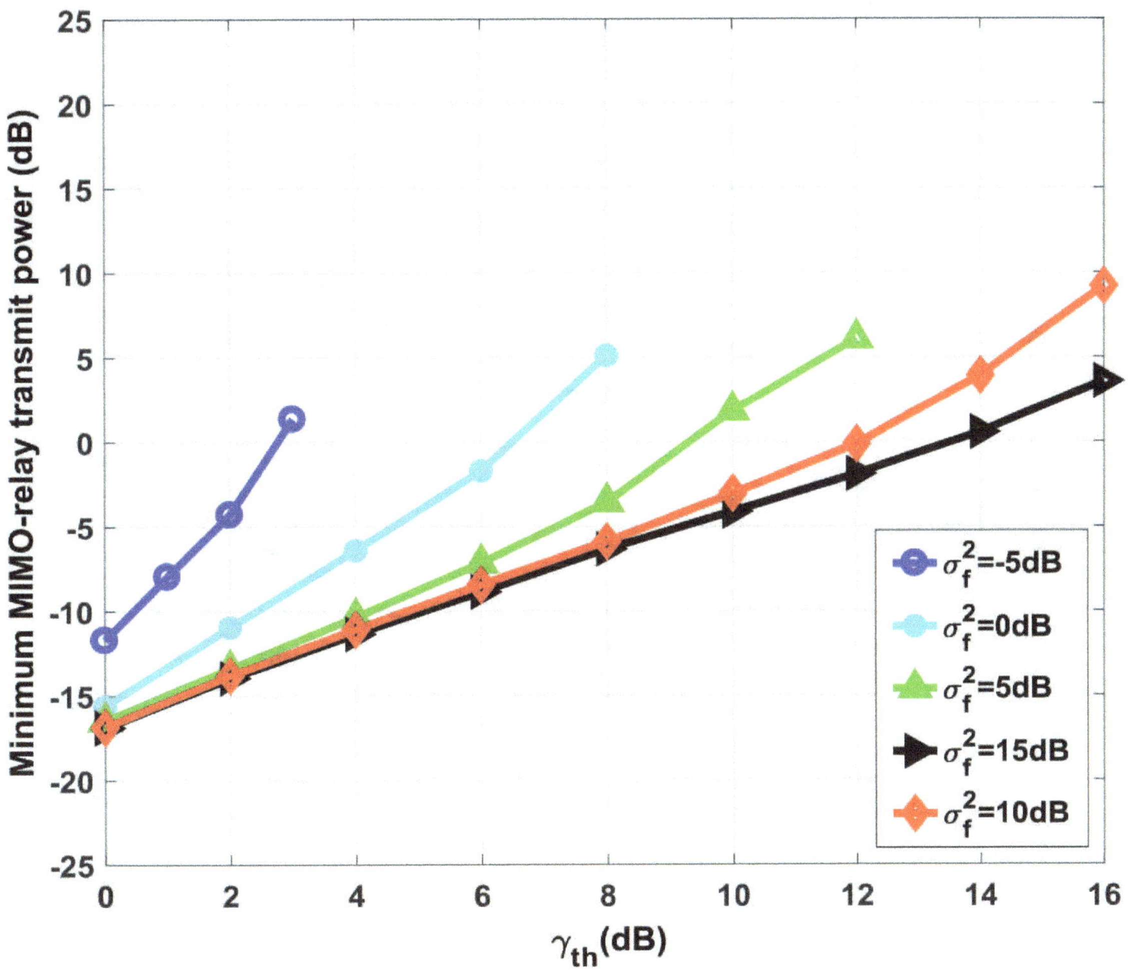


**Figure 2.** Minimum MIMO-relay transmit power $P_T^{\min}$ versus destination SINR threshold value $\gamma_{th}$, for different values of $\sigma_f^2$ and $\sigma_g^2 = 10dB$.

$$\begin{aligned} &\underset{\mathbf{X}}{\text{Minimize}}\ \beta \text{trace}\left(\mathbf{TX}\right) \\ &\text{Subject to } \beta \text{trace}\left(\mathbf{D}_k\mathbf{X}\right) \geq \gamma_{th}^{k}\sigma_{\varsigma_k}^{2}\ ,\ \ k\in\{1,\ldots,d\} \\ &\text{Rank}(\mathbf{X}) = 1\ ,\ \ \mathbf{X}\geq 0 \end{aligned} \tag{18}$$

In the above algorithm, the acceptable scaling factors are those that satisfy $\beta$ trace($\mathbf{T}_k\mathbf{X}$) ≥ 0. Thus, the maximum scaling factor should be selected as

$$\beta = \max_{k=1,\ldots,d}\left\{\frac{\gamma_{th}^{k}\sigma_{\varsigma_k}^{2}}{\text{trace}(\mathbf{D}_k\mathbf{X})}\right\} \tag{19}$$

Consequently, the approximate solution of problem (Eq. 16) is $\sqrt{\beta}x_c$. In our case, after an acceptable number of iterations (around 100 iterations), the solution of the randomization problem approached to its lower bound (the optimal value of relaxed problem). Therefore, $\mathbf{X}_{\mathbf{opt}}$ is an acceptable and a near optimal solution to the original non-convex problem. Another optimal solution of Eq. (16) can be found using a penalty function in the objective part of the problem and converting the objective function into the difference of two convex functions

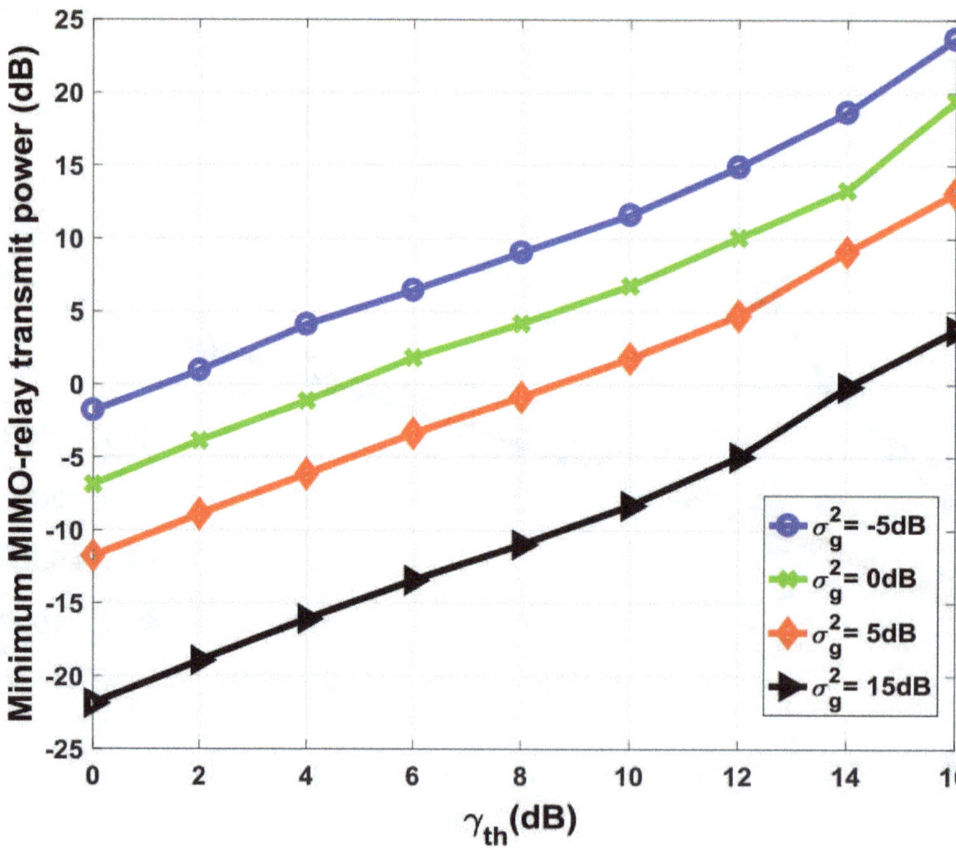


**Figure 3.** Minimum MIMO-relay transmit power $P_T^{\min}$ versus destination SINR threshold value $\gamma_{th}$, for different values of $\sigma_g^2$ and $\sigma_f^2 = 10dB$.

subject to current convex constraints [16], and applying an effective non-smooth optimization algorithm based on the sub-gradient of rank one constraint.

For examination, we assumed that channel state information is known at a processing center and the beamforming weights are optimized and spreaded to the nodes from this processing Center [17]. In each simulation snapshot, the channel coefficients $f_{rk}$, $g_{rk}$ are generated as i.i.d circularly symmetric complex Gaussian random variables with variances of $\sigma_f^2 = \sigma_g^2 = 10\text{dB}$. Also, it is assumed that we have the same output power at sources, that is, $\{P_k\}_{k=1}^{d} = 10\text{dB}$ and we set $\{\gamma_{th}^k\}_{k=1}^{d} = \gamma_{th}, \left\{\sigma_{\omega_i}^2\right\}_{i=1}^{R} = \left\{\sigma_{\varsigma_k}^2\right\}_{k=1}^{d} = 0\text{dB}$.

**Figures 2** and **3** show the minimum MIMO-relay transmit power $P_T^{\min}$ versus destination SINR threshold value $\gamma_{\text{th}}$, for different values of $\sigma_f^2$, $\sigma_g^2$. It can be seen from **Figures 2** and **3** that the better quality of uplink and/or downlink channels can decrease the minimum MIMO-relay transmit power for a certain threshold value.

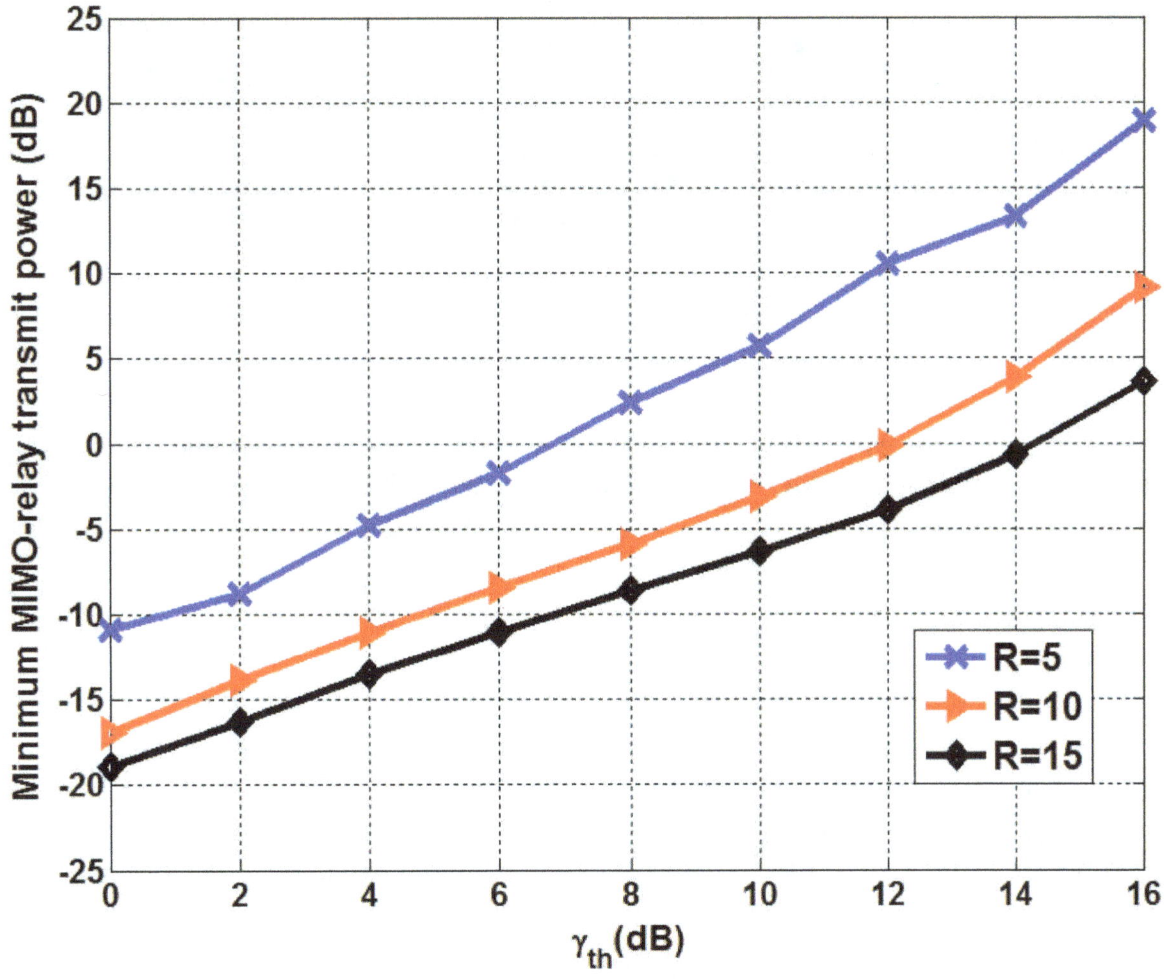


**Figure 4.** Minimum MIMO relay transmit power $P_T^{\min}$ versus destination SINR threshold value $\gamma_{th}$, for different number of antennas.

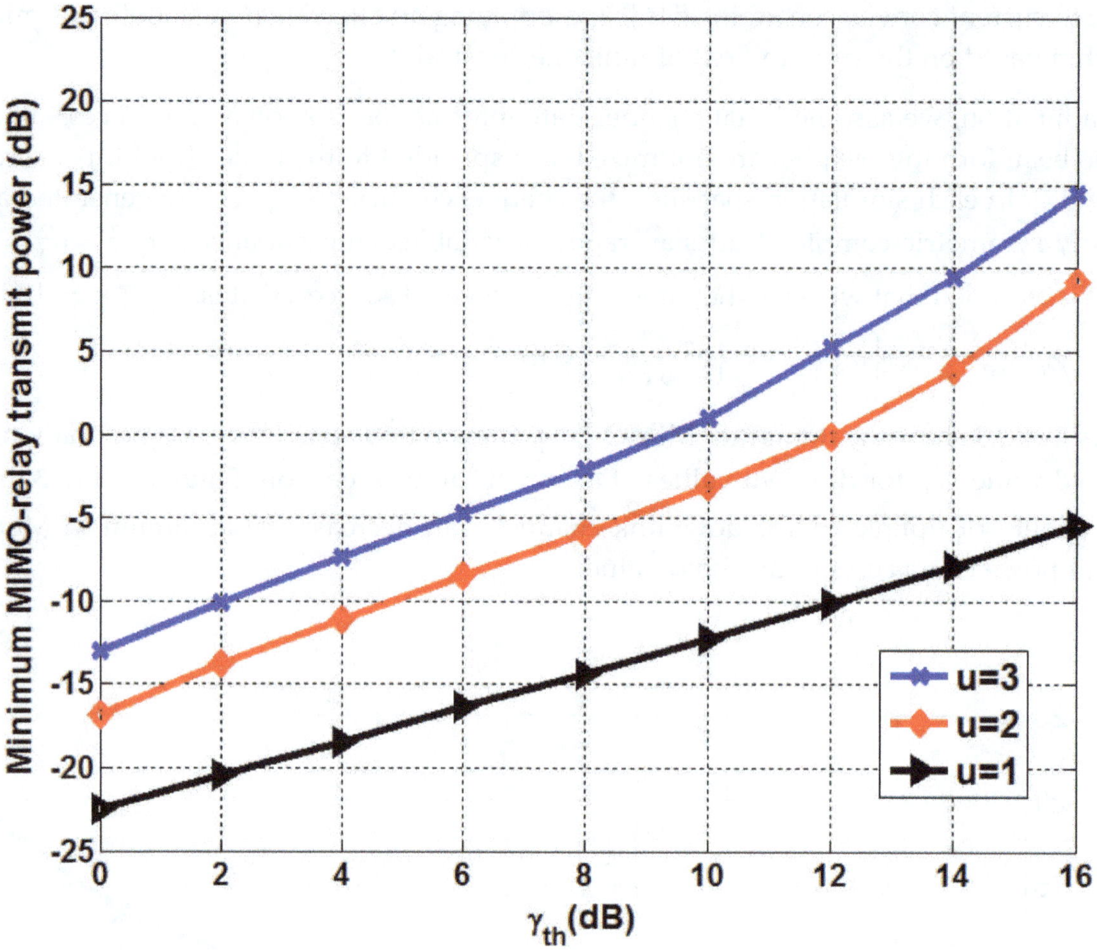


**Figure 5.** Minimum MIMO relay transmit power $P_T^{\min}$ versus destination SINR threshold value $\gamma_{th}$, for different number of source-destination pairs.

In **Figures 4** and **5**, we examine the network performance by changing the number of MIMO-relay antennas and number of source-destination pairs. As expected, more power saving will be obtained by increasing the number of MIMO antennas and/or decreasing the number of user nodes.

## 4. MIMO-CDMA relay networks

In the last section, we obtained the optimal beamforming weights for a MIMO relay network. Here, in addition to the multiple antenna technique, CDMA is applied to the network to increase the order of multiuser multiplexing. CDMA systems can share a fixed bandwidth among a large number of users without the need of frequency division or time division between nodes. CDMA introduces a diverse range of trade-off between receiver complexity and system performance.

As shown in **Figure 6**, a two-step AF protocol is used for this MIMO-relay network. In the first step, each source user broadcasts its precoded signal (i.e. $s_l u_l(t)$) at its maximum power $P_l$ toward the MIMO-relay. At the MIMO-relay, a matched filter is applied to retrieve the source's

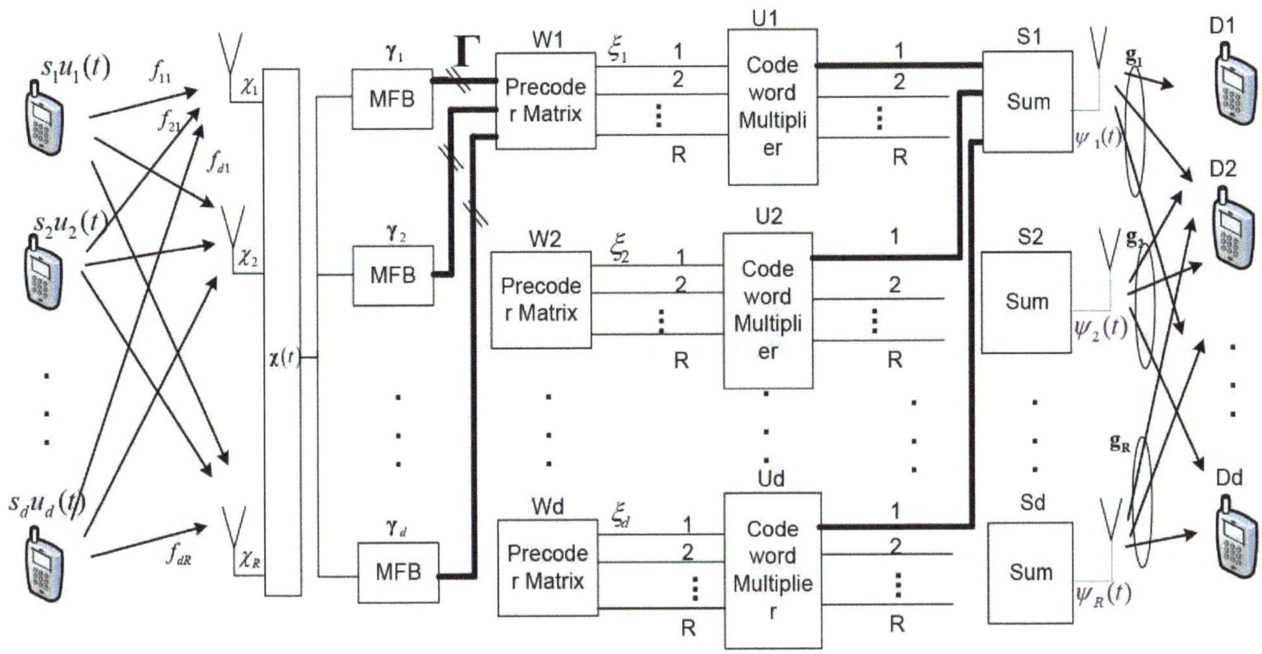


**Figure 6.** MIMO-relay multiuser network (from M.H. Golbon et al. [18]).

signals. In the second step, the adjusted and spreaded signals are transmitted from MIMO-relay to all destinations. Another matched filter is used at each destination to extract its corresponding symbols.

Let $u_k(t)$ denotes a signature waveform that is assigned to the $k^{\text{th}}$ source. Then, the received signal at the $r^{\text{th}}$ antennas of MIMO-relay is given by

$$\chi_r(t) = \sum_{l=1}^{d} f_{rl} s_l u_l(t) + \omega_r(t) \tag{20}$$

The vector form of Eq. (20) can be written as:

$$\boldsymbol{\chi}(t) = \sum_{l=1}^{d} \mathbf{f}_l s_l u_l(t) + \boldsymbol{\omega}(t) \in \mathbb{C}^{R\times 1} \tag{21}$$

where

$$\begin{aligned} &\boldsymbol{\chi}(t) \triangleq [\chi_1(t), \chi_2(t), \dots, \chi_R(t)]^T, \\ &\mathbf{v}(t) \triangleq [v_1(t), v_2(t), \dots, v_R(t)]^T, \\ &\mathbf{f}_l \triangleq [f_{1l}, f_{2l}, \dots, f_{Rl}]^T \end{aligned} \tag{22}$$

By denoting the cross correlation between $k$th user's codeword to the $l^{\text{th}}$ user's codeword as $\rho_{k,l} = u_k(t) * u_l(T_0 - t)|_{t=T_0}$, the output signal of the matched filter at the MIMO-relay can be expressed as

$$\begin{aligned}\gamma_n &= \chi(t)*u_n{}^*(T_0-t)|_{t=T_0}\\ &= \sum_{l=1}^{d}\mathbf{f}_l s_l u_l(t)*u_n{}^*(T_0-t)|_{t=T_0} + \omega(t)*u_n{}^*(T_0-t)|_{t=T_0}\\ &= \sum_{l=1}^{d}\mathbf{f}_l s_l \rho_{l,n} + \varsigma_n = \gamma_{n,-k} + \gamma_{n,k} + \varepsilon_n\end{aligned} \tag{23}$$

where $\rho_{l,n}$ is the cross correlation between $l^{th}$ user's code-word and $n^{th}$ user's code-word [19]:

$$\rho_{l,n} = u_l(t)*u_n{}^*(T_0-t)|_{t=T_0} = \langle u_l(t), u_n(t)\rangle \tag{24}$$

where $\gamma_{n,k}$, $\gamma_{n,-k}$, and $\varepsilon_n$ are defined as

$$\begin{aligned}\gamma_{n,-k} &\triangleq \sum_{l=1,\, l\neq k}^{d}\mathbf{f}_l s_l \rho_{l,n}\\ \gamma_{n,k} &\triangleq \mathbf{f}_k s_k \rho_{k,n}\\ \varepsilon_n &\triangleq \omega(t)*u_n{}^*(T_0-t)|_{t=T_0}\end{aligned} \tag{25}$$

The output of the matched filter in each relay has been processed by the beamforming weights $\mathbf{W}_l \in \mathbb{C}^{R\times Rd}$, which should be designed appropriately. We define the output of the matched filter bank as $\mathbf{\Gamma} = [\gamma_1{}^T, \gamma_2{}^T, \ldots, \gamma_\mathbf{d}{}^T]^T \in \mathbb{C}^{Rd\times 1}$, the adjusted MIMO-relay signals can be written as

$$\xi_l = \mathbf{W}_l\mathbf{\Gamma} \in \mathbb{C}^{R\times 1},\ \ l\in\{1,\ldots,d\} \tag{26}$$

Another filter bank is applied to the output of each MIMO antenna, which generates $R \times d$ filtered data. This data are processed in a processing center in the MIMO relay to achieve the proper symbol vector, which can be transmitted in each user's subspace. After beamforming by the above linear operation, the MIMO-relay transmits the following modulated and precoded signal to destination nodes:

$$\psi(t) = \sum_{l=1}^{d}\xi_l u_l(t) \in \mathbb{C}^{R\times 1} \tag{27}$$

The $r$th entry of $\psi(t)$ is the signal transmitted by $r$th relay antenna. Then, the received signal at the $k^{th}$ destination is given by

$$y_k(t) = \mathbf{g}_k{}^T\psi(t) + \zeta_k(t) \tag{28}$$

where $\zeta_k(t)$ is the noise at the $k$th receiver, which is also assumed to be $\mathbb{C}N(0, 1)$. Finally, each destination node convolves the received signals by its code-word to retrieve its corresponding data. So, the retrieved signal will be

$$\begin{aligned}
\lambda_k &= y_k(t)*u_k{}^*(T_0-t)\big|_{t=T_0} \\
&= \mathbf{g}_k{}^T\sum_{l=1}^{d}\xi_l\,\underbrace{u_l(t)*u_k^*(T_0-t)\big|_{t=T_0}}_{\rho l,k}+\underbrace{\zeta_k(t)*u_k^*(T_0-t)\big|_{t=T_0}}_{\varsigma_k} \\
&= \mathbf{g}_k{}^T\sum_{l=1}^{d}\xi_l\rho_{l,k}+\varsigma_k=\mathbf{g}_k{}^T\sum_{l=1}^{d}\mathbf{W}_l\boldsymbol{\Gamma}\rho_{l,k}+\varsigma_k \\
&= \mathbf{g}_k{}^T\left(\sum_{l=1}^{d}\rho_{l,k}\mathbf{I}_{R\times R}\mathbf{W}_l\right)\boldsymbol{\Gamma}+\varsigma_k=\mathbf{g}_k{}^T(\boldsymbol{\rho}_k\mathbf{W})\boldsymbol{\Gamma}+\varsigma_k \\
&= \mathbf{g}_k{}^T\boldsymbol{\rho}_k\mathbf{W}\left(\boldsymbol{\Gamma}_{\text{-}k}+\boldsymbol{\Gamma}_k+\boldsymbol{\Gamma}_{\varepsilon_n}\right)+\varsigma_k \\
&= \underbrace{\mathbf{g}_k{}^T\boldsymbol{\rho}_k\mathbf{W}\boldsymbol{\Gamma}_k}_{\text{desired received signal}}+\underbrace{\mathbf{g}_k{}^T\boldsymbol{\rho}_k\mathbf{W}\boldsymbol{\Gamma}_{\text{-}k}}_{\text{interference part}}+\underbrace{\mathbf{g}_k{}^T\boldsymbol{\rho}_k\mathbf{W}\boldsymbol{\Gamma}_{\varepsilon_n}+\varsigma_k}_{\text{noise part}}
\end{aligned} \tag{29}$$

where $\varsigma_k$ is the noise at the $k^{\text{th}}$ receiver, and the following notations are defined for simplicity:

$$\begin{aligned}
&\mathbf{r}_k\triangleq\left[\rho_{\mathbf{1,k}}\rho_{\mathbf{2,k}}\cdots\rho_{\mathbf{d,k}}\right]_{1\times d} \\
&\boldsymbol{\rho}_k\triangleq\mathbf{r}_k\otimes I_{R\times R}\in\mathbb{C}^{R\times Rd} \\
&\mathbf{W}\triangleq\left[\mathbf{W}_1{}^T,,\mathbf{W}_2{}^T,\ldots,,\mathbf{W}_d{}^T\right]^T\in\mathbb{C}^{Rd\times Rd} \\
&\boldsymbol{\Gamma}_{\text{-}k}\triangleq\left[\boldsymbol{\gamma}_{1,\text{-}k}{}^T,,\boldsymbol{\gamma}_{2,\text{-}k}{}^T,\ldots,,\boldsymbol{\gamma}_{d,\text{-}k}{}^T\right]^T \\
&\boldsymbol{\Gamma}_k\triangleq\left[\boldsymbol{\gamma}_{1,k}{}^T,,\boldsymbol{\gamma}_{2,k}{}^T,\ldots,,\boldsymbol{\gamma}_{d,k}{}^T\right]^T \\
&\boldsymbol{\Gamma}_{\varepsilon_n}\triangleq\left[\boldsymbol{\varepsilon}_1{}^T,,\boldsymbol{\varepsilon}_2{}^T,\ldots,,\boldsymbol{\varepsilon}_d{}^T\right]^T \\
&\boldsymbol{\Gamma}=\boldsymbol{\Gamma}_{\text{-}\mathbf{k}}+\boldsymbol{\Gamma}_{\mathbf{k}}+\boldsymbol{\Gamma}_{\boldsymbol{\varepsilon}_\mathbf{n}}\in\mathbb{C}^{Rd\times 1}
\end{aligned} \tag{30}$$

The object of the network beamforming is to minimize the total relay transmit power subject to maintaining every destination SINR above a pre-defined threshold value $\gamma_{th}$ (as a QoS parameter of the network).

First, using Eq. (27), the total MIMO-relay transmit power can be obtained as:

$$P_R = E(\langle \psi(t), \psi(t)\rangle) = E\left(\left(\sum_{l=1}^{d} \xi_l u_l(T_0-t)\right)^H * \left(\sum_{n=1}^{d} \xi_n u_n(t)\right)\Bigg|_{t=T_0}\right)$$
$$= E\left(\left(\sum_{l=1}^{d} \mathbf{W}_l u_l(T_0-t)\boldsymbol{\Gamma}\right)^H * \left(\sum_{n=1}^{d} \mathbf{W}_n u_n(t)\boldsymbol{\Gamma}\right)\Bigg|_{t=T_0}\right)$$
$$= E\left(\boldsymbol{\Gamma}^H \sum_{l=1}^{d} \mathbf{W}_l{}^H u_l(T_0-t) * \sum_{n=1}^{d} \mathbf{W}_n u_n(t)\boldsymbol{\Gamma}\Bigg|_{t=T_0}\right) \tag{31}$$
$$= E\left(\boldsymbol{\Gamma}^H \underbrace{\sum_{l=1}^{d}\sum_{n=1}^{d} \mathbf{W}_l{}^H \underbrace{u_l(T_0-t) * u_n(t)}_{\rho_{l,n}}\Bigg|_{t=T_0} \mathbf{W}_n}_{\mathbf{Q}}\boldsymbol{\Gamma}\right) = E\left(\boldsymbol{\Gamma}^H \mathbf{Q}\boldsymbol{\Gamma}\right)$$

where $\mathbf{Q} \triangleq \sum_{l=1}^{d}\sum_{j=1}^{d} \mathbf{W}_l{}^H \rho_{l,j}\mathbf{W}_j$ and the inner product of vectors $\mathbf{x}(t)$, $\mathbf{y}(t)$ is defined as

$$\langle \mathbf{x}(t), \mathbf{y}(t)\rangle \triangleq \int_{-\infty}^{\infty} \mathbf{x}^H(t)\mathbf{y}(t)dt = \mathbf{x}^H(T_0-t) * \mathbf{y}(t)\big|_{t=T_0} \tag{32}$$

For simplicity, Q can be represented by the following quadratic form:

$$\mathrm{Q} = \begin{bmatrix} \mathrm{W}_1 \\ \mathrm{W}_2 \\ \vdots \\ \mathrm{W}_d \end{bmatrix}^H \begin{bmatrix} \rho_{1,1}\mathbf{I}_{R\times R} & \rho_{1,2}\mathbf{I}_{R\times R} & \cdots & \rho_{1,d}\mathbf{I}_{R\times R} \\ \rho_{2,1}\mathbf{I}_{R\times R} & & & \\ \vdots & & \ddots & \rho_{d-1,d}\mathbf{I}_{R\times R} \\ \rho_{d,1}\mathbf{I}_{R\times R} & & \rho_{d,d-1}\mathbf{I}_{R\times R} & \rho_{d,d}\mathbf{I}_{R\times R} \end{bmatrix} \begin{bmatrix} \mathrm{W}_1 \\ \mathrm{W}_2 \\ \vdots \\ \mathrm{W}_d \end{bmatrix} \tag{33}$$

The kernel of the above form can be expressed as a Kronecker products as follows:

$$\mathrm{Q} = \mathbf{W}^H(\Upsilon \otimes \mathbf{I}_{R\times R})_{Rd\times Rd}\mathbf{W} \tag{34}$$

where $\Upsilon \triangleq \begin{bmatrix} \rho_{1,1} & \rho_{1,2} & \cdots & \rho_{1,d} \\ \rho_{2,1} & \ddots & & \vdots \\ \vdots & & & \\ \rho_{d,1} & \cdots & & \rho_{d,d} \end{bmatrix}$. Thus, Eq. (31) can be rewritten as:

$$\begin{aligned} P_R &= E\left(\boldsymbol{\Gamma}^H \left(\mathbf{W}^H(\Upsilon \otimes \mathbf{I}_{R\times R})\mathbf{W}\right)\boldsymbol{\Gamma}\right), \\ &= \mathrm{trace}\left(\mathbf{W}^{\mathbf{H}}(\Upsilon \otimes \mathbf{I}_{R\times R})\mathbf{W}\, E\left(\boldsymbol{\Gamma}\boldsymbol{\Gamma}^{\mathbf{H}}\right)\right), \\ &= \mathrm{vec}(\mathbf{W})^{\mathbf{H}} \underbrace{\left(E\left(\boldsymbol{\Gamma}\boldsymbol{\Gamma}^{\mathbf{H}}\right)^{\mathbf{T}} \otimes (\Upsilon \otimes \mathbf{I}_{R\times R})\right)}_{\mathbf{T}} \mathrm{vec}(\mathbf{W}), \\ &= \mathbf{w}^{\mathbf{H}}\mathbf{T}\mathbf{w} \end{aligned} \tag{35}$$

where $\mathbf{w} \triangleq \mathrm{vec}(\mathbf{W})$ and $\mathbf{T} \triangleq E(\boldsymbol{\Gamma}\boldsymbol{\Gamma}^{\mathbf{H}})^{\mathbf{T}} \otimes (\Upsilon \otimes \mathbf{I}_{R\times R})$.

Also, the instantaneous desired signal power at the $k^{\text{th}}$ destination is calculated as:

$$P_{s_k} = E\left[\mathbf{g_k}^{\mathbf{T}}\boldsymbol{\rho}_{\mathbf{k}}\mathbf{W}\boldsymbol{\Gamma}_{\mathbf{k}}\boldsymbol{\Gamma}_{\mathbf{k}}^{\mathbf{H}}\mathbf{W}^{\mathbf{H}}\boldsymbol{\rho}_{\mathbf{k}}^{\mathbf{T}}\mathbf{g_k}^{*}\right] \tag{36}$$

By defining $\boldsymbol{\Gamma}_{\mathbf{k}} \triangleq \mu_k S_k$ and $\mu_{\mathbf{k}} \triangleq \mathrm{r}_k^T \otimes \mathrm{f}_k$, Eq. (36) can be rewritten as

$$\begin{aligned} P_{s_k} &= P_{\mathrm{k}} E\left[\mathbf{g_k}^{\mathbf{T}}\boldsymbol{\rho}_{\mathbf{k}}\mathbf{W}\,\boldsymbol{\mu}_{\mathbf{k}}\boldsymbol{\mu}_{\mathbf{k}}^{\mathbf{H}}\mathbf{W}^{\mathbf{H}}\boldsymbol{\rho}_{\mathbf{k}}^{\mathbf{T}}\mathbf{g_k}^{*}\right] \\ &= P_{\mathrm{k}}\,\mathrm{trace}\left(\mathbf{W}^{\mathbf{H}}\boldsymbol{\rho}_{\mathbf{k}}^{\mathbf{T}}\underbrace{E\left(\mathbf{g_k}^{*}\mathbf{g_k}^{\mathbf{T}}\right)}_{\mathbf{R}_{\mathbf{g_k}}}\boldsymbol{\rho}_{\mathbf{k}}\mathbf{W}\,\underbrace{E\left(\boldsymbol{\mu}_{\mathbf{k}}\boldsymbol{\mu}_{\mathbf{k}}^{\mathbf{H}}\right)}_{\mathbf{R}_{\mu\mathbf{k}}}\right) \\ &= \mathrm{vec}(\mathbf{W})^{H}\left(\mathbf{R}_{\boldsymbol{\mu}_{\mathbf{k}}}^{\mathbf{T}}\otimes P_{\mathrm{k}}\left(\boldsymbol{\rho}_{\mathbf{k}}^{\mathbf{T}}\mathbf{R}_{\mathbf{g_k}}\boldsymbol{\rho}_{\mathbf{k}}\right)\right)\mathrm{vec}(\mathbf{W}) \\ &= \mathrm{vec}(\mathbf{W})^{H}\underbrace{\left(\mathbf{R}_{\boldsymbol{\mu}_{\mathbf{k}}}^{\mathbf{T}}\otimes\left(P_{\mathrm{k}}\boldsymbol{\tau}_{\mathbf{k}}\right)\right)}_{\mathbf{R}_k}\mathrm{vec}(\mathbf{W}) = \mathbf{w}^{H}\mathbf{R}_k\mathbf{w} \end{aligned} \tag{37}$$

where $\mathbf{R}_{\mu_k}$, $\mathbf{R}_{\mathbf{g}_k}$, $\boldsymbol{\tau}_{\mathbf{k}}$ and $\mathbf{R}_k$ are defined as

$\mathbf{R}_{\mu_k} \triangleq E\left(\boldsymbol{\mu}_{\mathrm{k}}\boldsymbol{\mu}_{\mathrm{k}}^{\mathrm{H}}\right)$, $\mathbf{R}_{\mathbf{g}_k} \triangleq E\left(\mathbf{g}_{\mathrm{k}}^{*}\mathbf{g}_{\mathrm{k}}^{\mathrm{T}}\right)$, $\boldsymbol{\tau}_{\mathrm{k}} \triangleq \boldsymbol{\rho}_{\mathrm{k}}^{\mathrm{T}}\mathbf{R}_{\mathbf{g}_k}\boldsymbol{\rho}_{\mathrm{k}}$ and $\mathbf{R}_k \triangleq \mathbf{R}_{\mu_{\mathbf{k}}}^{\mathbf{T}} \otimes P_k\boldsymbol{\tau}_{\mathbf{k}}$

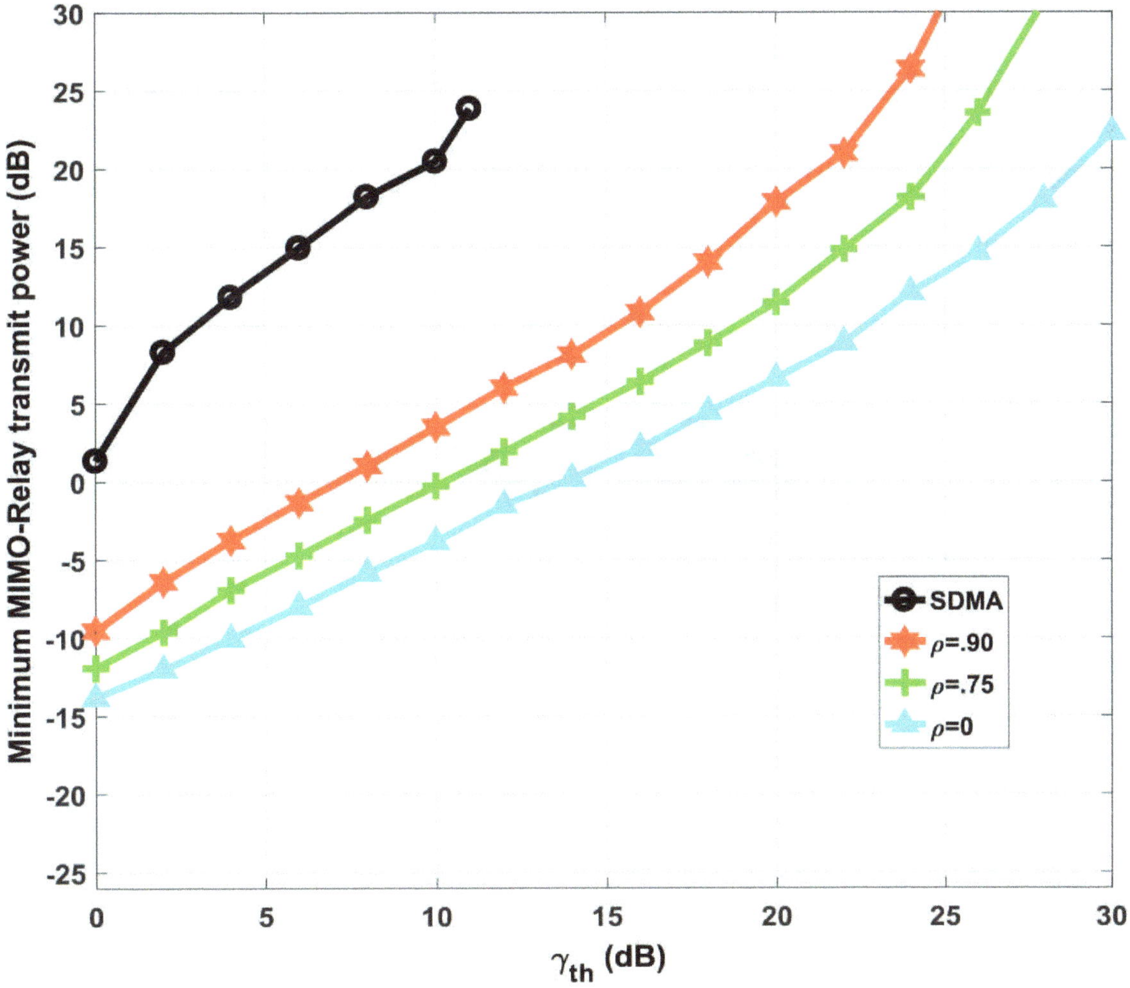


**Figure 7.** Minimum MIMO-relay transmit power $P_T^{\min}$ versus $\gamma_{th}$, for $R$=4, $u$=2.

Also, the received noise power at $k^{\text{th}}$ destination is given by:

$$\begin{aligned} P_{N_k} &= E\left[\mathbf{g_k}^{\mathbf{T}}\boldsymbol{\rho}_{\mathbf{k}}\mathbf{W}\boldsymbol{\Gamma}_{\varepsilon_n}\boldsymbol{\Gamma}_{\varepsilon_n}^{\mathbf{H}}\mathbf{W}^{\mathbf{H}}\boldsymbol{\rho}_{\mathbf{k}}{}^{\mathbf{T}}\mathbf{g_k}^{*}\right] + \sigma_{\varsigma_k}^2 \\ &= \text{trace}\left(\mathbf{W}^{\mathbf{H}}\boldsymbol{\rho}_{\mathbf{k}}{}^{\mathbf{T}}\mathbf{R}_{\mathbf{g_k}}\boldsymbol{\rho}_{\mathbf{k}}\mathbf{W}E\left(\boldsymbol{\Gamma}_{\varepsilon_n}\boldsymbol{\Gamma}_{\varepsilon_n}^{\mathbf{H}}\right)\right) + \sigma_{\varsigma_k}^2 P_{N_k} \\ &= \text{vec}(\mathbf{W})^H\left(\left(E\left(\boldsymbol{\Gamma}_{\varepsilon_n}\boldsymbol{\Gamma}_{\varepsilon_n}^{\mathbf{H}}\right)\right)^T \otimes \boldsymbol{\tau}_k\right)\text{vec}(\mathbf{W}) + \sigma_{\varsigma_k}^2 \\ &= \mathbf{w}^H\mathbf{N}_k\mathbf{w} + \sigma_{\varsigma_k}^2 \end{aligned} \tag{38}$$

where $\mathbf{N}_k \triangleq \left(E\left(\boldsymbol{\Gamma}_{\varepsilon_n}\boldsymbol{\Gamma}_{\varepsilon_n}^{\mathbf{H}}\right)\right)^T \otimes \boldsymbol{\tau}_k$. Also, it can be easily proved that:

$$E\left(\boldsymbol{\Gamma}_{\varepsilon_n}\boldsymbol{\Gamma}_{\varepsilon_n}^{H}\right) = \left[\int_{-\infty}^{\infty}\left(\mathbf{u}^*(t)\mathbf{u}^T(t)\right)dt\right] \otimes \sigma_{\omega}^2\mathbf{I}_{R\times R} \in \mathbb{C}^{Rd\times Rd} \tag{39}$$

Finally, the power of the received interference at the $k^{\text{th}}$ destination can be computed as

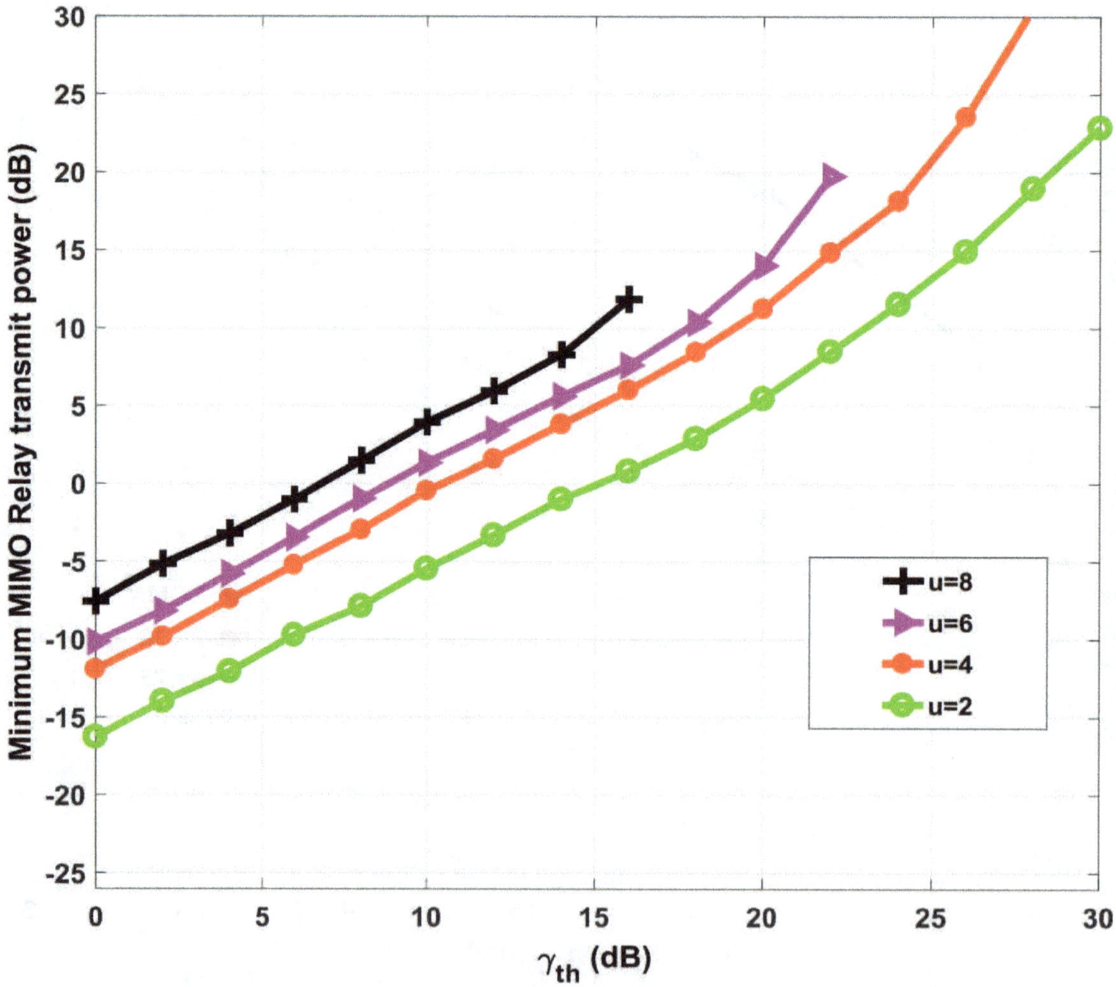


**Figure 8.** Minimum relay transmit power $P_T^{\min}$ versus $\gamma_{th}$, for $R$=2 and $\rho_{l,m}$ = 0.75.

$$\begin{aligned} P_{I_k} &= E\left[\mathbf{g_k}^{\mathbf{T}}\boldsymbol{\rho}_{\mathbf{k}}\mathbf{W}\boldsymbol{\Gamma}_{\text{-k}}\boldsymbol{\Gamma}_{\text{-k}}^{\mathbf{H}}\mathbf{W}^{\mathbf{H}}\boldsymbol{\rho}_{\mathbf{k}}^{\mathbf{T}}\mathbf{g_k}^{*}\right] \\ &= \text{trace}\left[\mathbf{W}^{\mathbf{H}}\boldsymbol{\rho}_{\mathbf{k}}^{\mathbf{T}}E\left(\mathbf{g_k}^{*}\mathbf{g_k}^{\mathbf{T}}\right)\boldsymbol{\rho}_{\mathbf{k}}\mathbf{W}\,E\left(\boldsymbol{\Gamma}_{\text{-k}}\boldsymbol{\Gamma}_{\text{-k}}^{\mathbf{H}}\right)\right] \\ &= \text{trace}\left[\mathbf{W}^{\mathbf{H}}\boldsymbol{\rho}_{\mathbf{k}}^{\mathbf{T}}\mathbf{R}_{\mathbf{g_k}}\boldsymbol{\rho}_{\mathbf{k}}\mathbf{W}\,E\left(\boldsymbol{\Gamma}_{\text{-k}}\boldsymbol{\Gamma}_{\text{-k}}^{\mathbf{H}}\right)\right] \\ &= \text{vec}(\mathbf{W})^{H}\underbrace{\left(\left(E\left(\boldsymbol{\Gamma}_{\text{-k}}\boldsymbol{\Gamma}_{\text{-k}}^{\mathbf{H}}\right)\right)^{T}\otimes\boldsymbol{\tau}_{\mathbf{k}}\right)}_{\mathbf{I}_{\mathbf{k}}}\text{vec}(\mathbf{W}) \\ &= \mathbf{w}^{H}\mathbf{I}_{\mathbf{k}}\mathbf{w} \end{aligned} \tag{40}$$

The instantaneous SINR for $k^{\text{th}}$ destination simply becomes the desired signal power of the desired signal to the power of interference plus noise. So, the optimization problem can now be written as

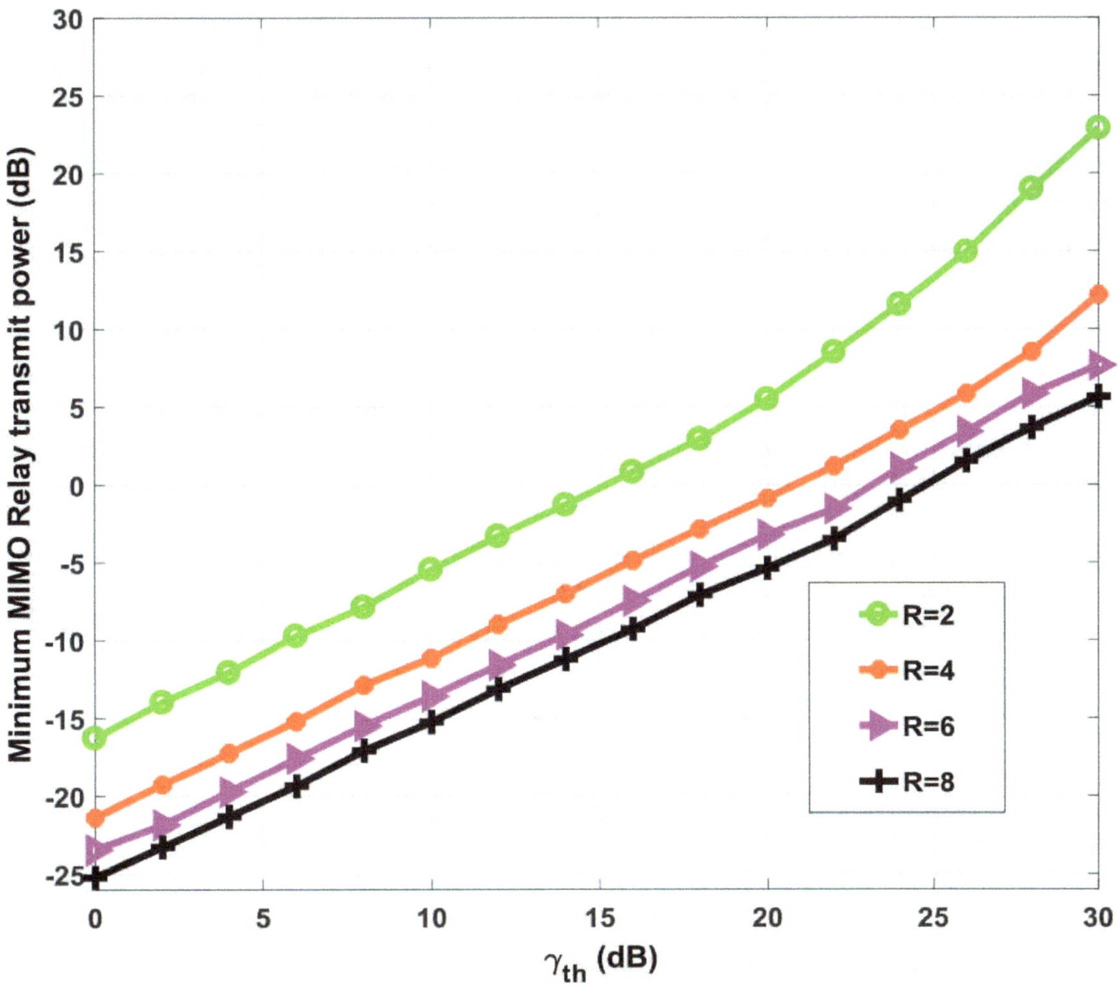


**Figure 9.** Minimum relay transmit power $P_T^{\min}$ versus $\gamma_{th}$, for $u$=2 and $\rho_{l,m} = 0.75$.

$$\begin{aligned} &\underset{\mathbf{w}}{\text{Minimize}}\ \mathbf{w}^H\mathbf{T}\mathbf{w} \\ &\text{Subject to}\ \text{SINR}_k = \frac{\mathbf{w}^H\mathbf{R}_k\mathbf{w}}{\mathbf{w}^H(\mathbf{N}_k+\mathbf{I}_k)\mathbf{w}+\sigma_{\varsigma_k}^2} \geq \gamma_{th}^k \\ &k \in \{1,2,\dots,d\} \end{aligned} \tag{41}$$

By defining $\mathbf{D}_k \triangleq \mathbf{R}_k - \gamma_{th}^k(\mathbf{N}_k+\mathbf{I}_k)$ , $\mathbf{X} \triangleq \mathbf{w}\mathbf{w}^H$, the optimization problem can recast to

$$\begin{aligned} &\underset{\mathbf{X}}{\text{Minimize}}\ \text{trace}(\mathbf{T}\mathbf{X}) \\ &\text{Subject to}\ \text{trace}(\mathbf{D}_k\mathbf{X}) \geq \gamma_{th}^k\sigma_{\varsigma_k}^2, \qquad k \in \{1,\dots,d\} \\ &\text{Rank}(\mathbf{X}) = 1\ , \mathbf{X} \geq 0 \end{aligned} \tag{42}$$

We solve this optimization problem in a same way as the previous section. The first simulation scenario was carried out to consider the total MIMO-relay transmit power versus destination SINR threshold value, for different values of users' correlation factors. The averaged results are shown in **Figure 7**. The network consists of two source-destination pairs and four MIMO-relay antennas. **Figure 7** shows that the total MIMO-relay transmit power in all cases increases by raising $\gamma_{th}$. Furthermore, **Figure 7** indicates that when the signature sequence correlation $\rho_{k,l}$ increases, more total transmit power is needed to ensure SINR constraints at destination nodes.

When $\rho_{k,l}$ approaching one, the problem downgrades to the SDMA network and the system loses the benefits of CDMA technique. Also, increasing the signal dependency by increasing the correlation factor results in the more infeasibility rate of the constraints. Therefore, when the correlation factor increases from 0 to 0.75, there is little difference between the curves, but when $\rho_{k,l}$ increases beyond 0.75, it can be seen that the difference becomes considerably larger. As a result, a large power gain can be achieved when moving from $\rho_{k,l} = 1$, by a small reduction of $\rho_{k,l}$. To study the effect of the number of relay nodes and the number of source-destination pairs in terms of quality of matched filter output, we have examined a network with $\rho_{k,l} = 0.75$.

**Figures 8** and **9** display the minimum relay transmit power versus $\gamma_{th}$, for different number of MIMO-relay antenna and different number of user pairs. As normally expected, more power saving can be achieved by increasing the number of relays or decreasing the number of users. Comparing **Figures 8** and **9** with **Figure 7** reveals that decreasing the correlation factor will be much more efficient for saving network power than increasing the number of relays.

## 5. Distributed relay networks

In this section, we considered a distributed relays network, instead of MIMO-relay. The optimization problem is defined to find the relay beamforming coefficients that minimize the total relay transmit power by keeping the SINR of all destinations above a certain threshold value.

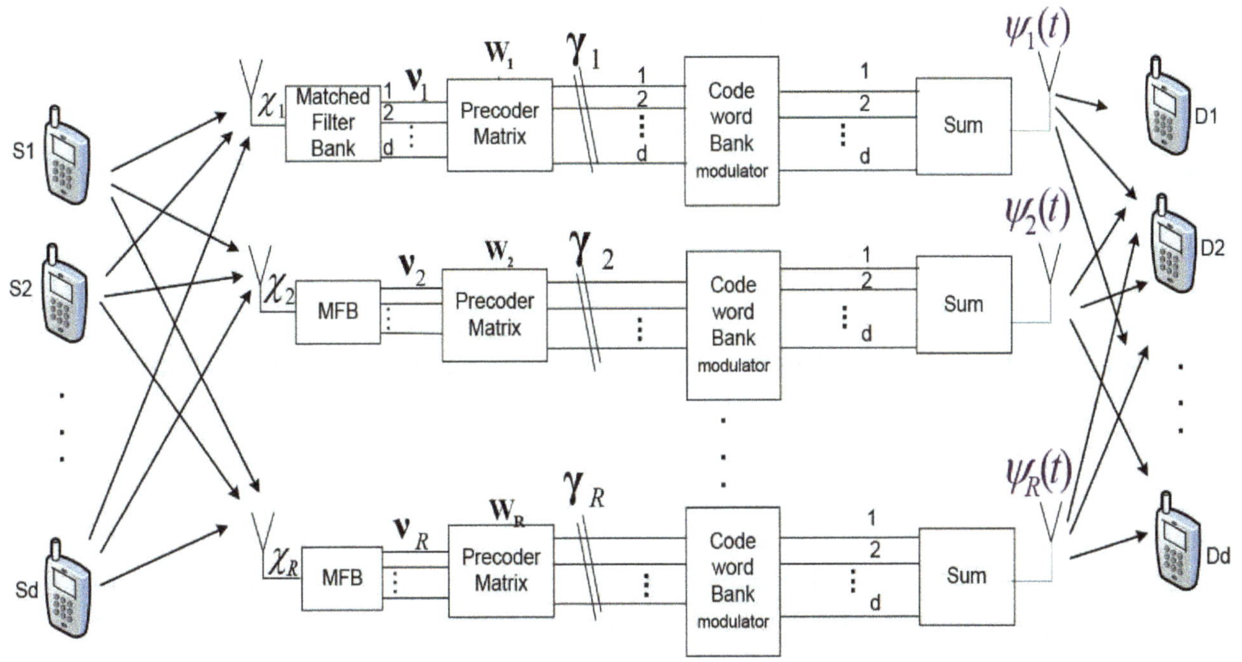


**Figure 10.** Distributed relay network.

Consider a wireless relay network with $d$ pairs of source-destination (peers) communicating without a direct link through $R$ single relay antennas, as shown in **Figure 10**. A two-step AF protocol is used. In the first step, each source user broadcasts its spread symbol toward the relays. A matched filter is applied in each relay in order to retrieve the source's signals. In the second step, the adjusted and spread signals by the relays are transmitted to destinations. Another matched filter is used at each destination to extract its corresponding symbols. Let $s_k$ stands for the $k^{th}$ source symbol that is assumed to be independent of the other sources, that is, $E\{s_k s_l^*\} = P_k\delta_{kl}$ and $u_k(t)$ denotes a signature waveform that is assigned to the $k^{th}$ source. Then, the received signal at the $r^{th}$ relay is given by:

$$\chi_r(t) = \sum_{l=1}^{d} f_{rl} s_l u_l(t) + \omega_r(t)\ , r\in\{1, \ldots, R\} \tag{43}$$

where $\omega_r(t)$ is the noise at the $r$th relay. By denoting the cross correlation between $k^{th}$ user's codeword to the $l^{th}$ user's codeword as $\rho_{k,l} = u_k(t)*u_l(T_0-t)|_{t=T_0}$, the output signal of the matched filter at the $r^{th}$ relay can be expressed as

$$\begin{aligned} \mathbf{v}_r &= \chi_r(t)*\mathbf{u}(T_0-t)|_{t=T_0} \\ &= \sum_{k=1}^{d} f_{rk} s_k \boldsymbol{\rho}_k + \boldsymbol{\varsigma}_r = \mathbf{v}_{r,k} + \mathbf{v}_{r,\text{-}k} + \mathbf{n}_r,\ \ r\in\{1, \ldots, R\} \end{aligned} \tag{44}$$

where the following definitions have been used:

$$\mathbf{u}(t) \triangleq [u_1(t), \ldots, u_d(t)]^T$$

$$\mathbf{n}_r \triangleq \omega_r(t) * \mathbf{u}(T_0 - t)\big|_{t=T_0}$$

$$\boldsymbol{\rho}_k \triangleq u_k(t) * \mathbf{u}(T_0 - t)\Big|_{t=T_0} = \left[\rho_{k,1}, \ldots, \rho_{k,d}\right]^T,$$

$$\mathbf{v}_{r,k} \triangleq f_{rk} s_k \boldsymbol{\rho}_k \ , \ \ \mathbf{v}_{r,-k} \triangleq \sum_{l=1,\, l \neq k}^{d} f_{rl} s_l \boldsymbol{\rho}_l \tag{45}$$

The output of the matched filter in each relay has been processed by the beamforming weights $\mathbf{W}_r \in \mathcal{C}^{d \times d}$, which should be designed appropriately. So, it can be expressed as

$$\gamma_r = \mathbf{W}_r \mathbf{v}_r \in \mathbb{C}^{d \times 1}, \ \ r \in \{1, \ldots, R\} \tag{46}$$

Another filter bank is applied to the output of each relay, which generates $d$ filtered data. These data are processed in the relay in order to achieve the proper symbol vector, which can be transmitted in each user's signal subspace. After beamforming by the above linear operation, the $r^{\text{th}}$ relay transmits the following modulated and precoded signal by a CDMA technique

$$\psi_r(t) = \gamma_r{}^T \mathbf{u}(t) \ , r \in \{1, \ldots, R\} \tag{47}$$

The vector forms of Eq. (47) can be written as

$$\begin{aligned} \boldsymbol{\psi}(t) &= \left[\psi_1(t), \psi_2(t), \ldots, \psi_R(t)\right]^T \\ &= \left[\gamma_1, \ldots, \gamma_R\right]^T \mathbf{u}(t) \\ &= \left[\mathbf{W}_1 \mathbf{v}_1, \ldots, \mathbf{W}_R \mathbf{v}_R\right]^T \mathbf{u}(t) \\ &= (\mathbf{W}\mathbf{H})^T \mathbf{u}(t) \end{aligned} \tag{48}$$

The $r^{th}$ entry of $\boldsymbol{\psi}(t)$ is the signal transmitted by $r^{th}$ relay and $\mathbf{W} \triangleq [\mathbf{W}_1, \ldots, \mathbf{W}_R] \in \mathcal{C}^{d \times Rd}$, $\mathbf{H} \triangleq \text{BD}(\mathbf{v}_1, \ldots, \mathbf{v}_R) \in \mathcal{C}^{Rd \times R}$, where BD(·) denotes the block diagonalization of matrices. Thus, the total received signal at the $k^{th}$ destination is given by

$$y_k(t) = \mathbf{g}_k{}^T \boldsymbol{\psi}(t) + \zeta_k(t) \tag{49}$$

where $\zeta_k(t)$ is the noise at the $k^{th}$ receiver and $\mathbf{g}_k \triangleq [g_{1k} g_{2k} \ldots g_{Rk}]^T$ is the vector of downlink channel coefficients. Finally, each destination node convolves the received signals by its codeword to retrieve its corresponding data. So, the retrieved signal will be

$$\begin{aligned}
\eta_k &= y_k(t) * u_k(T_0 - t)\big|_{t=T_0} \\
&= \mathbf{g}_k{}^T (\mathbf{WH})^T \mathbf{u}(t) * u_k(T_0 - t)\Big|_{t=T_0} + \zeta_k(t) * u_k(T_0 - t)\big|_{t=T_0} \\
&= \mathbf{g}_k{}^T \mathbf{H}^T \mathbf{W}^T \boldsymbol{\rho}_k + \varsigma_k \\
&= \underbrace{\mathbf{g}_k{}^T \mathbf{H}_k{}^T \mathbf{W}^T \boldsymbol{\rho}_k}_{\text{desired signal}} + \underbrace{\mathbf{g}_k{}^T \mathbf{H}_{-k}{}^T \mathbf{W}^T \boldsymbol{\rho}_k}_{\text{interference part}} + \underbrace{\mathbf{g}_k{}^T \mathbf{H}_{\mathbf{n}_k}{}^T \mathbf{W}^T \boldsymbol{\rho}_k + \varsigma_k}_{\text{noise part}}
\end{aligned} \tag{50}$$

where:

$$\begin{aligned}
&\varsigma_k \triangleq \zeta_k(t) * u_k(T_0 - t)\big|_{t=T_0} \\
&\mathbf{H}_k \triangleq \mathrm{BD}(\boldsymbol{\nu}_{1,k}, \ldots, \boldsymbol{\nu}_{R,k}) \\
&\mathbf{H}_{-k} \triangleq \mathrm{BD}(\boldsymbol{\nu}_{1,-k}, \ldots, \boldsymbol{\nu}_{R,-k}) \\
&\mathbf{H}_{\varsigma_n} \triangleq \mathrm{BD}(\boldsymbol{\varsigma}_1, \boldsymbol{\varsigma}_2, \ldots, \boldsymbol{\varsigma}_R) \\
&\mathbf{H} = \mathbf{H}_{-k} + \mathbf{H}_k + \mathbf{H}_{\varsigma_n}
\end{aligned} \tag{51}$$

The three last terms of Eq. (50) are the desired received signal, interference and noise at the $k^{th}$ destination, respectively.

The object of the network beamforming is to minimize the total relay transmit power subject to maintaining every destination SINR above a pre-defined threshold value $\gamma_{th}$ (as a QoS parameter of the network). First, using Eq. (48) the total relay transmit power can be obtained as

$$\begin{aligned}
P_R &= E\langle \boldsymbol{\psi}(t), \boldsymbol{\psi}(t) \rangle = E\left[ \left( (\mathbf{WH})^T \mathbf{u}(T_0 - t) \right)^T * (\mathbf{WH})^T \mathbf{u}(t) \Big|_{t=T_0} \right] \\
&= E\left[ \int_{-\infty}^{\infty} \left( \mathbf{u}(t)^T \mathbf{W}^* \mathbf{H}^* \right) \left( \mathbf{H}^T \mathbf{W}^T \mathbf{u}(t) \right) dt \right] \\
&= E\left[ \mathrm{Tr}\left( \mathbf{W}^* \mathbf{H}^* \mathbf{H}^T \mathbf{W}^T \int_{-\infty}^{\infty} \mathbf{u}(t) \mathbf{u}(t)^T dt \right) \right] \\
&= \mathrm{trace}\left\{ E\left( \mathbf{H}^* \mathbf{H}^T \right) \mathbf{W}^T \boldsymbol{\mu} \mathbf{W}^* \right\} \\
&= \mathrm{vec}(\mathbf{W})^T \left( E\left( \mathbf{H}^* \mathbf{H}^T \right)^T \otimes \boldsymbol{\mu} \right) \mathrm{vec}(\mathbf{W}^*) \\
&= \mathbf{w}^T \mathbf{T} \mathbf{w}^*
\end{aligned} \tag{52}$$

where

$$\mathbf{T} \triangleq E\left(\mathbf{H}^{*}\mathbf{H}^{T}\right)^{T} \otimes \boldsymbol{\mu},$$
$$\boldsymbol{\mu} \triangleq \mathbf{u}(t) * \mathbf{u}(T_0 - t)^{T}\Big|_{t=T_0} = \int_{-\infty}^{\infty} \mathbf{u}(t)\mathbf{u}^{T}(t)dt = \begin{bmatrix} \rho_{1,1} & \cdots & \rho_{1,d} \\ \vdots & \ddots & \vdots \\ \rho_{d,1} & \cdots & \rho_{d,d} \end{bmatrix} \tag{53}$$

Note that using Eqs. (48) and (44), $E(\mathbf{H}^{*}\mathbf{H}^{T})$ can be obtained as

$$E\left(\mathbf{H}^{*}\mathbf{H}^{T}\right) = \mathrm{BD}\left(E\left(\mathbf{v}_1{}^{*}\mathbf{v}_1{}^{T}\right), \ldots, E\left(\mathbf{v}_R{}^{*}\mathbf{v}_R{}^{T}\right)\right),$$
$$E\left(\mathbf{v}_r{}^{*}\mathbf{v}_r{}^{T}\right) = E\left[\left(\sum_{k=1}^{d} f_{rk}{}^{*} s_k{}^{*} \boldsymbol{\rho}_k{}^{*} + \mathbf{n}_r{}^{*}\right)\left(\sum_{k=1}^{d} f_{rk} s_k \boldsymbol{\rho}_k{}^{T} + \mathbf{n}_r{}^{T}\right)\right]$$
$$= P_k E\left(\sum_{k=1}^{d} f_{rk}{}^{*} f_{rk} \boldsymbol{\rho}_k{}^{*} \boldsymbol{\rho}_k{}^{T}\right) E\left(\mathbf{n}_r{}^{*}\mathbf{n}_r{}^{T}\right),$$
$$E\left(\mathbf{n}_r{}^{*}\mathbf{n}_r{}^{T}\right) = E\left[\left(\int_{-\infty}^{\infty} \omega_r{}^{*}(t)\mathbf{u}(t)dt\right)\left(\int_{-\infty}^{\infty} \omega_r{}^{*}(t)\mathbf{u}^{T}(t)dt\right)\right] \tag{54}$$
$$= E\left[\omega_r{}^{*}(t)\omega_r(t)\right]\left(\int_{-\infty}^{\infty} \mathbf{u}(t)\mathbf{u}^{T}(t)dt\right),$$
$$E\left[\omega_r{}^{*}(t)\omega_r(t)\right] = \sigma_{\omega}^{2},$$
$$\left(\int_{-\infty}^{\infty} \mathbf{u}(t)\mathbf{u}^{T}(t)dt\right) = \boldsymbol{\mu}$$

Using Eq. (50), the desired signal power at the $k^{th}$ destination can be obtained as

$$P_{S_k} = E\left(\boldsymbol{\rho}_k{}^{T}\mathbf{W}^{*}\,\mathbf{H}_k^{*}\,\mathbf{g}_k^{*}\,\mathbf{g}_k{}^{T}\,\mathbf{H}_k^{T}\mathbf{W}^{T}\boldsymbol{\rho}_k\right)$$
$$= \mathrm{trace}\left(E\left(\mathbf{H}_k^{*}\,\mathbf{g}_k^{*}\,\mathbf{g}_k{}^{T}\,\mathbf{H}_k^{T}\right)\mathbf{W}^{T}\boldsymbol{\rho}_k\boldsymbol{\rho}_k{}^{T}\mathbf{W}^{*}\right) \tag{55}$$
$$= \mathrm{vec}(\mathbf{W})^{T}\left(\boldsymbol{\tau}_k{}^{T} \otimes \boldsymbol{\rho}_k\boldsymbol{\rho}_k{}^{T}\right)\ \mathrm{vec}(\mathbf{W}^{*}) = \mathbf{w}^{T}\mathbf{R}_k\mathbf{w}^{*}$$

where $\boldsymbol{\tau}_k \triangleq E\left(\mathbf{H}_k^{*}\,\mathbf{g}_k^{*}\,\mathbf{g}_k{}^{T}\,\mathbf{H}_k^{T}\right) = P_k(\mathbf{F}_k \odot \mathbf{G}_k) \otimes \boldsymbol{\rho}_k\boldsymbol{\rho}_k^{T}$, $\mathbf{F}_k \triangleq \mathbf{f}_k^{*}\mathbf{f}_k^{T}$, $\mathbf{G}_k \triangleq \mathbf{g}_k^{*}\mathbf{g}_k^{T}$ and $\mathbf{f}_k \triangleq [f_{1k} \ldots f_{Rk}]^{T}$, $\mathbf{R}_k \triangleq \boldsymbol{\tau}_k{}^{T} \otimes \boldsymbol{\rho}_k\boldsymbol{\rho}_k{}^{T}$. Also, the received noise power at $k^{th}$ destination is given by

$$P_{N_k} = E\left(\boldsymbol{\rho}_k{}^{T}\mathbf{W}^{*}\mathbf{H}_{\mathbf{n}_k}{}^{*}\mathbf{g}_k{}^{*}\mathbf{g}_k{}^{T}\,\mathbf{H}_{\mathbf{n}_k}{}^{T}\mathbf{W}^{T}\boldsymbol{\rho}_k\right) + \sigma_{\varsigma_k}^{2}$$
$$= \mathrm{trace}\left\{E\left(\mathbf{H}_{\mathbf{n}_k}{}^{*}\mathbf{g}_k{}^{*}\mathbf{g}_k{}^{T}\,\mathbf{H}_{\mathbf{n}_k}{}^{T}\right)\mathbf{W}^{T}\boldsymbol{\rho}_k\boldsymbol{\rho}_k{}^{T}\mathbf{W}^{*}\right\} + \sigma_{\varsigma_k}^{2} \tag{56}$$
$$= \mathbf{w}^{T}\mathbf{N}_k\mathbf{w}^{*} + \sigma_{\varsigma_k}^{2}$$

where

$$\mathbf{N}_k \triangleq \Upsilon_k{}^{T} \otimes \boldsymbol{\rho}_k\boldsymbol{\rho}_k{}^{T},$$
$$\Upsilon_k \triangleq E\left(\mathbf{H}_{\mathbf{n}_k}{}^{*}\mathbf{g}_k{}^{*}\mathbf{g}_k{}^{T}\,\mathbf{H}_{\mathbf{n}_k}{}^{T}\right) = E\left(\mathbf{H}_{\mathbf{n}_k}{}^{*}\mathbf{G}_k\,\mathbf{H}_{\mathbf{n}_k}{}^{T}\right) = \sigma_{\omega}^{2}\mathbf{G}_k \otimes \boldsymbol{\mu} \tag{57}$$

The relay noises are assumed to be zero-mean and independent with the equal noise power. So, we have

$$\int_{-\infty}^{\infty} E\left(\boldsymbol{\omega}(t)\boldsymbol{\omega}^{H}(t)\right)dt = \sigma_{\omega}^{2} I_{R\times R},$$
$$\boldsymbol{\omega}(t) \triangleq [\omega_1(t), \ldots, \omega_R(t)]^{T} \tag{58}$$

Finally, the power of the received interference at the $k^{\text{th}}$ destination can be computed as

$$\begin{aligned} P_{I_k} &= E\left(\boldsymbol{\rho}_k{}^{T}\mathbf{W}^{*}\mathbf{H}_{\text{-}k}{}^{*}\mathbf{g}_k{}^{*}\mathbf{g}_k{}^{T}\mathbf{H}_{\text{-}k}{}^{T}\mathbf{W}^{T}\boldsymbol{\rho}_k\right) \\ &= \text{trace}\left\{E\left(\mathbf{H}_{\text{-}k}{}^{*}\mathbf{g}_k{}^{*}\mathbf{g}_k{}^{T}\mathbf{H}_{\text{-}k}{}^{T}\right)\mathbf{W}^{T}\boldsymbol{\rho}_k\boldsymbol{\rho}_k{}^{T}\mathbf{W}^{*}\right\} \\ &= \text{vec}(\mathbf{W})^{T}\left(\boldsymbol{\theta}_k{}^{T}\otimes\boldsymbol{\rho}_k\boldsymbol{\rho}_k{}^{T}\right)\text{vec}(\mathbf{W}^{*}) = \mathbf{w}^{T}\mathbf{I}_k\mathbf{w}^{*} \end{aligned} \tag{59}$$

where $\mathbf{I}_k \triangleq \boldsymbol{\theta}_k{}^{T} \otimes \boldsymbol{\rho}_k\boldsymbol{\rho}_k{}^{T}$ and

$$\boldsymbol{\theta}_k \triangleq E\left(\mathbf{H}_{\text{-}k}{}^{*}\mathbf{g}_k{}^{*}\mathbf{g}_k{}^{T}\mathbf{H}_{\text{-}k}{}^{T}\right) = \sum_{l=1,\, l\neq k}^{d}\left((\mathbf{F}_l \odot \mathbf{G}_l)\otimes\left(\boldsymbol{\rho}_l\boldsymbol{\rho}_l{}^{T}\right)\right)P_l \tag{60}$$

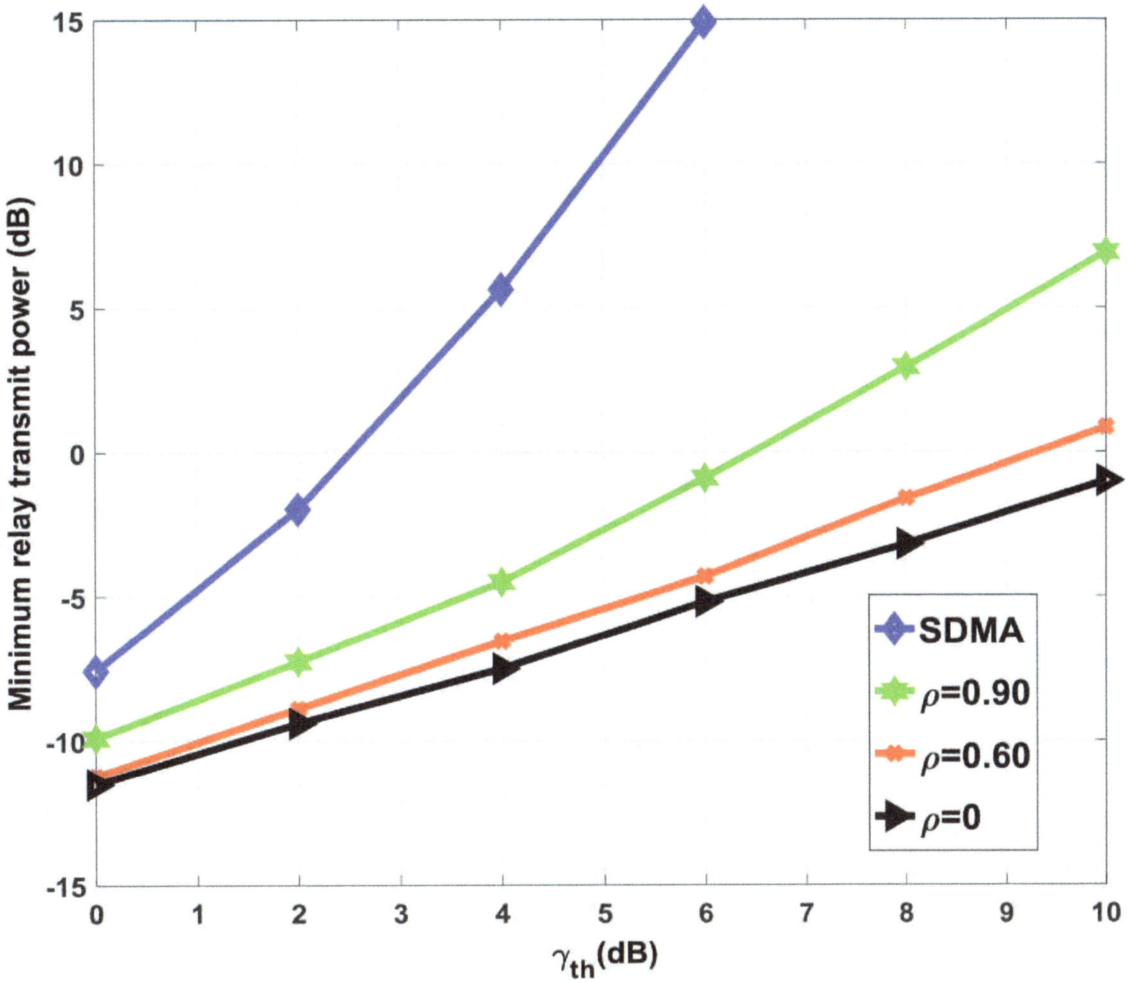


**Figure 11.** Minimum relay transmit power $P_T^{\min}$ versus $\gamma_{th}$, for $R$=4, $u$=2.

In this case, the instantaneous SINR for $k^{th}$ destination simply becomes the desired signal power of the desired signal to the power of interference plus noise. So, the optimization problem can now be written as

$$\begin{aligned} &\underset{\mathbf{w}}{\text{Minimize}}\ \mathbf{w}^T\mathbf{T}\mathbf{w}^* \\ &\text{Subject to}\ \text{SINR}_k = \frac{\mathbf{w}^T\mathbf{R}_k\mathbf{w}^*}{\mathbf{w}^T(\mathbf{N}_k+\mathbf{I}_k)\mathbf{w}^*+\sigma^2_{\varsigma_k}} \geq \gamma^k_{th} \\ &k \in \{1,2,\ldots,d\} \end{aligned} \tag{61}$$

Since $\mathbf{w}^T(\mathbf{N}_k+\mathbf{I}_k)\mathbf{w}^*+\sigma^2_{\varsigma_k} \geq 0$, the constraints of the optimization problem can be formulated as

$$\mathbf{w}^T\left(\mathbf{R}_k - \gamma^k_{th}(\mathbf{N}_k+\mathbf{I}_k)\right)\mathbf{w}^* \geq \gamma^k_{th}\sigma^2_{\varsigma_k} \tag{62}$$

In this problem, if all the matrices $\mathbf{R}_k - \gamma^k_{th}(\mathbf{N}_k+\mathbf{I}_k)$ are negative semi-definite for all $k$, the problem is convex and can be solved uniquely. However, the feasible set of our optimization

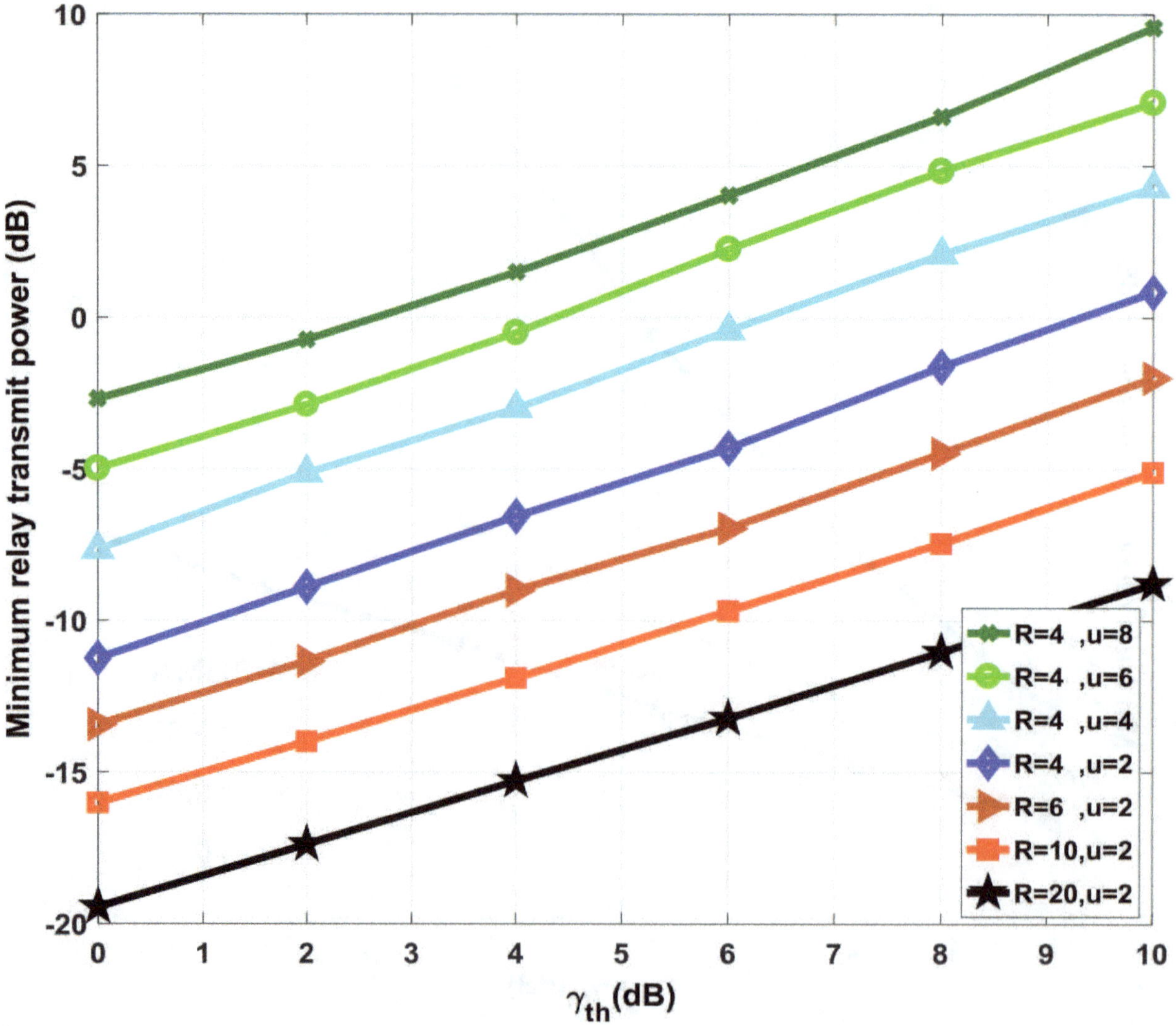


**Figure 12.** Minimum relay transmit power $P_T^{\min}$ versus $\gamma_{th}$.

problem is empty since $\mathbf{w}^T\left(\mathbf{R}_k-\gamma_{th}^k(\mathbf{N}_k+\mathbf{I}_k)\right)\mathbf{w}^*\leq 0$ for all $K$ and $W$. Therefore, $\mathbf{R}_k-\gamma_{th}^k(\mathbf{N}_k+\mathbf{I}_k)$ is non-negative definite matrix which results in non-convex inequality constraints, hence the QCQP problem is non-convex and NP-hard in general. However, we will show that a simple near optimal solution can be found in our problem. First, we replaced our QCQP problem with a semi-definite programming (SDP) problem. Let us define $\mathbf{D}_k \triangleq \mathbf{R}_k-\gamma_{th}^k(\mathbf{N}_k+\mathbf{I}_k)$ , $\mathbf{X}\triangleq\mathbf{w}^*\mathbf{w}^T$, the optimization problem can recast to

$$\begin{gathered}\underset{\mathbf{X}}{\text{Minimize}}\ \text{trace}\ (\mathbf{TX}) \\ \text{Subject to trace}\ (\mathbf{D}_k\mathbf{X}) \geq \gamma_{th}^k\sigma_{\varsigma_k}^2, k\in\{1,\ldots,d\} \\ \text{Rank}\ (\mathbf{X}) = 1\ , \mathbf{X}\succeq 0. \end{gathered} \tag{63}$$

The problem is non-convex, because the Rank($\mathbf{X}$) = 1 constraint is non-convex. We relax the problem by ignoring this non-convex constraint and convert it to a convex SDP problem. The following semi definite representation (SDR) form is the relaxed version of the problem (Eq. 63).

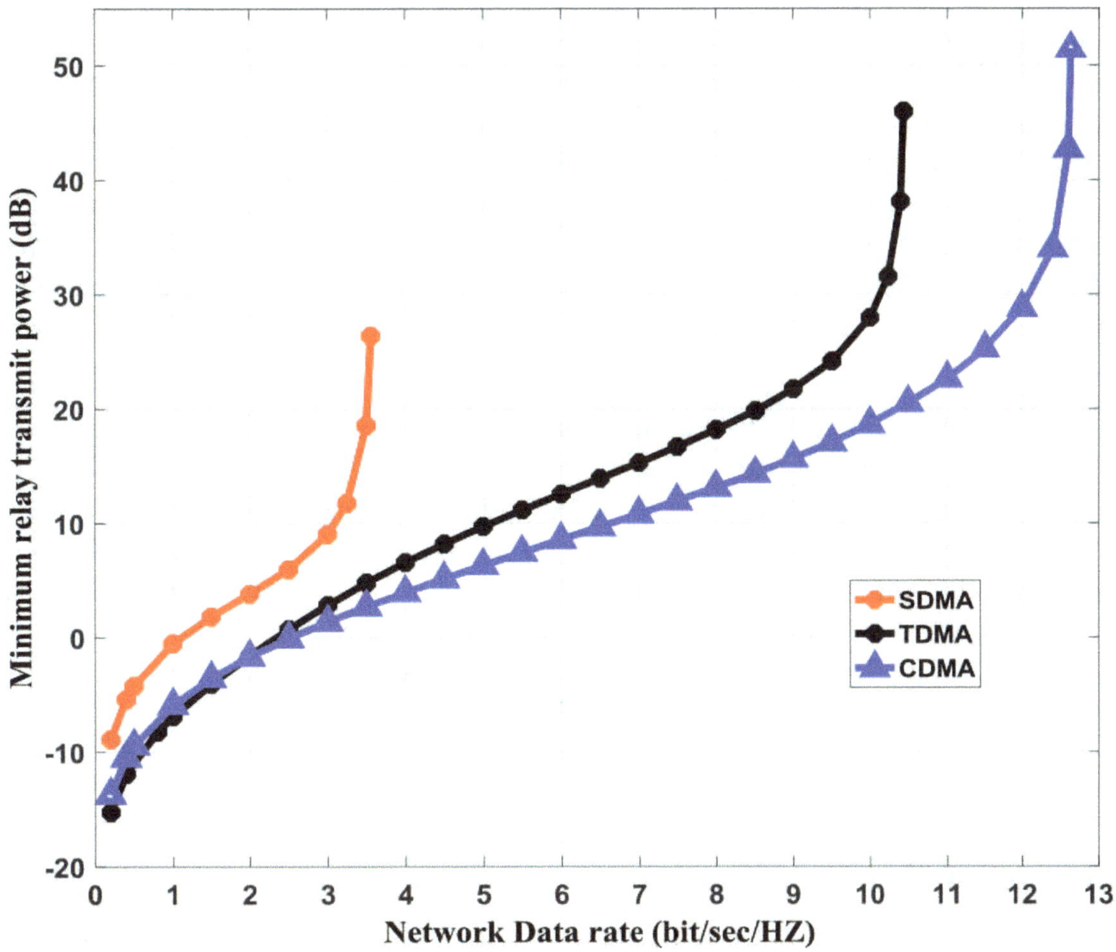


**Figure 13.** Minimum relay transmits power versus *D*, for *R*=4, *u*=4.

$$\begin{array}{l} \underset{\mathbf{X}}{\text{Minimize}}\ \text{trace}\ (\mathbf{TX}) \\ \text{Subject to}\ \text{trace}\ (\mathbf{D}_k\mathbf{X}) \geq \gamma_{th}^k \sigma_{\varsigma_k}^2, \quad k \in \{1, \dots, d\} \\ \quad \mathbf{X} \geq 0 \end{array} \tag{64}$$

This optimization problem has been solved in a same way as the previous sections. **Figure 11** shows the total relay transmit power versus destination SINR threshold value, for different values of users' correlation factors. The network consists of two source-destination pairs and four relays. **Figure 11** shows that the total relay transmit power in all cases increases by raising $\gamma_{th}$. Furthermore, **Figure 11** indicates that when the signature sequence correlation $\rho_{k,l}$ increases, more total transmit power is needed to ensure SINR constraints at destination nodes. When $\rho_{k,l}$ approaching one, the problem downgrades to the SDMA network and the system loses the benefits of CDMA technique. Also, increasing the signal dependency by increasing the correlation factor, results in the more infeasibility rate of the constraints.

**Figure 12** displays the minimum relay transmit power versus $\gamma_{th}$ for different number of relays and users. As normally expected, more power saving can be achieved by increasing the number of relays or decreasing the number of users. Comparing **Figure 2** with **Figure 3** reveals that decreasing the correlation factor will be much more efficient for saving network power than increasing the number of relays.

**Figure 13** shows the minimum relay transmit power versus the network data rate ($D$) for distributed CDMA, SDMA and TDMA schemes. In **Figure 13**, we consider a network with four relays and four source-destination pairs. For the sake of comparison fairness, we need to ensure that different schemes are compared with the same average source powers. So, we assume that the source power of CDMA and SDMA are one fourth of those in TDMA scheme. For **Figure 13**, the network data rate has the following relation to the SINR threshold value, $D = w \log_2(1 + \text{SINR}_{th})$. Signature sequences of the user are randomly generated for 100 trials and the best code in term of least maximum correlation is chosen for performance comparison.

Also, it can be seen from **Figure 13** that the minimum relay transmitted power increases with the increase of $D$. For the SDMA scheme, the problem quickly becomes infeasible due to the power of interference at destinations. So, for establishing connections between four users, SDMA-based networks should use at least 40 relays to overcome the TDMA scheme. Since the QoS constraints are less stringent in CDMA scheme, the network can establish the communication between source-destination pairs for a larger range of D. Consequently, it can be observed from **Figure 13** that the CDMA-based network can establish the source-destination connections with a significantly lower relay transmit power as compared to other schemes.

## 6. Computational complexity

Since the CDMA relay systems have a heavy computational complexity, the aim of this section is to analyze the computational form of related algorithms used in practice [20]. Here, the computational complexity of a standard SDP is introduced and extended to our case. The standard SDP problem with equality constraint is given as:

$$\begin{aligned} &\underset{\mathbf{X}}{\text{Minimize}}\ \text{trace}\ (\mathbf{CX}) \\ &\text{Subject to trace}\ (\mathbf{A}_\text{i}\mathbf{X}) = b_\text{i} \quad i\in\{1,\dots,d\} \\ &\mathbf{X}\geq 0 \end{aligned} \tag{65}$$

where $C$ and $\mathbf{A}_\text{i}$ are symmetric $n \times n$ matrices, and $b \in {}^d$.So for such a problem the complexity with large-update (or long-step) algorithm [21] based on the primal dual SDP algorithm is

$$O(\sqrt{n}\log n\log(n/\varepsilon)) \tag{66}$$

where $\varepsilon$ denotes the accuracy parameter of the algorithm, while this algorithm with small-update (or short-step) still has $O(\sqrt{n}\log(n/\varepsilon))$ iterations bound [22].

It is shown in Ref. [22] that small update interior point methods (IPMs) are restricted to unacceptably slow progress, while large-update IPMs are more efficient for faster. Also, large update IPMs perform much more efficiently in practice, however, they often have somewhat worse complexity bounds. The complexity order of solving standard SDP problem is polynomial time.

For evaluating the complexity of our SDP problem with inequality constraints, we have to calculate the dimension parameter $n$. Therefore, we should determine the dimensions of the matrices used in the objective and constraints of the problem Eq. (63). In the Kronecker product of two matrices, if $\mathbf{A} \in \mathbb{C}^{n \times n}$ and $\mathbf{B} \in \mathbb{C}^{m \times m}$, then $\mathbf{A} \otimes \mathbf{B}$ will be a $nm \times nm$ matrix. According to the new vectors definite in Eq. (65) and sizes of $\boldsymbol{\mu} \in \mathbb{C}^{d \times d}$ and $E(\mathbf{H}^*\mathbf{H}^T) \in \mathbb{C}^{Rd \times Rd}$, dimension of $\mathbf{T}$ will be $\mathbf{T}\in\mathbb{C}^{Rd^2\times Rd^2}$.

Similarly, we can obtain the above conclusion for $\mathbf{D}_k$ and $\mathbf{X}$, that is, $\mathbf{T}$, $\mathbf{D}_k, \mathbf{X}\in\mathbb{C}^{Rd^2\times Rd^2}$. It is notable that the constraints of our problem are not the same as the standard SDP form. Therefore, we have to equalize them so that they alter to a type similar to the standard format. In order to achieve this goal, first we have to eliminate the inequality constraints of Eq. (64) by defining $y_i$ as:

$$\text{trace}(\mathbf{D}_\text{i}\mathbf{X}) = \gamma_\text{i}\sigma_{\varsigma_k}^2 + y_\text{i}, \mathbf{X}\geq 0, y_\text{i}\geq 0 \quad \text{for i} = 1,\dots,d \tag{67}$$

Next, a new variable $\widehat{\mathbf{X}}$ should be defined in order to standardize the problem:

$$\widehat{\mathbf{X}} \triangleq \begin{bmatrix} \mathbf{X} & \mathbf{0}_{R^2d^2\times d} \\ \mathbf{0}_{d\times R^2d^2} & \begin{bmatrix} y_1 & \cdots & 0 \\ \vdots & \ddots & \vdots \\ 0 & \cdots & y_d \end{bmatrix} \end{bmatrix} \tag{68}$$

As a result, the following standard form will be attained.

$$\begin{aligned}&\underset{\mathbf{X}}{\text{Minimize}}\ \text{trace}\left(\widehat{\mathbf{T}}\widehat{\mathbf{X}}\right)\\&\text{Subject to}\ \ \text{trace}\left(\widehat{\mathbf{D}}_{\mathrm{i}}\widehat{\mathbf{X}}\right)=b_{\mathrm{i}}\,,\widehat{\mathbf{X}}\geq 0\ \ \text{for i}=1,\ldots,d\end{aligned} \tag{69}$$

where

$$\widehat{\mathbf{D}}\triangleq\begin{bmatrix}\mathbf{D} & \mathbf{0}_{Rd^2\times d}\\ \mathbf{0}_{d\times Rd^2} & \mathbf{0}_{d\times d}\end{bmatrix},\ \widehat{\mathbf{T}}_i\triangleq\begin{bmatrix}\mathbf{T}_i & \mathbf{0}_{Rd^2\times d}\\ \mathbf{0}_{d\times Rd^2} & \mathbf{0}_{d\times d}\end{bmatrix} \tag{70}$$

As a result of the above representation form, $n$ for Eq. (63) would be:

$$n_{Distributed\_Relay}=Rd^2+d\simeq Rd^2 \tag{71}$$

Also, we can use the same procedure to calculate $n$ for Eqs. (16) and (42):

$$\begin{aligned}n_{MIMO}&=R^2+d\simeq R^2\\ n_{MIMO\_CDMA}&=R^2d^2+d\simeq R^2d^2\end{aligned} \tag{72}$$

Therefore, the complexity for problems (16), (42) and (63) for MIMO, MIMO-CDMA, and distributed-relay networks are as follows:

$$\begin{aligned}&O\left(R\log\left(R^2\right)\log\left(R^2/\varepsilon\right)\right),\\&O\left(Rd\log\left(R^2d^2\right)\log\left(R^2d^2/\sqrt{\varepsilon}\right)\right),\\&O\left(\sqrt{Rd^2}\log\left(Rd^2\right)\log\left(Rd^2/\varepsilon\right)\right)\end{aligned} \tag{73}$$

while a SDMA relay network has the complexity order of $O\left(\sqrt{R}\log(R)\log(R/\varepsilon)\right)$.

## Author details

Mohammad-Hossein Golbon-Haghighi

Address all correspondence to: golbon@ou.edu

School of Electrical and Computer Engineering, University of Oklahoma, Norman, OK, USA

## References

[1] Laneman, J.N., D.N.C. Tse, and G.W. Wornell, *Cooperative Diversity in Wireless Networks: Efficient Protocols and Outage Behavior.* IEEE Trans. Inf. Theory, Dec. 2004. **50**: p. 3062–3080.

[2] Yindi, J. and H. Jafarkhani, *Network Beamforming Using Relays with Perfect Channel Information.* IEEE Trans. Inf. Theory, Jun. 2009. **55**: p. 2499–2517.

[3] Havary-Nassab, V., et al., *Distributed Beamforming for Relay Networks Based on Second-Order Statistics of the Channel State Information.* IEEE Trans. Signal Process., Sep. 2008. **56**: p. 4306–4316.

[4] Verdu, S., *MultiUser Detection.* 1998, New York, NY, USA: Cambridge University Press.

[5] Wan, C., J.G. Andrews, and R.W. Heath, *Multiuser Antenna Partitioning for Cellular MIMO-CDMA Systems.* IEEE Trans.Veh. Technol., Sep. 2007. **56**: p. 2448–2456.

[6] Assra, A., W. Hamouda, and A. Youssef, *EM-Based Joint Channel Estimation and Data Detection for MIMO-CDMA Systems.* IEEE Trans. Veh. Technol., Mar. 2010. **59**: p. 1205–1216.

[7] Dempster, A., N. Laird, and D.B. Rubin, *Maximum Likelihood from Incomplete Data via the EM Algorithm.* J. R. Stat. Soc. Ser. B (Methodological), 1977. 39(1): p. 1–38.

[8] Driouch, E. and W. Ajib, *Efficient Scheduling Algorithms for Multiantenna CDMA Systems.* IEEE Trans. Veh. Technol., Feb. 2012. **61**: p. 521–532.

[9] Yener, A., R.D. Yates, and S. Ulukus, *Combined Multiuser Detection and Beamforming for CDMA Systems: Filter Structures.* IEEE Trans. Veh. Technol., 2002. **51**: p. 1087–1095.

[10] Fazeli-Dehkordy, S., S. Shahbazpanahi, and S. Gazor, *Multiple Peer-to-Peer Communications Using a Network of Relays.* IEEE Trans. Signal Process., 2009. **57**(8): p. 3053–3062.

[11] Golbon-Haghighi, M.H., B. Mahboobi, and M. Ardebilipour, Multiple Antenna Relay Beamforming for Wireless Peer to Peer Communications. J. Inform. Syst. Telecommun. (JIST), Dec. 2013. **1**(4): p. 209–215.

[12] Karipidis, E., N.D. Sidiropoulos, and Z.-Q. Luo, *TransQuality of Service and Max-Min Fair Transmit Beamforming to Multiple Cochannel Multicast Groups.* IEEE Trans. Signal Process., Mar. 2008. **56**(3): p. 1268–1279.

[13] Luo, Z.-Q., et al., *TransApproximation Bounds for Quadratic Optimization with Homogeneous Quadratic Constraints.* SIAM J. Optim., Feb. 2007. **18**: p. 1–28.

[14] Chang, T.-H., Z.-Q. Luo, and C.-Y. Chi, *Approximation Bounds for Semidefinite Relaxation of Max-Min-Fair Multicast Transmit Beamforming Problem.* IEEE Trans. Signal Process., Aug. 2008. **56**(8): p. 3932–3943.

[15] Ma, W.-K., et al., *Quasi-Maximum-Likelihood Multiuser Detection Using Semi-Definite Relaxation with Application to Synchronous CDMA.* IEEE Trans. Signal Process., Apr. 2002. **50**(4): p. 912–922.

[16] Phan, A.H., et al., *Beamforming Optimization in MultiUser AF Wireless Relay Networks.* IEEE Trans. on W.Commun., Apr. 2012. **11**: p. 1510–1520.

[17] Golbon-Haghighi, M.H., B. Mahboobi, and M. Ardebilipour, Optimal Beamforming in Wireless Multiuser MIMO-Relay Networks, in 21st Iranian Conference on Electrical Engineering (ICEE). 2013, IEEE. p. 1–5.

[18] Golbon-Haghighi, M.H., B. Mahboobi, and M. Ardebilipour, *Linear Pre-coding in MIMO-CDMA Relay Networks.* Wireless Pers. Commun. (Springer), Jul. 2014. **79**(2): p. 1321–1341.

[19] Golbon-Haghighi, M.-H., et al., *Detection of Ground Clutter from Weather Radar Using a Dual-Polarization and Dual-Scan Method.* Atmos. J., 2016. **7**(6).

[20] Nesterov, Y.E. and A.S. Nemirovskii, Interior Point Polynomial Algorithms in Convex Programming. SIAM Studies in Applied Mathematics, 1994, Philadelphia, PA: SIAM. p. 13.

[21] Wolkowicz, H., R. Saigal, and L. Vandenberghe, *Handbook of Semidefinite Programming: Theory, Algorithms, and Applications.* 2000, New York, USA: Springer Science & Business Media.

[22] Peng, J., C. Roos, and T. Terlaky, *Self-Regularity: A New Paradigm for Primal-Dual Interior-Point Algorithms.* 2009, Princeton, NJ: Princeton University Press.

4

# Waveform Design Considerations for 5G Wireless Networks

Evren Çatak and Lütfiye Durak-Ata

**Abstract**

In this chapter, we first introduce new requirements of 5G wireless network and its differences from past generations. The question "Why do we need new waveforms?" is answered in these respects. In the following sections, time-frequency (TF) lattice structure, pulse shaping, and multicarrier schemes are discussed in detail. TF lattice structures give information about TF localization of the pulse shape of employed filters. The structures are examined for multicarrier, single-carrier, time-division, and frequency-division multiplexing schemes, comparatively. Dispersion on time and frequency response of these filters may cause interference among symbols and carriers. Thus, effects of different pulse shapes, their corresponding transceiver structures, and trade-offs are given. Finally, performance evaluations of the selected waveform structures for 5G wireless communication systems are discussed.

**Keywords:** waveform design, orthogonal frequency division multiplexing (OFDM), filtered multitone (FMT), time-frequency lattice, pulse shaping, multicarrier modulation, generalized frequency division multiplexing (GFDM)

## 1. Introduction

In communication systems waveforms enable the allocation of data on the joint time-frequency (TF) domain by transmitting and receiving proper signals. As the waveform design deals with the methods to generate transmitted signals at the transmitter, and receive at receiver side through a channel, the design criteria depend on demands of users, channel conditions, system, and technology criteria. Therefore, the design criteria change with respect to the advancement of technologies. The waveform techniques in 2G/3G/4G mobile technologies cannot meet the demands of next-generation wireless networks. To overcome problems stemming from the new demands, either it is required to design new waveform techniques, or propose improved versions of the waveform used in 4G, i.e., the orthogonal frequency division multiplexing (OFDM) [1, 2] at least.

The answer to the question "Why do we need new waveforms?" reveals important issues. The state-of-the-art radio access technology is summarized in **Figure 1**. Accordingly, the ambitious performance goals for 5G networks are 10–100 times higher typical user data rates, 10–100 times more connected devices, 10 times lower network energy consumption, less than 1 ms end-to-end latency, and 10000 times higher mobile data traffic per geographical area [1, 3]. The 5G communication systems that are expected to have a heterogeneous network structure are planned to design in such a way that they provide service not only for people as real users but also for various kinds of equipment. While designing the system in this way, we should keep in mind that, features for each user, such as transmission packet lengths, data rates, data transmission frequencies, and capacities would be different. These various requests of users, lead to lots of issues, such as synchronization in time and frequency. To overcome these problems, it is required to design new techniques capable of utilizing the spectrum more efficiently, with higher data rates, with lower energy consumption, and latency [4, 5].

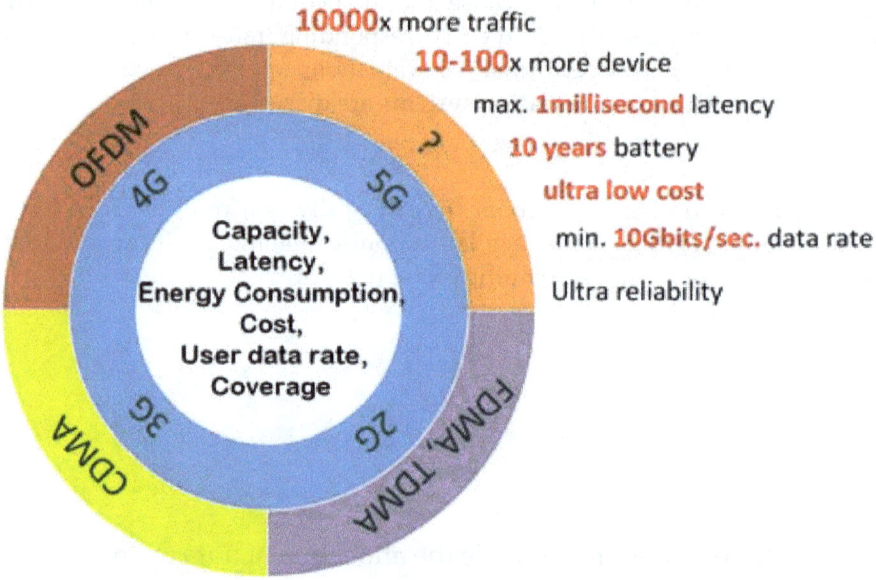


**Figure 1.** The state-of-the-art radio access technology: moving from voice to 5G.

An ideal waveform shall fulfill the following requirements (i) low power consumption, (ii) high data rates, (iii) spectrum efficient, (iv) low latency, (v) easy to implement, and (vi) low out-of-band emission. Additionally, a well-designed waveform must be robust to disruptive features of communication channels, and be able to easily extract these effects at the receiver side. It must be compliant with massive multiple-input multiple-output (MIMO) systems, and adaptive for users with different access requirements on heterogeneous networks. Absolutely, it is not possible to find a waveform that supplies to all requirements perfectly. However, the accurate waveform design procedure meets most of these features at optimum ways.

OFDM is the dominant technology for today's broadband multicarrier communications. However, it is considered as an undesirable solution for 5G wireless networks due to its shortcomings on some channel effects [6]. The other shortcomings are the out-of-band (OOB) emission [7] and peak-to-average power ratio (PAPR) problems [8]. Rectangular pulse shaping of OFDM introduces the nonnegligible out-of-band emissions, which cause interferences among adjacent bands, whereas usage of independent phases for subcarriers causes PAPR problem.

In literature, up to now several candidate waveforms are proposed to achieve 5G communication system requirements. The multicarrier waveforms based on filtering operations are good candidate waveforms to overcome OOB emission problems. Filter bank-based multicarrier (FBMC) and its varieties, generalized frequency division multiplexing (GFDM), and universal filtered multicarrier (UFMC) are among these candidate waveforms.

FBMC is one of the multicarrier waveforms using filtering operation. Filtered multitone (FMT), staggered multitone (SMT), and cosine-modulated multitone (CMT) modulations are variants of the FBMC transmission scheme [9]. The main differences of these schemes are their TF domain allocations. Contrary to FMT, the subcarriers of SMT and CMT are overlapping. So, FMT is not spectrally efficient.

GFDM can be considered as a type of filter bank-based multicarrier modulation scheme with transmission filters that are shifted in time and frequency domains. The novelty of GFDM is in its flexibility, which can address the different applications. On the other hand, most of the real-time applications (i.e., tactile Internet) need lower latency. Low latency can be obtained with small symbol durations and less complex transceiver structures. It is possible to reduce signal durations for GFDM by designing appropriate TF structures [10]. The complexity that is caused by filtering operations can be reduced by using polyphase structures of filters [11]. OOB emission can be reduced via these using filters that have low side lobe levels at their frequency responses.

UFMC is another waveform with low OOB emission [12, 13]. The distinguishing feature of UFMC is in filtering the group of subcarriers instead of filtering each subcarrier. The filters used for UFMC have large bandwidth and short impulse response. It makes short burst transmission. This scheme is not suitable for applications that need time synchronization.

The purpose of this chapter is to present the basics of waveform design for 5G networks. To achieve this, the rest of the chapter is organized as follows. In Section 2, the fundamentals of waveform design that includes TF lattice structures and pulse shaping are explained. In Section 3, the concept of multicarrier waveforms and transceiver structures such as OFDM, FBMC, and FMT with nonuniformly divided bandwidth allocations and GFDM are discussed. In Section 4, the performance comparisons of the waveforms are evaluated. Conclusion and future directions remarks are given in Section 5.

## 2. Fundamentals of waveform design

Forming TF lattice structures and pulse shaping are the essential steps for waveform design. Time and frequency allocation of transmitted and also received signals are performed through TF lattice structures. The pulse shaping is also an important step to avoid interferences among the symbols in both time and frequency domains.

### 2.1. TF lattice structures

TF lattice structures contain information about the relationship between time and frequency support information for all symbols. TF lattice structures depend on transmission schemes, i.e., single-carrier, multicarrier, time-division, and frequency-division transmission schemes.

**Figure 2** shows the TF lattice structures of time and frequency division multiplexing (TDMA and FDMA, respectively). If frequency spectrum is divided into subbands, the waveform is called multicarrier waveform. Each carrier in a subband is called a subcarrier. Each grid in TF lattice structure indicates a subsymbol. The symbols are transmitted at every $T$ seconds.

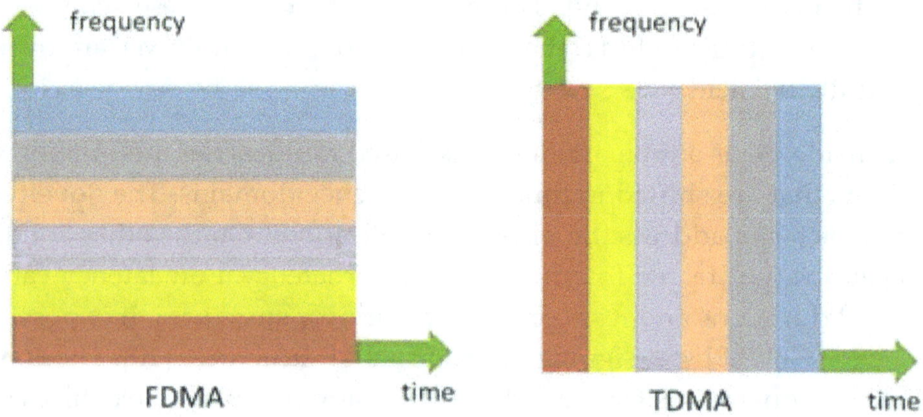


**Figure 2.** Frequency division and time division multiplexing as a TF lattice structure.

Data rate depends on the transmission bandwidth, channel capacity, signal-to-noise ratio (SNR), and the receiver capacity. Data rate is related to the frequency resolution that is expressed by

$$\Delta f = \frac{1}{T} = \frac{f_s}{N} \tag{1}$$

where $f_s$ is the sampling frequency and $\Delta f$ is the difference between two adjacent frequency bins. In order to resolve frequencies, it needs to make $\Delta f$ sufficiently small and that is referred to as increasing the frequency resolution.

A signal $s(t)$ can be represented in the frequency domain by its Fourier transform $S(f)$ as

$$S(f) = \mathcal{F}\{s(t)\} = \int_{-\infty}^{\infty} s(t)e^{-j2\pi ft}dt. \quad (2)$$

Time-domain signal $s(t)$ has a finite duration. Finite time duration implies infinite bandwidth. On the contrary, finite bandwidth implies infinite time duration. In practice, time duration and bandwidth are limited. A time-limited signal $s_T(t)$ can be expressed by multiplying a rectangular pulse of duration $T$ as

$$s_T(t) = s(t)\ \mathrm{rec}(t / T). \quad (3)$$

The Fourier transform of the time-limited signal in Eq. (3) is

$$S_T(f) = S(f) * T\mathit{sinc}(fT) \quad (4)$$

where * is the convolution operation in the frequency domain. Because of the convolution operation, bandwidth of $S_T(f)$ becomes unlimited.The time and frequency domain representations of the rectangular pulse are given in **Figure 3**. Time domain is limited, but frequency response spreads over a large range of bandwidth.

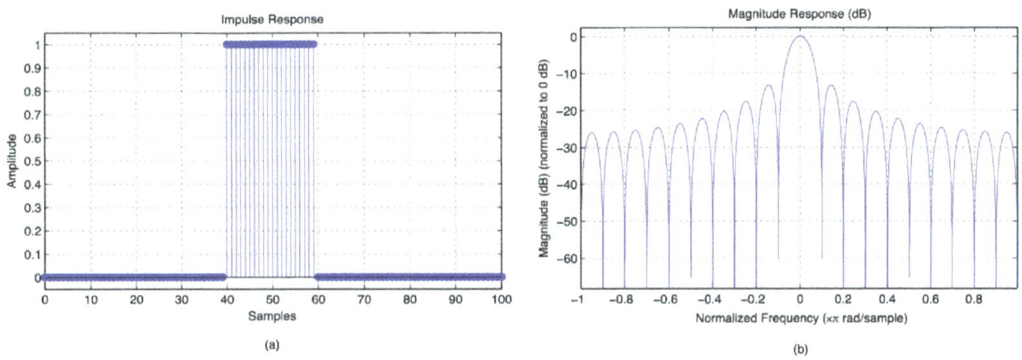


**Figure 3.** (a) The impulse response and (b) frequency response of a rectangular pulse: The impulse response is limited; frequency response spreads over the frequency domain and includes high-level side lobes.

Such infinite bandwidth information is not realistic. For that reason, a bandwidth that contains most of the signal energy can be used. The extreme frequencies ($f_{min}, f_{max}$) can be defined from the desired signal energies, and the bandwidth is $B = f_{max} - f_{min}$.

Time-bandwidth product is a design parameter of TF lattice structure. Time-bandwidth product is expressed by $B \times T$ that measures localization in time and frequency domain. The aim is to minimize the unit area of TF lattice structures. But there is a lower limit that is obtained from the uncertainty principle [14, 15]. The time domain representation of a Gaussian pulse is

$$s(t) = e^{-\propto^2 t^2} \tag{5}$$

with time duration $T = 1/2 \propto$ and bandwidth $B = \propto /2\pi$. The time-bandwidth product of Gaussian pulse becomes

$$B \times T = \frac{1}{4\pi}. \tag{6}$$

Time-bandwidth product of Gaussian pulses in Eq. (6) is the lower limit. For all other signals, time-bandwidth product is limited below $B \times T > \frac{1}{4\pi}$ based on the celebrated uncertainty principle.

The TF lattice structures of several waveforms are shown in **Figure 4**. These structures give information about the rules of frequency division and time division of waveforms. TF lattice structure of OFDM is shown in **Figure 4(a)** for a transmission bandwidth, $B$. The transmission bandwidth is divided into $N$ subbands through IFFT operations. On the other hand, according to the TF lattice structure of GFDM, the time domain is also divided into time slots.

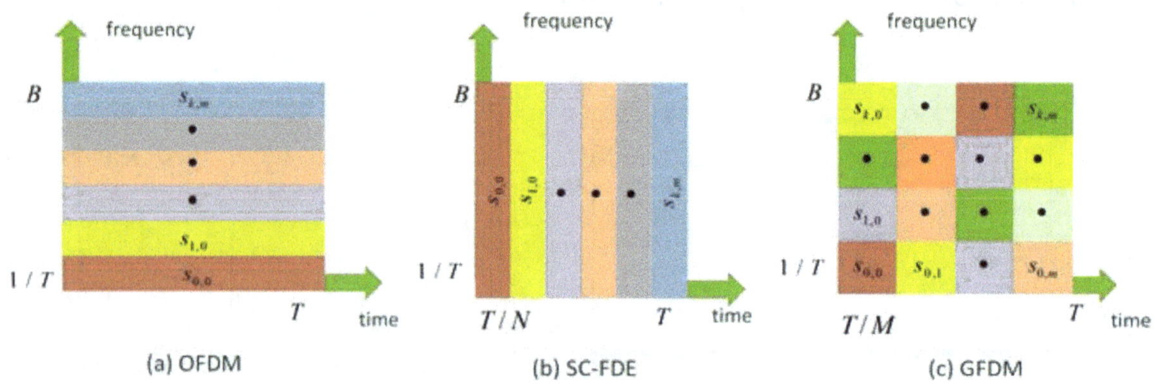


**Figure 4.** (a) OFDM, (b) single carrier-FDE, and (c) GFDM.

The transmitted signal with proper time and frequency shifts can be expressed as

$$\boldsymbol{x}(n) = \sum_{m=0}^{M-1} \sum_{k=0}^{K-1} s_{k,m}\, \boldsymbol{g}_m(n) e^{-j2\pi n \frac{k}{K}} \tag{7}$$

where $s_{k,m}$ is the data symbol with a subcarrier subscript $k$ and subsymbol subscript $m$ where $k = 0, 1, \ldots, K-1$ and $m = 0, 1, \ldots, M-1$, respectively. $g_m(n)$ is a time-shifted version of a prototype filter $g(n)$. In OFDM, prototype filter $g(n)$ is replaced with 1 and each subcarrier contains one subsymbol, which means $M = 1$. Thus, the OFDM symbol is simply

$$x(n) = \sum_{k=0}^{K-1} s_{k,m}\, e^{-j2\pi n \frac{k}{K}}. \quad (8)$$

In the same approach, single carrier transmission is obtained by replacing $K = 1$ and $g(n)$ with Dirichlet pulse [16]. The symbols are transmitted by dividing into time slots and each sub-symbol contains all frequency components of the transmission bandwidth.

TF lattice structures of GFDM waveform are the combination of the frequency-division and time-division based waveforms that are defined in Eq. (7). The transmitted signal is obtained by convolution of data with filter $g_m(n)$ that is the time- shifted and frequency-shifted version of prototype filter $g(n)$. The projection of filters $g_m(n)$ on time-frequency domain is not rectangular as indicated in **Figure 3**.

Toroidal lattice [17] and hexagonal lattice [18] are other lattice structures proposed in the literature. Hermite-Gaussian functions are well-localized in both time and frequency domains and the time-bandwidth product of its zeroth-order function equals to the lowest time-bandwidth product, i.e., $1/4\pi$. The time- and frequency-domain representation of the third-order Hermite-Gaussian pulse and a toroidal rectangular TF lattice structure are given in **Figure 5**.

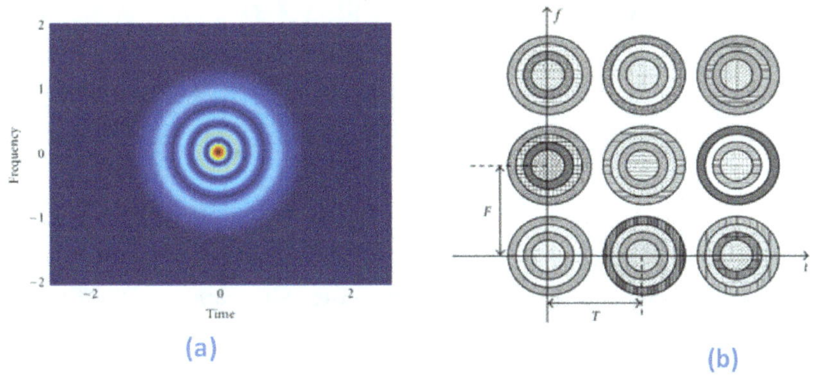


**Figure 5.** Toroidal lattice structure. (a) Third order of Hermite pulse and (b) rectangular lattice with Hermite pulses [17].

Toroidal rectangular lattice structure provides more data rate as indicated in [17]. On the other hand, the hexagonal lattice structure is more robust for inferences and channel effects [18, 19].

Briefly, the symbol durations and bandwidths are important parameters of TF lattice structures. These parameters are chosen according to the requirements of the users and channel conditions. The details are given in Section 4. The next step of the waveform design is pulse shaping. The pulse shaping is the determination of time and frequency limits of a pulse to fill in each grid in the TF lattice. The methods and constraints of pulse shaping are given in the following section.

## 2.2. Pulse shaping

In a communication system, pulse shaping is important to generate band- and time-limited transmitted signal. Limiting the signals of symbols in time and frequency domains is important to avoid interferences.

The definition of pulse shaping is the filtering process that maps modulated signals to the TF lattice to control the interferences. The main problem of pulse shaping is the reciprocal relation between time and frequency domains. It means that a narrow pulse in the time domain has wider spectrum in the frequency domain. If the width of a pulse is increased in the time domain, the width of the spectrum in the frequency domain will be decreased. Of course, the pulse cannot be widened to infinity as in the ideal case. This causes out-of-band emission in the frequency domain. Well-designed filters according to design requirements can prevent or at least decrease out-of-band emission and also interference.

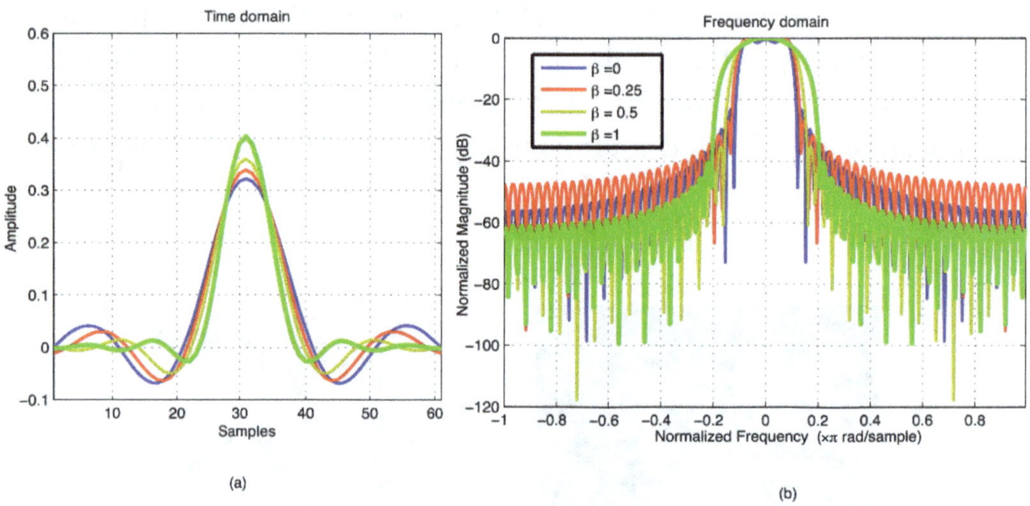


**Figure 6.** Raised-cosine filter: (a) time and (b) frequency responses with various roll-off factors. If roll-off factor is $\beta = 0$, the impulse response is similar to the rectangular pulse.

The Fourier transform of the rectangular pulse is a *sinc* function that has very large bandwidth because of the side lobes. The problems of reducing the level of side lobes and the signal power out of the transmitted band can be solved by windowing. The windowing operation limits the out-of-band energy by smoothing the time-domain function. So, in order to mask to spectrum, pulse shaping, i.e., time-domain windowing is used. Raised cosine filter and Gaussian filter are the famous pulse shaping filters. The impulse response of these filters are given by

$$\boldsymbol{h}_{RC}(t) = \frac{\sin(\pi t/T)}{\pi t/T} \frac{\cos(\pi \beta t/T)}{1 - 4\beta^2 t^2/T^2} \tag{9}$$

and

$$h_{Gaussian}(t) = \sqrt{\frac{2\pi}{In2}}\,(BT)e^{-\frac{2\pi^2}{In2}(BT)^2t^2} \tag{10}$$

respectively. Here $\beta$ is called the roll-off factor that is in the range of $0 \leq \beta \leq 1$. The frequency responses are

$$H_{RC}(f) = \begin{cases} T & 0 \leq |f| \leq \frac{1-\beta}{2T} \\ \frac{T}{2}\left\{1 + cos\left[\frac{\pi T}{\beta}\left(|f| - \frac{1-\beta}{2T}\right)\right]\right\} & \frac{1-\beta}{2T} \leq |f| \leq \frac{1+\beta}{2T} \\ 0 & |f| \geq \frac{1+\beta}{2T} \end{cases} \tag{11}$$

$$H_{Gaussian}(f) = e^{-\frac{In2}{2}\left(\frac{f}{BT}\right)^2}. \tag{12}$$

The time and frequency responses of the raised-cosine filter for different $\beta$ values are given in **Figure 6**. The roll-off factor $\beta$ is the measure of the excess bandwidth of the filter. If $\beta = 0$, the impulse response approaches to *sinc(t/T)* function and the frequency response approaches to *rect(fT)* rectangular function.

The famous windowing functions and their time-domain sequences are given in **Table 1**.

| | **Window** | **Time domain sequence** <br> ***h(n)* for *0 ≤ n ≤ L − 1* length of filter** |
|---|---|---|
| 1 | Blackman | $0.42 - 0.5\cos\frac{2\pi n}{L-1} + 0.08\cos\frac{4\pi n}{L-1}$ |
| 2 | Hamming | $0.54 - 0.46\cos\frac{2\pi n}{L-1}$ |
| 3 | Hanning | $\frac{1}{2}\left(1 - \cos\frac{2\pi n}{L-1}\right)$ |
| 4 | Kaiser | $\frac{I_0\left[\alpha\sqrt{\left(\frac{L-1}{2}\right)^2 - \left(n - \frac{L-1}{2}\right)^2}\right]}{I_0\left[\alpha\left(\frac{L-1}{2}\right)\right]}$ <br> $I_0$: zeroth-order Bessel, $\alpha$: positive real number |

**Table 1.** Common window functions.

## 3. Transceiver schemes for 5G wireless networks

Multicarrier transmission is the best way to fix the problems due to frequency-selective channel conditions. Contrary to the single-carrier modulation techniques, that use only one carrier at all times, multicarrier modulation divides the band into more subcarriers. The ideal equalizer has a frequency response that is the inverse of the frequency response of the channel. So, the equalization of multicarrier transmission is easier for the frequency-selective channel. OFDM is an orthogonal multicarrier transmission scheme that has subcarriers with sinc-shaped spectra. The transceiver structure of the OFDM is given in **Figure 7**.

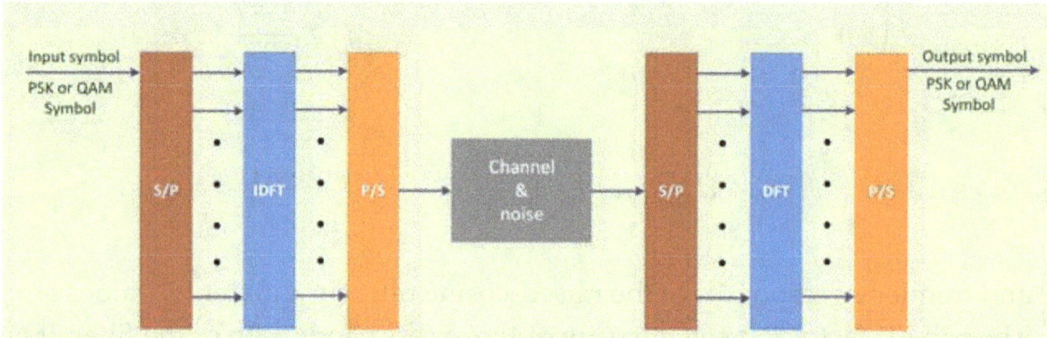


**Figure 7.** OFDM transmission scheme implemented using IDFT/DFT.

Accordingly, a sequence of PSK or QAM symbols is converted into $N$ parallel streams before the $N$-point inverse DFT (IDFT) operation. Parallel streams are converted to a serial form after the IDFT operation. The same operations are done at the receiver sides that include DFT operations instead of the IDFT operation.

The advantages and disadvantages of OFDM are as follows:

Advantages:

- Resilience to frequency selective fading: by dividing the channel into narrow flat fading channels.
- Spectrum efficiency: by allowing overlap.
- Resilience to interference: by using acyclic prefix (CP) to avoid intersymbol and interframe interferences.
- Channel equalization: by using multiple subchannels.
- Computationally efficient: by using fast Fourier transform (FFT) and inverse FFT (IFFT) operations to implement modulation and demodulation.

Disadvantages:

- High peak-to-average power ratio (PAPR): because of using independent phases for the subcarriers.

- Sensitive to carrier frequency offset (CFO): because of small subcarrier spacing and the necessity of good receiver synchronization.
- Out-of-band interference: because of the rectangular pulse shape.
- Loss of efficiency: because of using the cyclic prefix (CP) and guard intervals (GIs).

Therefore, OFDM is a very useful multicarrier modulation scheme because of its advantages. On the other hand, new modulation schemes are needed to overcome the drawbacks of OFDM.

### 3.1. Filter bank-based multicarrier

FBMC is the set of filtering operations that separate the input signal to the subbands with the frequency-shifted versions of low-pass prototype filters. The differences of FBMC from OFDM are: (i) CP extension is not required, (ii) having low side lobe and low spectral leakage depends on the filter type, (iii) more complex, and (iv) less sensitive to CFO. The benefits of FBMC are allowing to pulse shaping of filters that produce well-localized subbands in both time and frequency domain. FBMC is a candidate waveform of 5G communication networks to overcome some problems. The features such as lower side-lobes, lower sensitivity to CFO, and higher bandwidth efficiency—because of the absence of CP—makes FBMC a possible replacement of OFDM in 5G wireless communications. Furthermore, frequency allocations of subbands become more flexible with benefits of filtering operations.

FBMC modulation-based systems are more complex than OFDM due to exchange of FFT/IFFT operations by the filter banks. The CFO is caused by Doppler shift due to mobility. Orthogonality between adjacent subcarriers is destroyed by CFO and it introduces intercarrier interference (ICI) and intersymbol interference (ISI). Besides, the *sinc*-shape frequency response of each subcarrier causes large ICI in presence of CFO. Using the windows with smooth edges reduces the sensitivity of CFO, thus FBMC satisfies this condition.

In the conventional FBMC system, the frequency spectrum is divided into equal subbands and each symbol in subbands is filtered after upsampling operations. The upsampling value ($K$) and the number of the subbands ($M$) determine the overlapping of subbands [20] and the allocations of subbands of FBMC are given in **Figure 8**. When the $K$ equals to the $M$, the filter bank is said to be critically sampled; otherwise, it is noncritically sampled.

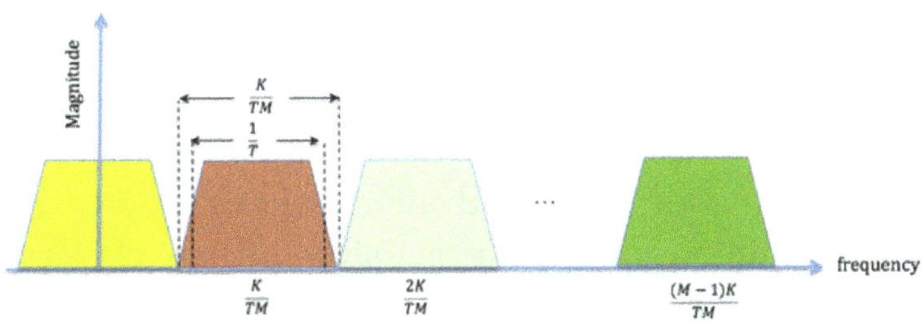


**Figure 8.** Frequency allocation of FBMC: the channel is uniformly divided by subbands.

According to the FMT modulation, each user symbols in subbands are filtered by the frequency-shifted versions of a low-pass prototype filter after upsampling operations. The transceiver scheme of FMT is given in **Figure 9**. Here, symbols with the same data rates share frequency spectrum equally.

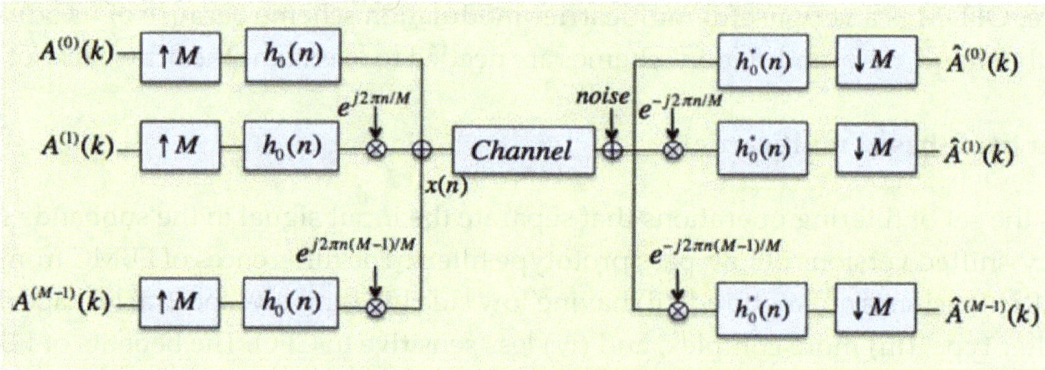


**Figure 9.** The transceiver structure of FMT: symbols are transmitted with multicarrier modulation by filtering. If the low-pass prototype filters $h_0(n)$ are symmetric finite impulse response (FIR) filters, then the transceiver filters are their complex conjugates.

The transmitted signal of the FMT scheme in **Figure 9** is given by

$$x(n) = \sum_{m=0}^{M-1} \sum_{k=-\infty}^{\infty} A^{(m)}(k)\, \mathbf{h}_0(n - kM) e^{j2\pi mn/M} \tag{13}$$

where $\mathbf{h}_0(n)$ is the prototype filter. The transmitted signal $x(n)$ is the sum of the convolutions of upsampled of data and the frequency-shifted versions of a low-pass prototype filter.

Generally, the bandwidth allocations of users need not be equal to each other because of different data rates. Especially, some users in 5G communication channel may upload their video streams, while some users are a part of internet-of-things/machine-type communications (IoT/MTC). The bandwidth requirements of these users are not the same and may change according to the applications of users. Hence, it is not advantageous to use traditional multi-carrier structures for the users that need different transmission bandwidths. In LTE (long-term evaluation), the frequency spectrum is shared by users with predefined bandwidths (i.e., 1.4, 3, 5, 10, 15, and 20 MHz), which is not a flexible solution for users having different data rate demands. Recent studies on FBMC modulation have not provided an effective remedy for such users. For that reason, FMT modulation can be modified for user demands on different data rates to allow nonuniformly divided bandwidth allocations as proposed by Çatak and Durak-Ata in [21]. The main contributions of [21] are as follows: (i) the classical FBMC modulation schemes are modified for user demands on data rates. (ii) The assignments of user bandwidths are done at the physical layer. (iii) The bandwidth allocations become adaptive for user requirements instead of system orders.

### 3.2. FMT with nonuniformly divided bandwidth allocation

The nonuniformly divided bandwidth allocation is important for users with different data rate demands. Data-rate demands of users depend on their applications. For instance, video streaming applications require higher data rates. On the other hand, machine-type communications (MTC), sensors, etc., need lower data rates. FMT with nonuniformly divided bandwidth allocation structures can serve to such heterogeneous users and applications in the same transceiver structure and assign users on bandwidth on the physical layer.

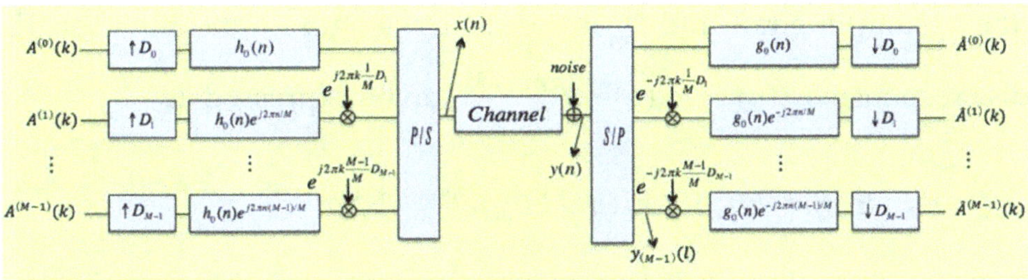

**Figure 10.** The block diagram of the FMT with nonuniformly divided bandwidth allocation.

The transceiver structure of the FMT multicarrier system for nonuniformly divided bandwidth allocations is given in **Figure 10**. Each user symbols ($A^{(m)}(k)$) in subbands are filtered by the frequency-shifted versions of a low-pass prototype filter after upsampling operations. The upsampling values and the filter lengths may be different for all subbands.

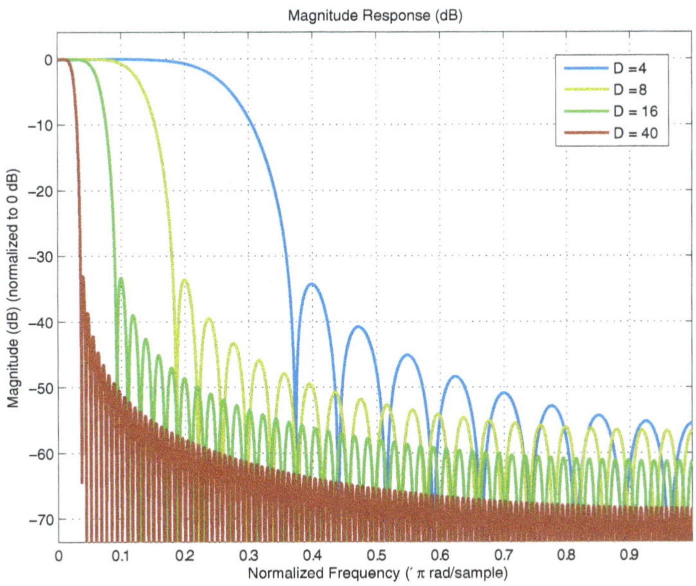


**Figure 11.** Frequency responses of raised cosine filters for different upsampling rates.

In **Figure 10**, the upsampling operation $\uparrow D$ is inserting $D-1$ zeros between consecutive samples. The frequency responses of the raised cosine filter for different upsampling numbers are given in **Figure 11**. Accordingly, if the sampling rate increases, the frequency resolution will be increased. Thus, the users need less bandwidth. According to the limit of time-bandwidth product, less bandwidth means longer symbol duration and also high latency.

The transmitted signal for FMT with nonuniformly divided bandwidth allocation is given by

$$\boldsymbol{x}(n) = \sum_{m=0}^{M-1} \sum_{k \epsilon T_m} \boldsymbol{A}^{(m)}(k)\, \boldsymbol{h}_0(n - kD_m) e^{j2\pi mn/M} \tag{14}$$

where $D_m$ is the upsampling rate and $T_m$ is the symbol length for the $m$th user. The prototype filter of impulse response $\boldsymbol{h}_m(n) = \boldsymbol{h}_0(n)e^{j2\pi mn/M}$ can be expressed as

$$\boldsymbol{h}_m(n - kD_m) = \boldsymbol{h}_0(n - kD_m)e^{j2\pi m(n-kD_m)/M} \tag{15}$$

And the transmitted signal in **Figure 10** becomes

$$\boldsymbol{x}(n) = \sum_{m=0}^{M-1} \sum_{k \epsilon T_m} \boldsymbol{A}^{(m)}(k)\, \boldsymbol{h}_m(n - kD_m) e^{j2\pi kmD_m/M}. \tag{16}$$

In the same way, the received signal is obtained by

$$\widehat{\boldsymbol{A}}^{(m)}(k) = \sum_{l=-\infty}^{\infty} \boldsymbol{y}_m(l) \boldsymbol{g}_m(lD_m - k)e^{-j2\pi mlD_m/M} \tag{17}$$

where $g_m(lD_m - k) = g_0(lD_m - k)e^{j2\pi m(lD_m - k)/M}$. If the transmitter filter $h_0(n)$ is assumed to be symmetric, the receiver filter $g_0(n)$ equals complex conjugate of $h_0(n)$. Finally, the received signal becomes

$$\widehat{\boldsymbol{A}}^{(m)}(k) = \sum_{l=-\infty}^{\infty} \boldsymbol{y}_m(l) \boldsymbol{h}_m^*(lD_m - k)e^{-j2\pi mlD_m/M} \tag{18}$$

### 3.3. Generalized frequency division multiplexing

GFDM can be considered as type of filter bank-based multicarrier modulation scheme with transmission filters that are shifted in time and frequency domains. This scheme offers more flexible pulse shaping for individual subcarriers [22]. However, GFDM has complicated receiver designs and needs high-order filtering and tail biting. To simplify transceiver structures, polyphase filters can be employed [10].

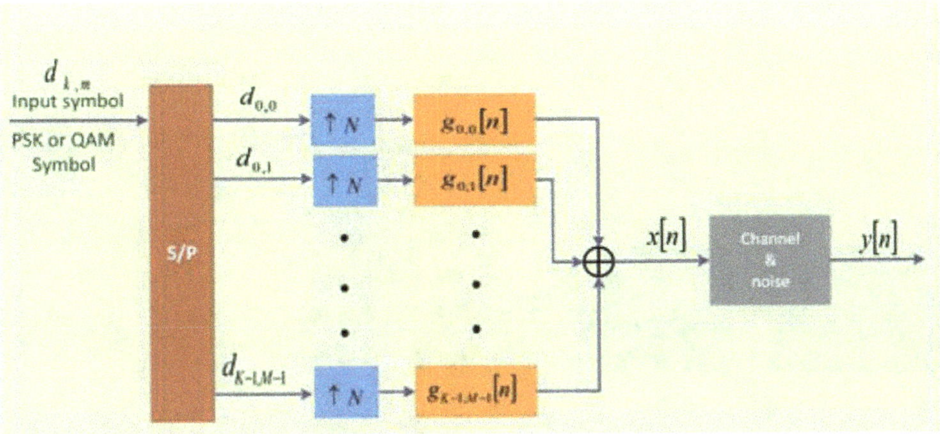


**Figure 12.** The transmitter structure of GFDM: the transmission filters are shifted in time and frequency domains.

The receiver structure of GFDM is given in **Figure 12**. Accordingly, the data is transmitted with $K$ subcarriers that carry $M$ subsymbols. Data is mapped into the complex valued QAM symbols. The mapped data are upsampled by the factor $N$, where $N = MK$. The transmitter filter $g_{k,\ m}[n]$ with $N$ samples is the time- and frequency-shifted version of $\boldsymbol{g}(n)$ that is expressed by

$$g_{k,m}[n] = \boldsymbol{g}[(n - mK)\mathrm{mod}N]\exp(-j2\pi\frac{k}{K}n) \tag{19}$$

where $k$ and $m$ are the subcarrier and subsymbol indices where $k = 0, 1, ..., K - 1$ and $m = 0, 1, ..., M - 1$, respectively. The transmitted signal is given by

$$x\left[n\right] = \sum_{m=0}^{M-1}\sum_{k=0}^{K-1} d_{k,m} g_{k,m}\left[n\right] \tag{20}$$

and Eq. (20) can be expressed with modulation matrix as

$$x = \mathbf{A}d \tag{21}$$

where $x$ is a vector that contains transmitted samples of $x[n]$ and $\mathbf{A}$ is the $KM \times KM$ modulation matrix that contains samples of transmitter filter $g_{k,\ m}[n]$ where

$$\mathbf{A} = \begin{bmatrix} \overbrace{\begin{matrix} g_{0,0}[0] & \cdots & g_{K-1,0}[0] \\ g_{0,0}[1] & \cdots & g_{K-1,0}[1] \\ \vdots & \ddots & \vdots \\ g_{0,0}[N-1] & \cdots & g_{K-1,0}[N-1] \end{matrix}}^{Subsymbol\ 0} & \begin{matrix} \cdots \\ \cdots \\ \\ \cdots \end{matrix} & \overbrace{\begin{matrix} g_{0,M-1}[0] & \cdots & g_{K-1,M-1}[0] \\ g_{0,M-1}[1] & \cdots & g_{K-1,M-1}[1] \\ \vdots & \ddots & \vdots \\ g_{0,M-1}[N-1] & & g_{K-1,M-1}[N-1] \end{matrix}}^{Subsymbol\ M-1} \end{bmatrix} \tag{22}$$

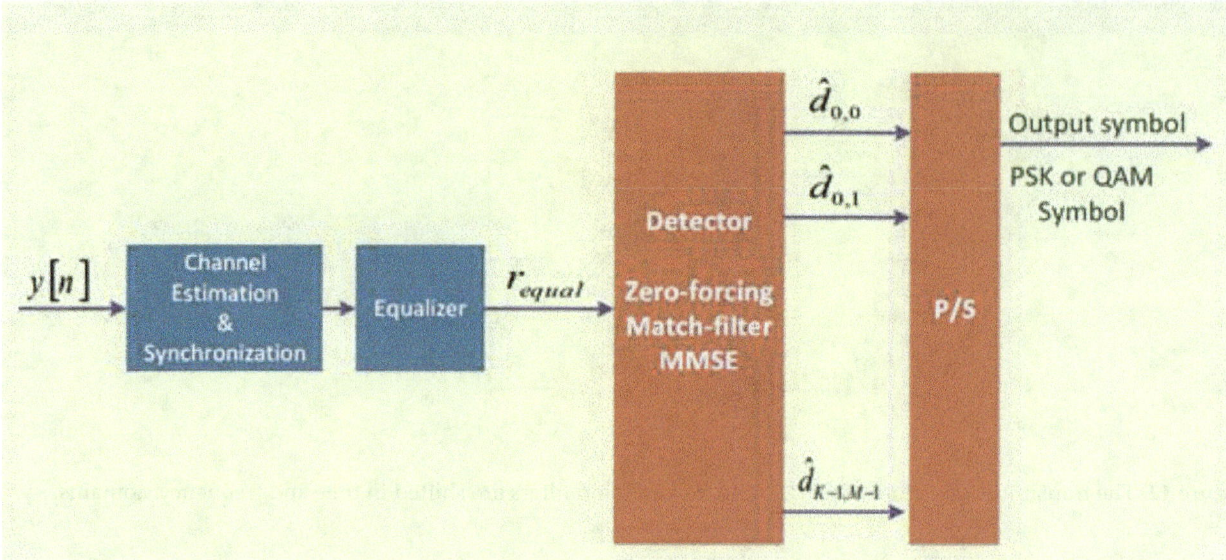


**Figure 13.** The receiver structure of GFDM with equalizer and detector.

The transmitter diagram of GFDM is given in **Figure 13**. The signal that passes through the channel must be equalized to clarify from the channel effects. If the number of subcarriers is high enough, the channel frequency response can be flat for each subcarrier. Thus, subcarrier bandwidths become smaller than the coherence bandwidth. In such a case, the received signal can be equalized with a zero-forcing equalizer. According to the zero-forcing equalizer, inverse of the frequency response of the channel is applied to the received signal. The implementation is simple for flat channels; otherwise, it becomes very hard due to inversing operations. The signal that passes through the channel is

$$\boldsymbol{y}[n] = \boldsymbol{h}[n] * \boldsymbol{x}[n] + \boldsymbol{w}[n] \tag{23}$$

where $w[n]$ is the additive noise and $c[n]$ is the impulse response of channel. The equalized signal with zero-forcing equalizer is given by

$$\mathbf{r}_{equal}[n] = IDFT\left\{\frac{Y(e^{j\omega})}{C(e^{j\omega})}\right\}, \tag{24}$$

where $Y\left(e^{j\omega}\right)$ and $C\left(e^{j\omega}\right)$ are the corresponding frequency responses. After equalization procees, the received signal can be estimated by a detection process. Zero-forcing receiver, matched-filter receiver, and minimum mean square error (MMSE) receiver structures are common detection methods.

Zero-forcing receiver is based on the inverse of modulation matrix in Eq. (21). Accordingly, the detected signal is

$$\hat{\boldsymbol{d}}_{\text{zero-forcing}} = \mathbf{A}^{-1}\boldsymbol{r}_{\text{equal}} \tag{25}$$

where $\mathbf{A}^{-1}$ is the inverse matrix of the modulation matrix $\mathbf{A}$ and $\boldsymbol{r}_{\text{equal}}$ is the equalized signal. The pseudo-inverse matrix can be used for the nonsquare case of $\boldsymbol{A}$. The pseudo-inverse of $\boldsymbol{A}$ can be evaluated by

$$\mathbf{A}^{+} = \mathbf{A}^{H}\left(\mathbf{A}\mathbf{A}^{H}\right)^{-1} \tag{26}$$

where $\boldsymbol{A}^{H}$ is Hermitan matrix of $\boldsymbol{A}$. Then, the detected signal by zero-forcing receiver in Eq. (25) becomes

$$\hat{\boldsymbol{d}}_{\text{zero-forcing}} = \mathbf{A}^{H}\boldsymbol{r}_{\text{equal}}. \tag{27}$$

Matched-filter receiver maximizes the SNR per subcarrier. The detected signal by the matched-filter receiver is given by

$$\hat{\boldsymbol{d}}_{\text{match-filtering}} = \mathbf{A}^{H}\boldsymbol{r}_{\text{equal}} \tag{28}$$

According to MMSE receiver, the detected signal is given by

$$\hat{\boldsymbol{d}}_{MMSE} = \mathbf{A}^{\dagger}\boldsymbol{r}_{equal} \text{ with } \mathbf{A}^{\dagger} = \left(\frac{\sigma^2{}_n}{\sigma^2{}_d}\boldsymbol{I} + \mathbf{A}^{H}\mathbf{A}\right)^{-1}\mathbf{A}^{H}. \tag{29}$$

where $\sigma^2{}_n$ and $\sigma^2{}_d$ are the variance of the noise and data symbol.

Briefly, zero-forcing receiver extracts the channel effects from the transmitted signal and removes all ISI for ideal noiseless channel condition. It amplifies the noise for noisy channels. The matched-filter receivers overperform the zero-forcing receiver in low SNR regime.

Matched-filter receiver suffers from self-interference. On the other hand, MMSE receiver is successful at high and low SNR similar to zero-forcing receiver and matched-filter receiver, respectively [23].

## 4. Performance evaluation

The waveform design issues depend on the requirements of users, communication types, and communication networks. These requirements are changing and evolving every year. Today, on the verge of 5G communication technology, most important requirements are data rate, latency, power, efficiency, complexity, and robustness to the channel [24]. Also, there are some design issues to execute these technology requirements. PAPR, OOB emission, interferences, and complexity issues are investigated and their importance is verified.

The PAPR is the ratio of peak power to the average power of a transmitted signal. A multicarrier signal consists of lots of modulated signals in each subcarrier, which can cause large PAPR value after addition. The comparisons of GFDM and OFDM on PAPR performances are given in **Figure 14**. Accordingly, the PAPR values of GFDM are better than OFDM. Low PAPR is important to reduce hardware cost and power consumption. One advantage of GFDM over OFDM is obviously in reducing the OOB radiation.

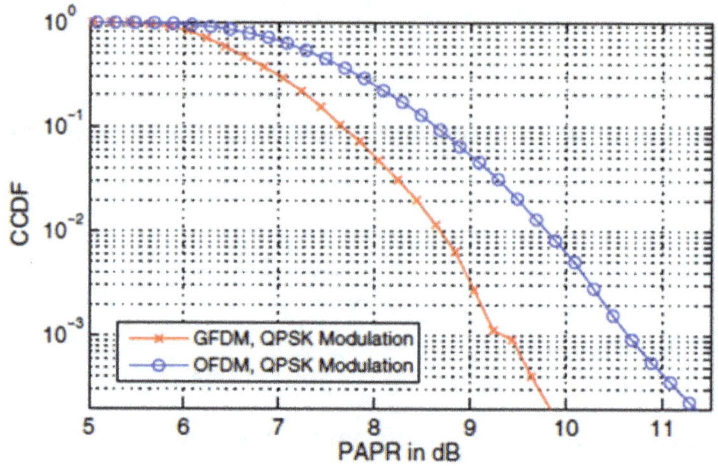


**Figure 14.** The comparison of PAPR of GFDM and OFDM: the PAPR of GFDM is less than OFDM. If multicarrier signals are summed up with same phases, the PAPR values increase [25].

The out-of-band (OOB) emission is the emission outside the necessary bandwidths. It causes waste of spectral resources and serious interference problems to adjacent wireless channels. These redundant emissions cause interference. Interference between carriers (ICI) and symbols (ISI) are two issues of waveform design. ICI is caused by channel frequency offsets and it is one of the major problems of OFDM. It can be avoided by frequency domain equalization, time domain windowing, and using redundant subcarrier between carriers. ISI is caused by the dispersion of the channel. It can be avoided by leaving enough space in between the transmitted symbols.

In **Figure 15**, OOB emissions of OFDM symbol and FBMC symbol are given comparatively. Here, OFDM suffers from high-level OOB emission. Conversely, filter bank-based operations allow lower out-of-band emissions.

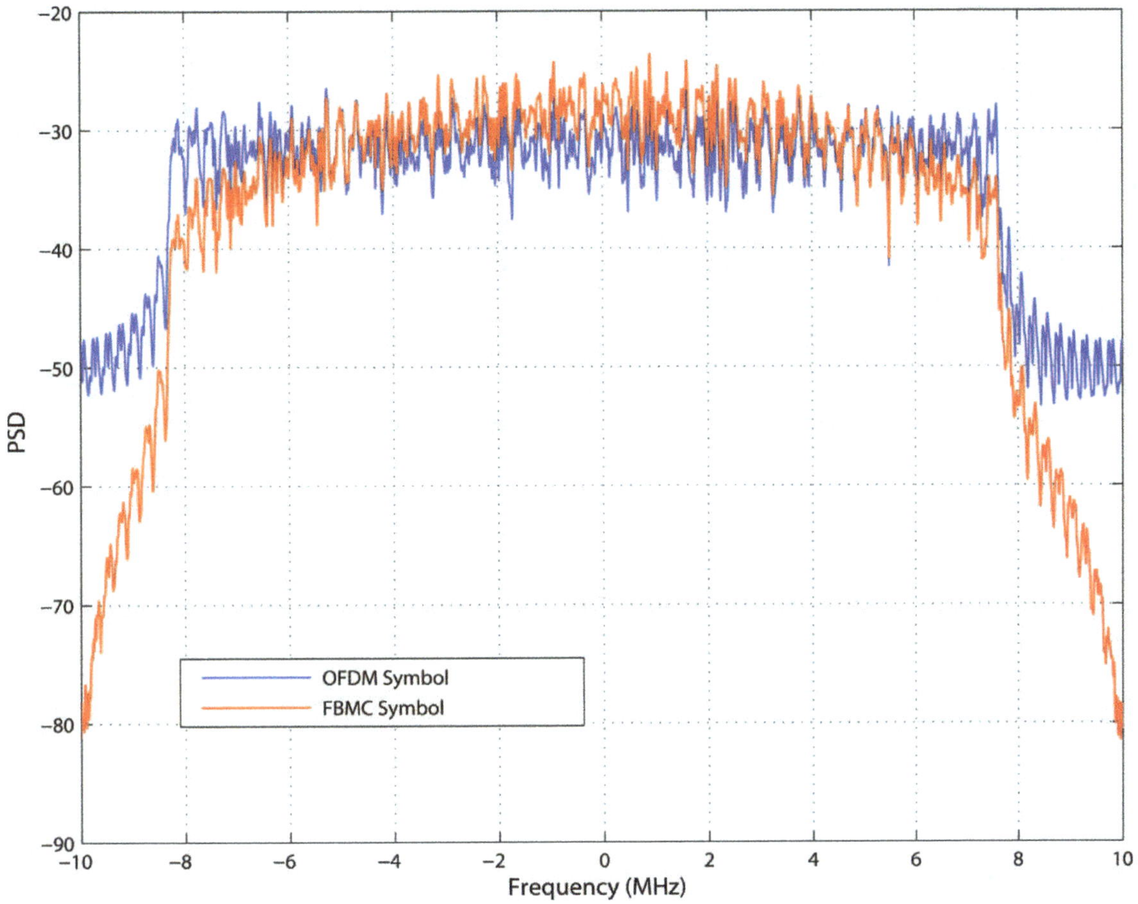


**Figure 15.** Power spectrum density of OFDM and FBMC symbols: FBMC scheme allows lower out-of-band emissions.

Complexity is defined by the total number of operations in the transmitters and receivers. The transmitter structures must be adapted to channel conditions and provide easy detection. Filtering operations make the systems more complex. Polyphase filter structures are used to overcome these problems. Another issue is channel equalization at the receivers while taking the inverse of a matrix. The performance evaluations are summarized in **Table 2**.

| | Complexity | OOB | PAPR | Spectral efficiency |
|---|---|---|---|---|
| OFDM | Low | High | High | Good |
| FBMC [6, 9] | High | Low | Low | Bad |
| GFDM [6, 25] | High | Low | Low | Good |
| UFMC [12, 13] | High | Low | High | Good |

**Table 2.** Pros and cons summary of waveforms.

## 5. Conclusion and future directions

This chapter presents the requirements of 5G communication systems and the fundamentals of waveform design to cover them for 5G wireless communication networks. According to the report of 5G PPP Architecture Working Group, the 5G network will "operate in a wide spectrum range with a diverse range of characteristics" [26]. Accordingly, the 5G communication channel will be heterogeneous and will provide users with different demands. The waveform design part of the physical layer is a critical issue in meeting the new demands and requirements, such as low latency, low power consumption, high data rates, and spectrum efficiency. TF lattice structures and pulse shaping must be determined. The transmission scheme, time and frequency allocation of symbols, resolution in time and frequency, and time-bandwidth product are the design criteria of time frequency lattice structures. Also, pulse shaping is the filtering process that maps the modulated signals to the TF lattice to control the interferences. Besides, transceiver scheme of some candidate waveforms and performance evaluations are given. Accordingly, OFDM has an easy implementation, but the high level of OOB emission and PAPR value. The waveforms that include filtering have lower OOB emission but high complexity.

In this chapter, the waveform design is assumed to be performed at baseband. On the other hand, one of the potential of 5G communication technologies under consideration is the use of millimeter wave frequencies. In this way, signals allocate more bandwidths to faster transmission, high-resolution video broadcasting, etc. Massive-MIMO and advanced beamforming technologies will allow high data rate.

## Author details

Evren Çatak[1] and Lütfiye Durak-Ata[2*]

*Address all correspondence to: lutfiye@ieee.org

1 Department of Electronics and Communications Engineering, Yıldız Technical University, Istanbul, Turkey

2 Informatics Institute, Istanbul Technical University, Istanbul, Turkey

## References

[1] Wunder G et al.: 5GNOW transceiver and frame structure concept. 5th generation non-orthogonal waveforms for asynchronous. Signalling (5GNOW) Project Report. 2015:D3.3

[2] Schaich F, Ringset V, Bellanger M, Zhang D, Ruyet D L: Compatibility of OFDM and FBMC systems and reconfigurability of terminals. Physical Layer for DYnamic AccesS and Cognitive Radio (PHYDYAS) Project Report. 2009:D7.1

[3] Chen T, Matinmikko M, Chen X, Zhou X, Ahokangas P: Software defined mobile networks: concept, survey, and research directions. IEEE Communications Magazine. 2015;53(11):126–133

[4] Andrews J G, Buzzi S, Choi W, Hanly S, Lozano A, Soong A C K, Zhang J C: What will 5G be? IEEE Journal on Selected Areas in Communications. 2014;32(6):1065–1082

[5] Wunder G, Jung P, Kasparick M, Wild T, Schaich F, Chen Y, Brink S, Gaspar I, Michailow N, Festag A, Mendes L, Cassiau N, Ktenas D, Dryjanski M, Pietrzyk S, Eged B, Vago P, Wiedmann F: 5GNOW: Non-orthogonal, asynchronous waveforms for future mobile applications. IEEE Communications Magazine. 2014:52(2):97–105

[6] Siohan P, Lin H. An advanced multi-carrier modulation for future radio systems. In IEEE International Conferance on Acoustics, Speech and Signal Processing(ICASSP); 2014.

[7] Van De Beek J, Berggren F: Out-of-band power suppression in OFDM. IEEE Communications Letters. 2008:12(9):609–611

[8] Hee Han S, Hong Lee J: An overview of peak-to-average power ratio reduction techniques for multicarrier transmission. IEEE Wireless Communications. 2005:12(2): 56–65

[9] Farhang-Boroujeny B: Filter bank multicarrier modulation: a waveform candidate for 5G and beyond. Advances in Electrical Engineering. 2014. 25 pages. Article ID 482805–25.

[10] N. Michailow et al.: Generalized frequency division multiplexing for 5th generation cellular networks. IEEE Transactions on Communications. 2014:62(9); pp. 3045–3061.

[11] Farhang A, Marchetti N, Doyle L E: Low complexity transceiver design for GFDM. In arXiv: 1501.02940 [Online] available: http://arxiv.org/abs/1501.02940, 2015

[12] Vakilian V, Wild T, Schaich F, Brink S T, Frigon J F: Universal-filtered multi-carrier technique for wireless systems beyond LTE. In IEEE Globecom Workshops (IEEE GC Wkshps); 2013:pp. 223–228

[13] Schaich F, Wild T. Waveform contenders for 5G – OFDM vs. FBMC vs. UFMC. In IEEE, Communications, Control and Signal Processing (ISCCSP); 2014; p. 457–460

[14] Gabor D: Theory of communication. Part 1: the analysis of information electrical engineers—Part III: radio and communication engineering. Journal of the Institution of Electrical Engineers; 1946:93(26):429–441

[15] Boashash B. Time-frequency signal analysis and processing: a comprehensive reference. First ed. Hungary:Academic Press; 2003.

[16] Michailow N, Fettweis G. Low peak-to-average power ratio for next generation cellular systems with generalized frequency division multiplexing. In International Symposium on Intelligent Signal Processing and Communications Systems (ISPACS); 2013; pp. 651–655.

[17] Aldirmaz S, Serbes A, Durak-Ata L: Spectrally efficient OFDMA lattice structure via toroidal waveforms on the time-frequency plane. EURASIP Journal on Advances in Signal Processing. 2010:14 pages. DOI: 10.1155/2010/684097

[18] Han F M, Zhang X D: Hexagonal multicarrier modulation: a robust transmission scheme for time-frequency dispersive channels. IEEE Transactions on Signal Processing. 2007:55(5):1955–1961

[19] Senay S, Durak L, Chaparro L F: A time-frequency division multiplexing communications system with hexagonal lattice structure. In Signal Processing Conference; 2009: pp. 1186–1189

[20] Amini P, Farhang-Boroujeny B: Packet format design and decision directed tracking methods for filter bank multicarrier systems. EURASIP Journal on Advances in Signal Processing. 2010:13 pages. DOI: 10.1155/2010/307983

[21] Catak E, L. Durak-Ata L: Filtered multitone system for users with different data rates at 5G wireless networks. In IEEE Signal Processing and Communication Application Conference (SIU); 2016; pp. 741–744

[22] Michailow N, Datta R, Krone S, Lentmaier M, Fettweis G: Generalized frequency division multiplexing: A flexible multi-carrier modulation scheme for 5th generation cellular networks. in German Microwave Conference (GeMiC); 2012

[23] Michailow N, Krone S, Lentmaier M, Fettweis G: Bit error rate performance of generalized frequency division multiplexing. In IEEE Vehicular Technology Conference (VTC); 2012; pp. 1–5

[24] Long Bao Le et al.: Enabling 5G mobile wireless technologies. EURASIP Journal on Wireless Communications and Networking. 2015;218:14 pages. DOI:10.1186/s13638-015-0452-9

[25] Fettweis G, Krondorf M, Bittner S: GFDM-Generalized frequency division multiplexing. In IEEE Vehicular Technology Conference (VTC); 2009; pp. 1–4

[26] "View on 5G architecture," 5G PPP Architecture Working Group, Version 1.0 July 2016.

# 5

# Superallocation and Cluster-Based Cooperative Spectrum Sensing in 5G Cognitive Radio Network

Md Sipon Miah, Md Mahbubur Rahman and Heejung Yu

**Abstract**

Consequently, the research and development for the 5G systems have already been started. This chapter presents an overview of potential system network architecture and highlights a superallocation technique that could be employed in the 5G cognitive radio network (CRN). A superallocation scheme is proposed to enhance the sensing detection performance by rescheduling the sensing and reporting time slots in the 5G cognitive radio network with a cluster-based cooperative spectrum sensing (CCSS). In the 4G CCSS scheme, first, all secondary users (SUs) detect the primary user (PU) signal during a rigid sensing time slot to check the availability of the spectrum band. Second, during the SU reporting time slot, the sensing results from the SUs are reported to the corresponding cluster heads (CHs). Finally, during CH reporting time slots, the CHs forward their hard decision to a fusion center (FC) through the common control channels for the global decision. However, the reporting time slots for the SUs and CHs do not contribute to the detection performance. In this chapter, a superallocation scheme that merges the reporting time slots of SUs and CHs by rescheduling the reporting time slots as a nonfixed sensing time slot for SUs to detect the PU signal promptly and more accurately is proposed. In this regard, SUs in each cluster can obtain a nonfixed sensing time slot depending on their reporting time slot order. The effectiveness of the proposed chapter that can achieve better detection performance under –28 to –10 dB environments and thus reduce reporting overhead is shown through simulations.

**Keywords:** 5G, software-defined network, cognitive radio, superallocation technique, cluster head, fusion center

## 1. Introduction

Around 2020, the promising 5G technology in cognitive radio networks is expected to be developed 5G networks that will have to support advanced services and multimedia applications with a wide variety of requirements, including higher peak and user data rates, reduced latency, enhanced indoor and outdoor coverage, improved energy efficiency, capacity and throughput, network densification, autonomous applications and network management, and Internet of things [1, 2].

The primary technologies and approaches to address the requirements for the 5G systems can be classified as follows [1, 2]:

- Network densification of existing mobile cellular networks (e.g., peer-to-peer [P2P], machine-to-machine [M2M], device-to-device [D2D], and heterogeneous networks);
- Full-duplex (FD) communication (e.g., simultaneous transmission and reception);
- Improvement of capacity and throughput (e.g., massive multiple-input multiple-output [massive MIMO]);
- Improvement of energy efficiency by wireless charging and energy harvesting;
- Advanced services and applications by a cloud-based radio access network (C-RAN) (e.g., smart city and service-oriented communication);
- Multiple network operators to share common resources by cooperation and network virtualization (e.g., network infrastructure, backhaul, licensed spectrum, core and radio access network, energy/power, etc.).

In this chapter, the main objectives of the beyond 2020 5G cognitive radio network by providing the technical support that needed to address the very challenging requirements foreseen for this time frame are proposed. A 5G system (i) has to be significantly more efficient in terms of energy, cost, and resource utilization (e.g., licensed spectrum utilization) than today's system (e.g., 4G); (ii) has to be significantly more versatile to support a significant diversity of requirements; and (iii) should provide better scalability in terms of the number of connected devices, densely deployed access points, spectrum usage, energy, and cost. In CRN, both higher data volume and higher data rates are required to access more spectrum band. As mentioned before, in 4G, it is clearly expected that more spectrum will be released for licensed wireless mobile communications. This new spectrum lies in the frequency range between 300 MHz and 6 GHz. However, for the future 5G system, these new spectrum opportunities will not be sufficient. Moreover, wireless local area networks operating in the unlicensed bands, such as the ISM and U-NII bands at 2.4 and 5 GHz, as well as the 60 GHz band, can be more tightly integrated. The present chapter discusses the superallocation and cluster-based cooperative spectrum sensing in the 5G CRN (e.g., highlights the number (i)) to provide more efficient spectrum utilization.

Cognitive radio (CR) is a new promising technology in the wireless communication era that has changed the policy of spectrum allocation from a static to a more flexible paradigm [3]. Recently, CRs that enable opportunistic access to underutilized licensed bands have been proposed as a promising technology for the improvement of spectrum operations. In an overlay cognitive radio network, an overlay waveform is used to exploit idle spectra and transmit information data within these unused regions. On the other hand, in an underlay cognitive radio network an underlay waveform with low transmit power is used to transmit data without harmful effects on the primary network [4]. In this chapter, we focus on overlay networks where secondary users find the idle channel with spectrum sensing. A precondition of secondary access is that there shall be no interference with the primary system [5]. This means spectrum sensing plays a vital role in the 5G CRN.

There are a number of spectrum sensing techniques, including matched filter detection, cyclostationary detection, and energy detection [6–8]. Matched filter detection is known as the optimum method for detection of the primary users when the transmitted signal is known. The main advantage of matched filtering is that it takes a short time to achieve spectrum sensing below a certain value for the probability of false alarm or the probability of detection compared to the other methods. However, it requires complete knowledge of the primary user's signaling features, such as bandwidth, operating frequency, modulation type and order, pulse shaping, and packet format. Cyclostationary detection is especially appealing because it is capable of differentiating the primary signal from the interference and noise. Due to noise rejection property, it works even in a very low signal-to-noise ratio (SNR) region, where the traditional signal detection method such as the energy detection is used. It offers good performance but requires knowledge of the PU cyclic frequencies and also requires a long time to complete sensing. On the other hand, the energy detection senses spectrum holes by determining whether the primary signal is absent or present in a given frequency slot. It operates without the knowledge of the primary signal parameters. Its key parameters, including detection threshold, number of samples, and estimated noise power, determine the detection performance. Also, it is an attractive and suitable method due to its easy implementation and low computation complexity. However, it is vulnerable to the uncertainty of noise power and cannot distinguish between noise and signal. Conversely, its major limitation is that the received signal strength can be dangerously weakened at a particular geographic location due to multipath fading and the shadow effect [9].

In order to improve the reliability of spectrum sensing, cooperative spectrum sensing was proposed [10–13]. Each SU performs local spectrum sensing independently and then forward the sensing results to the fusion center (FC) through the noise-free reporting channels between the SUs and the FC. In Zarrin and Lim [13], basic methods including AND, OR, and k-out-of-N logic are used to take hard decisions for a final decision at the FC. However, the reporting channels are always subject to fading effects in real environments [14]. When reporting channels become very noisy, cooperative sensing offers no advantages [15–16]. To overcome this problem, Sun et al. [17] and Xia et al. [18] proposed a cluster-based cooperative sensing scheme by dividing all the SUs into a number of clusters and selecting the most favorable SU in each cluster as a CH to report the sensing results, which can dramatically reduce the performance deterioration caused by fading of the wireless channels. In these schemes, the SU selected as the CH has to fuse sensing data from all cluster members (the SUs in this cluster). However, in these schemes, each SU's reporting time slot and the CH reporting time slot offer no contribution to spectrum sensing, while SU sensing and reporting times and CH reporting time are in different time slots.

Jin et al. [19] proposed a superposition-based cooperative spectrum-sensing scheme that increases the sensing duration by super positing the SUs' reporting duration into the sensing duration. However, this scheme adopts various individual reporting durations. In this case, synchronization problems occur at the FC. Moreover, the data processing burden at the FC increases for a large CR network.

In this chapter, we propose a superallocation and cluster-based cooperative spectrum sensing 5G scheme to provide more efficient spectrum sensing. In this scheme, each SU achieves a nonfixed and longer sensing time for sensing the PU signal bandwidth because both the SUs and the CHs are superallocated to different reporting time slots. On the other hand, both the SU and the CH reporting time slots are of fixed length because the synchronization problem for the FC is relieved. In addition, this proposed scheme decreases the data processing burden of the FC while all the SUs in the CRN are divided into fewer clusters such that each SU reports its local decision to the corresponding CH, which then reports to the FC. Simulation results show that the proposed 5G scheme can improve sensing performance in a low signal-to-noise ratio environment (i.e., –28 dB) and also greatly reduces reporting overhead in comparison with cluster-based cooperative spectrum sensing in 4G CRNs.

The remainder of this chapter is organized as follows. Section 2 describes the system model. Section 3 offers an overview of energy detection. Section 4 describes the cluster-based cooperative spectrum sensing in the 4G CRN. The proposed superallocation and cluster-based cooperative spectrum sensing in the 5G CRN is presented in Section 5 that addresses the spectrum utilization goal of this chapter for the 5G CRN. Some simulations and comparisons are presented in Section 6. We finally present the main conclusion of this chapter in Section 7.

## 2. Cognitive radio network system model

In CRN, the detection performance of the PU signal might be degraded when the sensing decisions are forwarded to an FC through fading channels. **Figure 1** shows the CRN deployment where SUs are grouped into a cluster governed by a CH based on low-energy adaptive clustering hierarchy-centralized (LEACH-C) protocol [20] and the CHs of the clusters report their decisions to an FC through a common control channel. Here, HDF will be applied to obtain a final decision on the presence of the PU activities. The process of the LEACH-C protocol is made up of several rounds, and each round consists of two phases: a setup phase when the CHs and clusters are organized and a steady-state phase when the cluster members begin to send their measurements to CH and CHs send their decision to the FC. In the setup phase, each SU sends information about its current location and SNR of reporting channel to the FC. Based on this information, the FC determines CHs among all CRUs, while the remaining CRUs will act as cluster members. After the CHs are determined, the FC broadcasts a message that contains not only the CH ID for each SU but also the information of time synchronization. If an SU's CH ID matches its own ID, the SU is a CH; otherwise, the SU is a cluster member and goes to sleep. In the steady-state phase, the SUs start to forward the measurement of the received PU's signal to the CH, and then the CH collects the measurements from the cluster members and makes the cluster decision about the presence of the PU and sends it to the FC during their allocated reporting time slots. Afterward, the FC combines

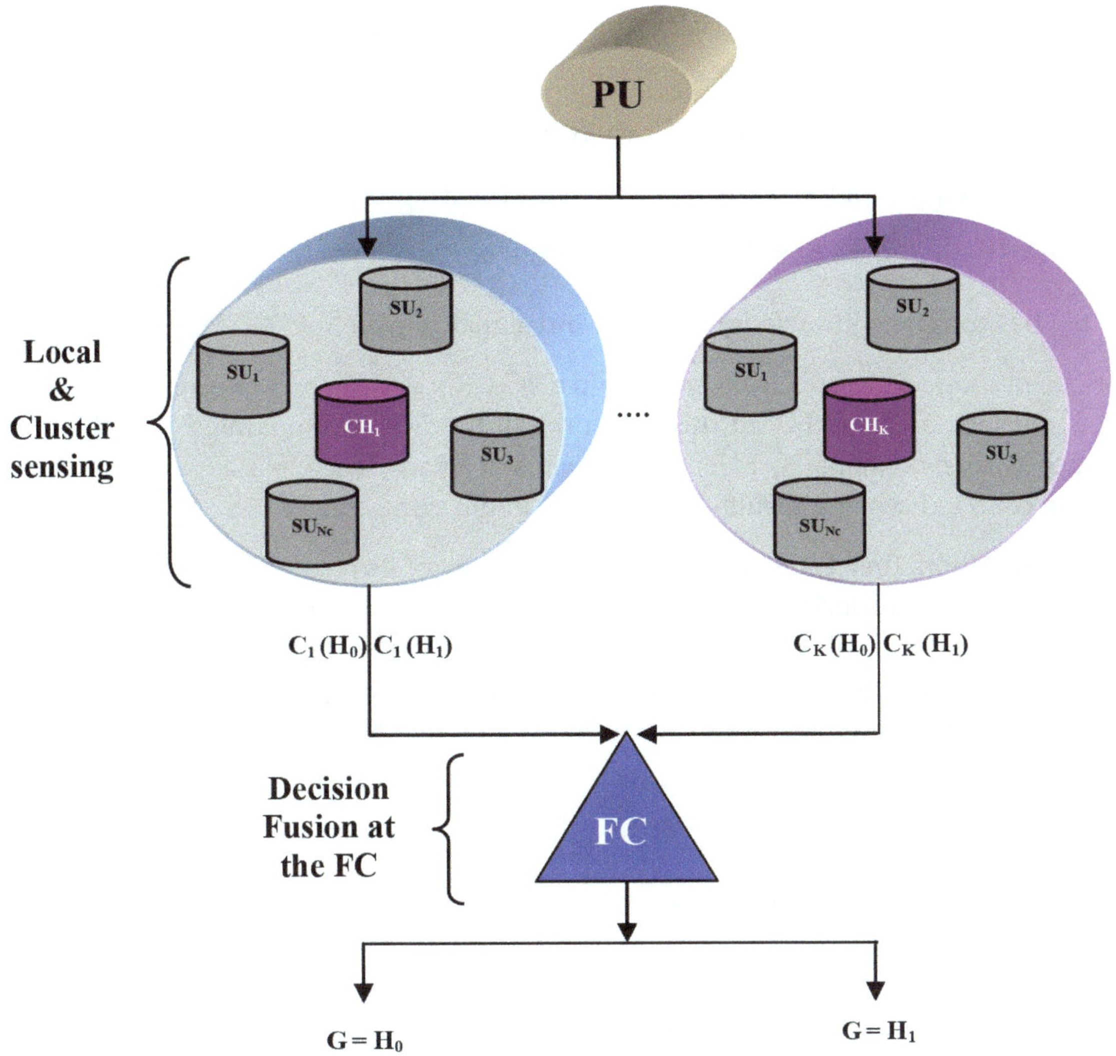


**Figure 1.** Cluster-based cooperative spectrum sensing in the 5G cognitive radio network.

the received clustering decision to make the final decision, then broadcasts it back to all CHs and the CHs send it to their cluster members.

Spectrum sensing can be formulated as a binary hypothesis-testing problem as follows:

$$\begin{cases} H_1: & \text{PU signal is present,} \\ H_0: & \text{PU signal is absent.} \end{cases} \tag{1}$$

Each SU implements a spectrum sensing process that is called local spectrum sensing to detect the PU's signal. According to the status of the PU, the received signal of an SU can be formulated as follows:

$$y_j(t) = \begin{cases} \eta_j(t), & H_0 \\ h_j(t)x(t) + \eta_j(t), & H_1 \end{cases} \tag{2}$$

where $y_j(t)$ represents the received signal at the $j$th SU, $h_j(t)$ denotes the gain of the channel between the $j$th SU and the PU, $x(t)$ with variance of $\sigma_x^2$ represents the signal transmitted by the

PU, and $\eta_j(t)$ is a circularly symmetric complex Gaussian (CSCG) with variance of $\sigma^2_{\eta,j}$ at the $j$th SU.

In addition, we make the following assumptions [21]:

- $x(t)$ is a binary phase shift keying (BPSK) modulated signal.
- $x(t)$ and $\eta_j(t)$ are mutually independent random variables.
- The SU has complete knowledge of noise and signal power.

Cluster-based cooperative spectrum sensing in a 5G CRN is shown in **Figure 1**, which contains N SUs, $K$ clusters, and one FC. In this network, all the SUs are separated into $K$ clusters, in which each cluster contains $N_c$ SUs; and the cluster head $CH_k$, $k = 1,2, \ldots, K$, is selected to process the collected sensing results from all SUs in the same cluster.

For sensing duration, first, each SU calculates the energy of its received signal in the frequency band of interest. Local decisions are then transmitted to the corresponding CH through a control channel, which will combine local decisions to make a cluster decision. Second, all cluster decisions will be forwarded to the FC through a control channel. At the FC, all cluster decisions from the CHs will be combined to make a global decision about the presence or the absence of the PU signal.

## 3. Overview of energy detection

The energy detection method has been demonstrated to be simple, quick, and able to detect primary signals, even if prior knowledge of the signal is unknown [22–25]. A block diagram of the energy detection method in the time domain is shown in **Figure 2**. To measure the energy of the signal in the frequency band of interest, a band-pass filter is first applied to the received signal, which is then converted into discrete samples with an analog-to-digital (A/D) converter.

An estimation of the received signal power is given by each SU with the following equation:

$$E_j = \frac{1}{L}\sum_{t=1}^{L} |y_j(t)|^2 \tag{3}$$

where $y_j(t)$ is the $t$th sample of a received signal at the $j$th SU and $L$ is the total number of samples. $L = T_s F_s$, where $T_s$ and $F_s$ are the sensing time and signal bandwidth in hertz, respectively. According to the central limit theorem, for a large number of samples, e.g., $L > 250$, the probability distribution function (PDF) of $E_j$, which is a chi-square distribution

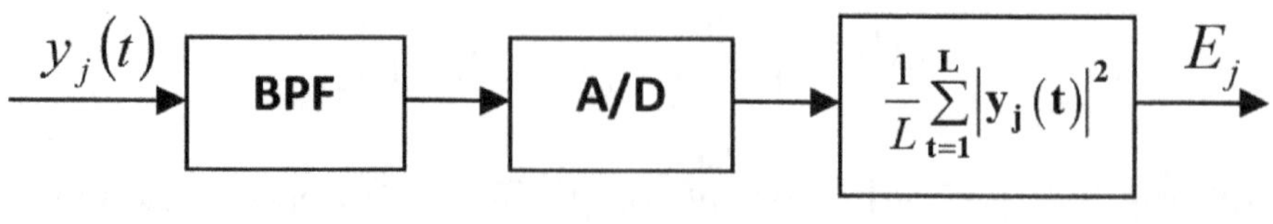


**Figure 2.** Block diagram of the energy detection scheme.

under both hypothesis $H_0$ and hypothesis $H_1$, can be well approximated as a Gaussian random variable such that

$$E_j = \begin{cases} N(\mu_{0,j},\ \sigma_{0,j}^2) \\ N(\mu_{1,j},\ \sigma_{1,j}^2) \end{cases} \tag{4}$$

where $N(\mu,\ \sigma^2)$ denotes a Gaussian distribution with mean of $\mu$ and variance of $\sigma^2$, $\mu_{0,j}$ and $\sigma_{0,j}^2$ represent the mean and variance, respectively, for hypothesis $H_0$, and $\mu_{1,j}$ and $\sigma_{1,j}^2$ represent the mean and variance for hypothesis $H_1$.

**Lemma 1.** When the primary signal is a BPSK-modulated signal and noise is a CSCG, the decision rule in Eq. (4) is modified as follows:

$$E_j = \begin{cases} N\left(\sigma_\eta^2,\ \frac{1}{L}\sigma_\eta^4\right) \\ N\left(\sigma_\eta^2(1+\gamma),\ \frac{1}{L}(1+2\gamma)\sigma_\eta^4\right) \end{cases} \tag{5}$$

where $\gamma = \frac{\sigma_x^2}{\sigma_\eta^2}$ that is the SNR of the primary signal at the $j$th SU. The SNR is a constant in the nonfading additive white Gaussian noise environment [25]. Here, we omit the subscript of $j$ in $\sigma_{\eta,j}^2$, which denotes that index of SU, to simplify the notation.

**Proof.** For hypothesis $H_1$, the mean $\mu_{1,j}$ is expressed as

$$\begin{aligned} \mu_{1,j} &= \sigma_x^2 + \sigma_\eta^2 = \sigma_\eta^2\left(1 + \frac{\sigma_x^2}{\sigma_\eta^2}\right) \\ &= (1+\gamma)\sigma_\eta^2 \end{aligned} \tag{6}$$

From Boyed and Vandenberghe [26], variance $\sigma_{1,j}^2$ is

$$\sigma_{1,j}^2 = \frac{1}{L}[E|x(t)|^4 + E|\eta(t)|^4 - (\sigma_x^2 - \sigma_\eta^2)^2] \tag{7}$$

For a complex $M$-array quadrature amplitude modulation signal [27], $E|x(t)|^4$ is given as

$$E|x(t)|^4 = \left(3 - \frac{2}{5}\frac{(4M-1)}{(M-1)}\right)\sigma_x^4 \tag{8}$$

For the BPSK signal [27], then we set $M = 4$. By substituting the value $M = 4$ into Eq. (8), we obtain

$$E|x(t)|^4 = \sigma_x^4 \tag{9}$$

For the CSCG noise signal [26], $E|\eta(t)|^4$ is given as

$$E|\eta(t)|^4 = 2\sigma_\eta^4 \tag{10}$$

Substituting the values $E|x(t)|^4$ and $E|\eta(t)|^4$ into Eq. (7), we obtain

$$\begin{aligned}\sigma_{1,j}^2 &= \frac{1}{L}[\sigma_x^4 + 2\sigma_\eta^4 - (\sigma_x^4 - 2\sigma_x^2\sigma_\eta^2 + \sigma_\eta^4)]\\ &= \frac{1}{L}[\sigma_\eta^4 + 2\sigma_x^2\sigma_\eta^2] = \frac{1}{L}\left[1 + 2\frac{\sigma_x^2}{\sigma_\eta^2}\right]\sigma_\eta^4\\ &= \frac{1}{L}[1 + 2\gamma]\sigma_\eta^4.\end{aligned} \tag{11}$$

For hypothesis $H_0$, substituting the value $\sigma_x^2 = 0$ into Eq. (6), mean $\mu_{0,j}$ is expressed as

$$\mu_{0,j} = \sigma_\eta^2 \tag{12}$$

Again, substituting the value $\sigma_x^2 = 0$ into Eq. (7), variance $\sigma_{0,j}^2$ is expressed as

$$\begin{aligned}\sigma_{0,j}^2 &= \frac{1}{L}[E|\eta(t)|^4 - (\sigma_\eta^2)^2]\\ &= \frac{1}{L}[2\sigma_\eta^4 - \sigma_\eta^4]\\ &= \frac{1}{L}\sigma_\eta^4\end{aligned} \tag{13}$$

Then, we can have distributions of a decision statistic under null and alternative hypotheses as in Eq. (5).

By the definition of a false alarm probability in a hypothesis testing with a decision statistic of $E_j$ depending on $T_s$, and a decision threshold of $\lambda_j$, the probability of false alarm for the $j$th SU is given by

$$\begin{aligned}P_f^j(T_s, \lambda_j) &= \Pr[E_j \geq \lambda_j | H_0]\\ &= Q\left(\frac{\lambda_j - \mu_{0,j}}{\sqrt{\sigma_{0,j}^2}}\right)\end{aligned} \tag{14}$$

where $Q(x)$ is the Gaussian tail function given by $Q(x) = \frac{1}{\sqrt{2\pi}}\int \exp\left(-\frac{t^2}{2}\right)dt$. Form Lemma 1, the probability of false alarm under a CSCG noise is given by

$$P_f^j(T_s, \lambda_j) = Q\left(\left(\frac{\lambda_j}{\sigma_\eta^2} - 1\right)\sqrt{T_s F_s}\right) \tag{15}$$

By the definition of a probability of detection in hypothesis testing and Lemma 1, the detection probability for the BPSK-modulated primary signal under a CSCG noise for the $j$th SU is given by

$$P_d^j(T_s, \lambda_j) = \Pr[E_j \geq \lambda_j | H_1] = Q\left(\frac{\lambda_j - \mu_{1,j}}{\sqrt{\sigma_{1,j}^2}}\right) = Q\left(\left(\frac{\lambda_j}{\sigma_\eta^2} - \gamma - 1\right)\sqrt{\frac{T_s F_s}{(1+2\gamma)}}\right) \quad (16)$$

The last equality is obtained by using Eq. (5).

With Eqs. (15) and (16), the probabilities of false alarm and the detection of the PU signal can be calculated when the duration of sensing time $T_s$ is given.

## 4. Cluster-based cooperative spectrum sensing in the 4G CRN

A general frame structure for the cluster-based cooperative spectrum sensing in the 4G CRN is shown in **Figure 3**. With this frame structure, all local decisions are forwarded to the CHs in the scheduled SU reporting time slots and are then forwarded to the FC in the scheduled CH reporting time slots.

**Lemma 2.** In the cluster-based cooperative spectrum sensing in the 4G CRN, the $N$ SUs in the network adopted fixed sensing time slot $T_s^{con}$ are given by

$$T_s^{con} = \frac{1}{F_s \gamma^2}\left[Q^{-1}(P_f^j) - Q^{-1}(P_d^j)\sqrt{(1+2\gamma)}\right]^2 \quad (17)$$

to sense the PU's signal with false alarm and detection probabilities of $P_f^j$ and $P_d^j$, respectively. Here, the superscript "con" means the conventional or 4G CRN.

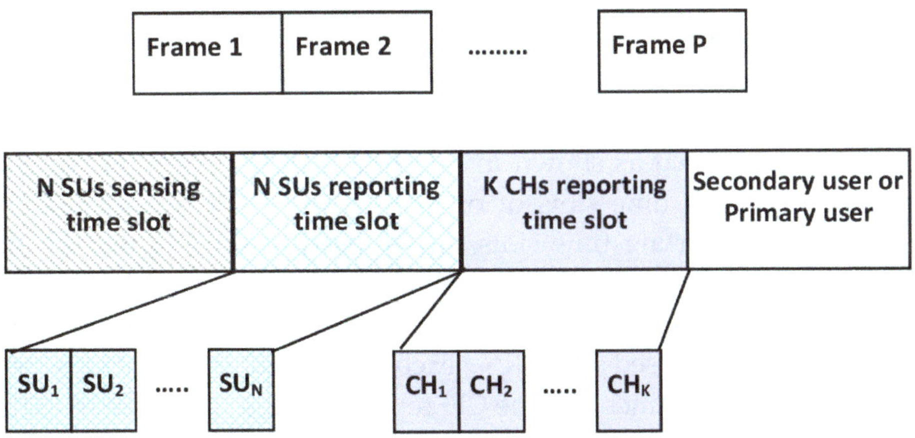


**Figure 3.** A cluster-based cooperative spectrum sensing in a 4G CRN [18].

**Proof:** We focus on the BPSK signal and CSCG noise. The probability of detection can be obtained with Eq. (18) by using Eq. (17):

$$\left(\frac{\lambda_j}{\sigma_\eta^2}-\gamma-1\right)\sqrt{\frac{T_sF_s}{(1+2\gamma)}}=Q^{-1}(P_d^j) \tag{18}$$

From Eq. (15), the probability of false alarm can be obtained by

$$\left(\frac{\lambda_j}{\sigma_\eta^2}-1\right)\sqrt{T_sF_s}=Q^{-1}(P_f^j)\,. \tag{19}$$

By substituting Eq. (19) into Eq. (18) and rewriting this equation, we have

$$\begin{aligned}
&\left(\frac{Q^{-1}(P_f^j)}{\sqrt{T_sF_s}}-\gamma\right)\sqrt{T_sF_s}=Q^{-1}(P_d^j)\sqrt{(1+2\gamma)}\\
&Q^{-1}(P_f^j)-\gamma\sqrt{T_sF_s}=Q^{-1}(P_d^j)\sqrt{(1+2\gamma)}\\
&\sqrt{T_sF_s}=\frac{1}{\gamma}\left[Q^{-1}(P_f^j)-Q^{-1}(P_d^j)\sqrt{(1+2\gamma)}\right]\\
&T_s=\frac{1}{F_s\gamma^2}\left[Q^{-1}(P_f^j)-Q^{-1}(P_d^j)\sqrt{(1+2\gamma)}\right]^2
\end{aligned} \tag{20}$$

Defining the sensing time with the last equation in Eq. (20), i.e., $T_s^{con}=T_s$, we can meet the requirement on false alarm and detection probabilities.

Because all SUs in $k$ clusters have the same fixed sensing time slot, $T_s^{con}$, the sensing performances, i.e., false alarm and detection probabilities, depend on the SNR of an SU. Therefore, sensing performance is not improved with a fixed sensing time slot. In addition, the reporting time slots for the SU and the CH are not utilized by the 4G CRN.

## 5. Proposed superallocation and cluster-based cooperative spectrum sensing in the 5G CRN

In the 4G CRN approach, sensing time slots, reporting time slots of SUs, and reporting time slots of CHs are strictly divided as shown in **Figure 3**. Due to this rigid structure in the 4G CRN approach, the reporting time slots of other SUs and CHs are not used for spectrum sensing. However, these reporting time slots can be used in sensing the spectrum by other SUs by scheduling sensing and reporting time slots effectively. To this end, a superallocation and cluster-based cooperative spectrum sensing in the 5G CRN is proposed by increasing the sensing time slot. In the proposed 5G CRN, each SU can obtain longer sensing time slot because the other SU reporting times and the CH reporting times are merged to the SU sensing time. Therefore, the sensing time slots for SUs in the proposed 5G CRN can be longer than those in the 4G CRN.

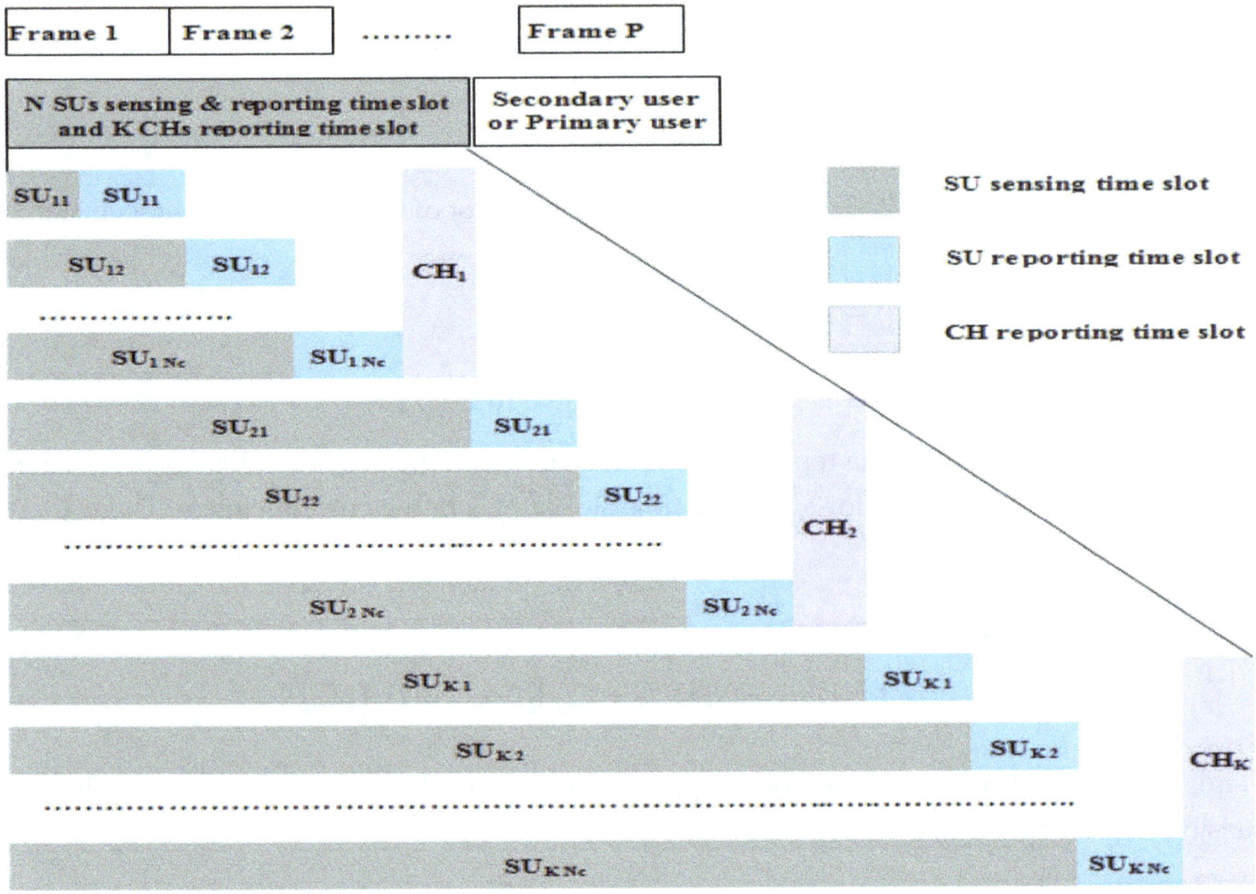


**Figure 4.** A superallocation and cluster-based cooperative spectrum sensing in the 5G CRN.

**Figure 4** shows the proposed scheduling method of sensing and reporting time slots in the superallocation for cluster-based cooperative spectrum sensing in the 5G CRN. In the figure, $SU_{nk}$ means the $k$th SU in the $n$th cluster in the network. To explain the duration of sensing time slot for $SU_{nk}$, we can define the durations of the sensing and reporting time for $SU_{nk}$ with $T_s^{nk}$ and $T_r^{nk}$, respectively.

In this proposed scheme, the sensing time slot for the first SU in the first cluster, i.e., $SU_{11}$, is equal to the sensing time slot in the 4G CRN, i.e., $T_s^{11} = T_s^{con} = T_s$. Except for $SU_{11}$, other SUs can obtain longer sensing time slots by scheduling SU reporting slots followed by the reporting slot for the CH of that cluster. With such a scheduling method, SUs can sense the spectrum during the reporting time slots of other SUs and CHs. For example, the sensing time slot of $SU_{12}$, $T_s^{12}$ is equal to the total duration of sensing time slot and the reporting time slot of the $SU_{11}$, i.e., $T_s^{12} = T_s + T_r^{11}$. Similarly, $T_s^{13}$ becomes the sum of the sensing duration of $SU_{12}$ and the reporting duration of $SU_{12}$, i.e., $T_s^{13} = T_s^{12} + T_r^{12} = T_s + \sum_{i=1}^{2} T_r^{1i}$. Obviously, the relationship of the sensing time slot $T_s^{1(j+1)}$ of the $SU_{1\ (j+1)}$ with the sensing time slot and the reporting time slot of the previous SUs can be given by

$$T_s^{1(j+1)} = T_s^{1j} + T_r^{1j} = T_s + \sum_{i=1}^{j} T_r^{1i} \tag{21}$$

for $j = 1, 2, 3, \ldots N_c$.

When $T_r^{\mathrm{prop}} = T_r^{1j}$ for $j = 1, 2, 3, \ldots, N_c$, the sensing time slot of the $j$th SU in the first cluster is written as

$$T_s^{1j} = T_s + (j{-}1)T_r^{\mathrm{prop}} \tag{22}$$

Therefore, $T_s^{1j}$ in the first cluster is greater than or equal to $T_s^{\mathrm{con}}$.

For SU in the other clusters, the reporting time slots of SUs in the previous clusters and that of the previous CH can be used for a sensing time slot of SUs in the current cluster. Thus, $T_s^{nj}$ is given by

$$\begin{aligned} T_s^{nj} &= \sum_{i=1}^{n-1} T_s^{iN_c} + \sum_{i=1}^{k} T_r^{ni} \\ &= (n{-}1)(T_s + N_c T_r^{\mathrm{prop}} + T_{r,CH}^{\mathrm{prop}}) + T_s + (j{-}1)T_r^{\mathrm{prop}} \end{aligned} \tag{23}$$

Here, $T_{r,CH}^{\mathrm{prop}}$ is the duration of the reporting time slot of a CH. Therefore, we can obtain longer sensing time as the index of CH increases.

### 5.1. Local sensing

As shown in Eq. (16), the detection probability $P_d^j$ is a function of parameters $\lambda_j$, $\gamma$, and $T_s F_s$. For fixed $F_s$, $\gamma$ and $\lambda_j$, $P_d^j$ is a function of $T_s$, which can be represented as $P_d^j(T_s)$.

**Lemma 3.** In the proposed cluster-based cooperative spectrum sensing in the 5G CRN, the $N$ SUs in the network adopts nonfixed sensing time slot $T_s^{nk}$ ($\geq T_s^{\mathrm{con}}$) in Eq. (23) to sense the PU's signal. Therefore, the sensing performance in the 5G CRN is improved over the 4G CRN.

**Proof:** Let $P_{d(\mathrm{con})}^j$ and $P_{d(\mathrm{prop})}^{1j}$ denote the probability of detection for the conventional and proposed schemes, respectively. When SU belongs to the first cluster, the CH reporting time slot is not included in its sensing time. Here, the subscript "prop" means the proposed scheme in the 5G CRN.

Substituting the values of $T_s$ and $T_s^{1j}$ into Eq. (16), we have

$$P_{d(\mathrm{con})}^j(T_s, \lambda_j) = Q\left(\left(\frac{\lambda_j}{\sigma_\eta^2} - \gamma - 1\right)\sqrt{\frac{T_s F_s}{(1+2\gamma)}}\right) \tag{24}$$

$$P_{d(\mathrm{prop})}^{1j}(T_s^{1j}, \lambda_j) = Q\left(\left(\frac{\lambda_j}{\sigma_\eta^2} - \gamma - 1\right) \times \sqrt{\frac{\left(T_s + (j{-}1) \times T_r^{\mathrm{prop}}\right) \times F_s}{(1+2\gamma)}}\right) \tag{25}$$

When the sensing time $T_s^{1j}$ becomes longer, then obviously the detection probability $P^j_{d(\text{prop})}$ increases. Hence, we show that

$$P^{1j}_{d(\text{prop})} \geq P^j_{d(\text{con})} \tag{26}$$

Because $\left(T_s + (j-1) \times T_r^{\text{prop}}\right) \geq T_s^{\text{con}}$ for $j = 1, 2, 3, \ldots, N_c$. When $j = 1$, then we obtain $P^{1j}_{d(prop)} = P^j_{d(con)}$.

If SU is not included in the first cluster, $P^{nj}_{d(\text{prop})}$ denotes the probability of detection for the proposed scheme. In this case, the sensing time slot includes the CH reporting time slots. Substituting the value of $T_s^{nj}$ into Eq. (16), we obtain

$$P^{nj}_{d(\text{prop})}\left(T_s^{nj}, \lambda_j\right) = Q\left(\left(\frac{\lambda_j}{\sigma_\eta^2} - \gamma - 1\right) \times \sqrt{\frac{\left((n-1)\left(T_s + N_c T_r^{\text{prop}} + T_{r,CH}^{\text{prop}}\right) + T_s + (k-1) T_r^{\text{prop}}\right) \times F_s}{(1+2\gamma)}}\right) \tag{27}$$

Therefore, $P^{nj}_{d(\text{prop})}\left(T_s^{nj}, \lambda_j\right) > P^{(n-1)N_c+j}_{d(\text{con})}\left(T_s, \lambda_j\right)$.

Each SU makes a local hard decision $d_j^{hd}$ as follows.

$$d_{nj}^{hd} = \begin{cases} 1, & if\ P^{nj}_{d(\text{prop})} > P^{nj}_{f(\text{prop})} \\ 0, & \text{Otherwise} \end{cases} \tag{28}$$

### 5.2. Cluster decision

At the $n$th CH, all local decisions $d_{nj}^{hd}$ received from the SUs will be combined to make a cluster decision $Q_{d,n}^{\text{prop}}$ as follows:

$$Q_{d,n}^{\text{prop}} = \begin{cases} 1, & \sum_{j=1}^{N_c} d_{nj}^{hd} > \xi \\ 0, & \text{Otherwise} \end{cases} \tag{29}$$

where $\xi$ is the threshold for the cluster decision.

### 5.3. Global decision

At the FC, all cluster decisions $(Q_{d,n}^{\text{prop}})$ received will be combined to make a global decision $(G)$ about the presence or the absence of the PU signal by using a $\tau$-out-of-$K$ rule as follows:

$$G = \begin{cases} 1, & if\ \sum_{n=1}^{K} Q_{d,n}^{\text{prop}} \geq \tau \quad : H_1 \\ 0, & \text{Otherwise} \quad : H_0 \end{cases} \tag{30}$$

where $\tau$ is the threshold for the global decision.

## 6. Simulation and result analysis

To evaluate the performance of the proposed cluster-based cooperative spectrum sensing in the 5G CRN, Monte Carlo simulations were carried out under following conditions:

The number of SUs is 12.

The number of clusters is 3.

The number of SUs in each cluster is 4.

The durations of sensing, SU reporting, and CH reporting time slots are 1 ms.

Average SNR of each SU in a cluster is –17 dB.

The PU signal is a BPSK signal.

The noise in SUs is CSCG.

The number of samples is 300.

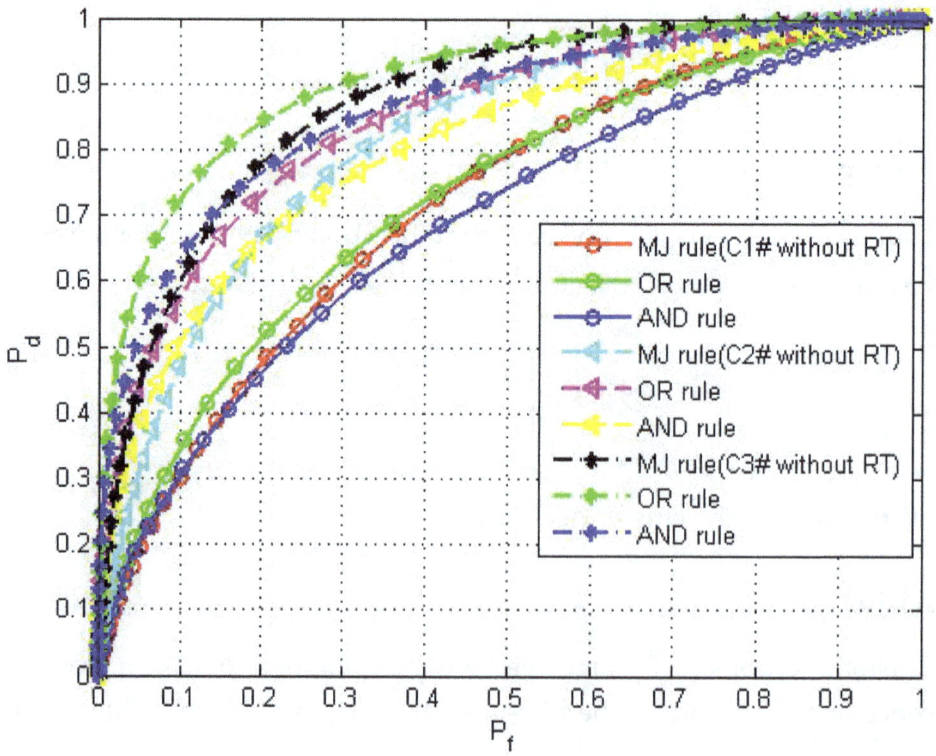


**Figure 5.** ROC curves of the proposed 5G scheme without cluster reporting time where C1#, C2#, and C3# mean the first, second, and third clusters, respectively.

First, the sensing performance of the proposed 5G and 4G cluster-based schemes, in terms of receiver operating characteristic (ROC), was evaluated under a CSCG channel. In this simulation, each SU conducts local sensing using equal gain combining (EGC).

**Figures 5** and **6** show ROC curves for the proposed 5G cluster-based schemes without and with cluster reporting time (RT), respectively. The proposed 5G scheme outperforms in the detection of the PU compared with the 4G scheme because the proposed superallocation technique can have longer sensing time than the 4G one. Test statistics (Eq. (25)) was considered for the proposed 5G scheme without reporting time for the cluster decision. In addition, test statistics (Eq. (27)) was considered for the proposed 5G scheme with reporting time for the cluster decision. When the index of the cluster increases from one to three, the detection probability increases (**Figures 7** and **8**).

From the detection efficiency of cooperative spectrum sensing, the probability of detection is 0.8 and the probability of false alarm is 0.2. However, in the worst environment, we need the probability of detection to be more than 0.9 and the probability of false alarm to be less than 0.1. In the 4G scheme, we can achieve these sensing performances with a longer sensing time slot but the throughput of the 4G cognitive radio network decreases. In the proposed 5G CRN, we can easily achieve more than 0.9 and less than 0.1 for the probabilities of detection and false alarm, respectively, because SU reporting time and CH reporting time merge to sense the PU signal without decreasing system throughput.

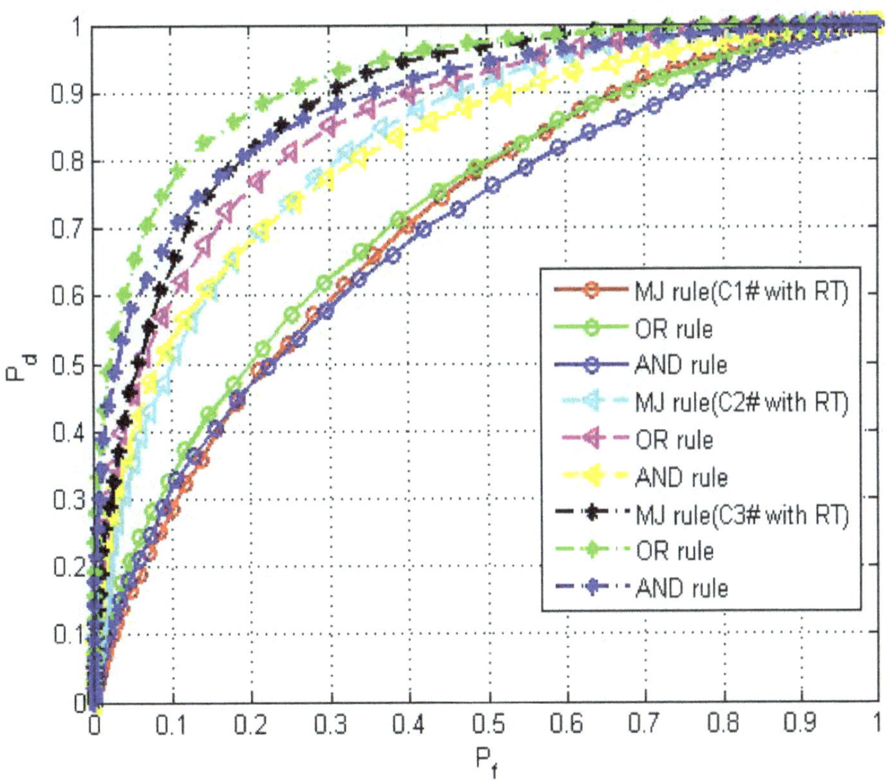


**Figure 6.** ROC curves of the proposed 5G scheme with cluster reporting time.

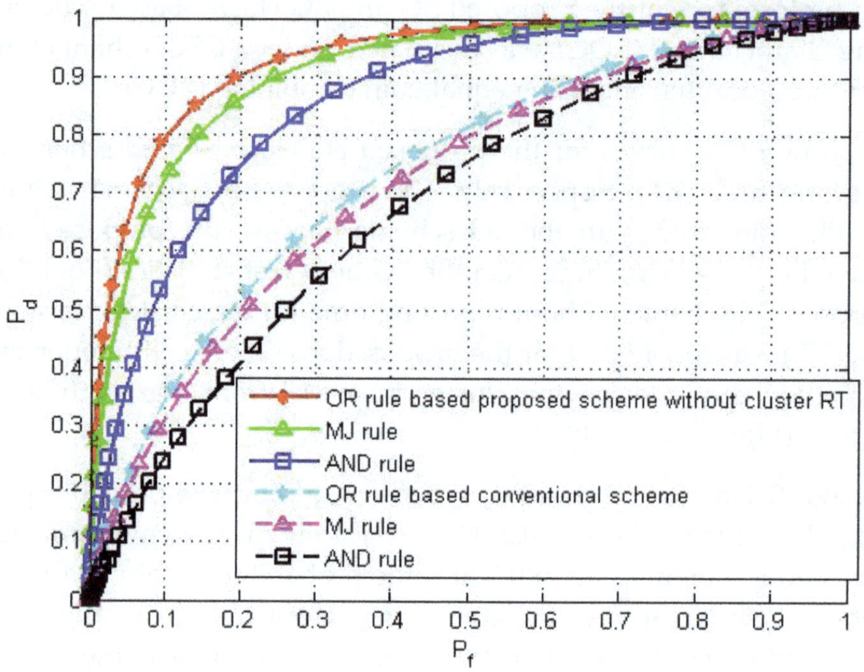


**Figure 7.** ROC curves of the proposed 5G scheme without cluster reporting time and the 4G scheme.

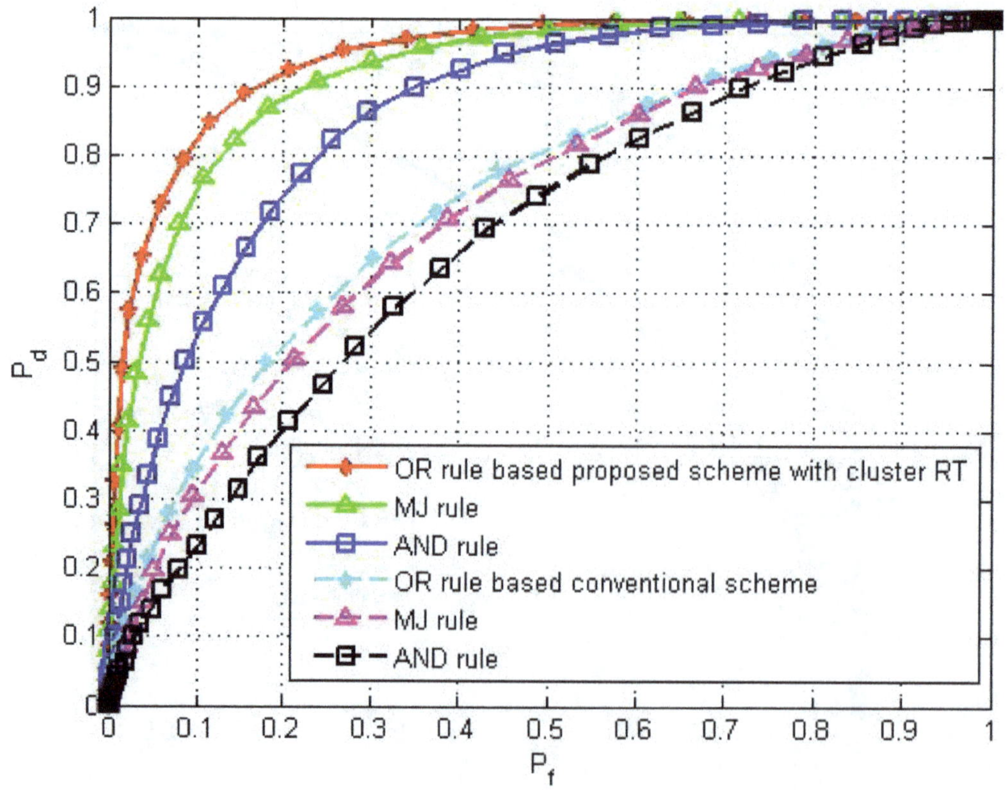


**Figure 8.** ROC curves of the proposed 5G scheme with cluster reporting time and the 4G scheme.

Second, the simulation was carried out under conditions whereby the SNRs of the PU's signal at the nodes are from –28 to –10 dB. The ROC curves of the proposed 5G scheme without

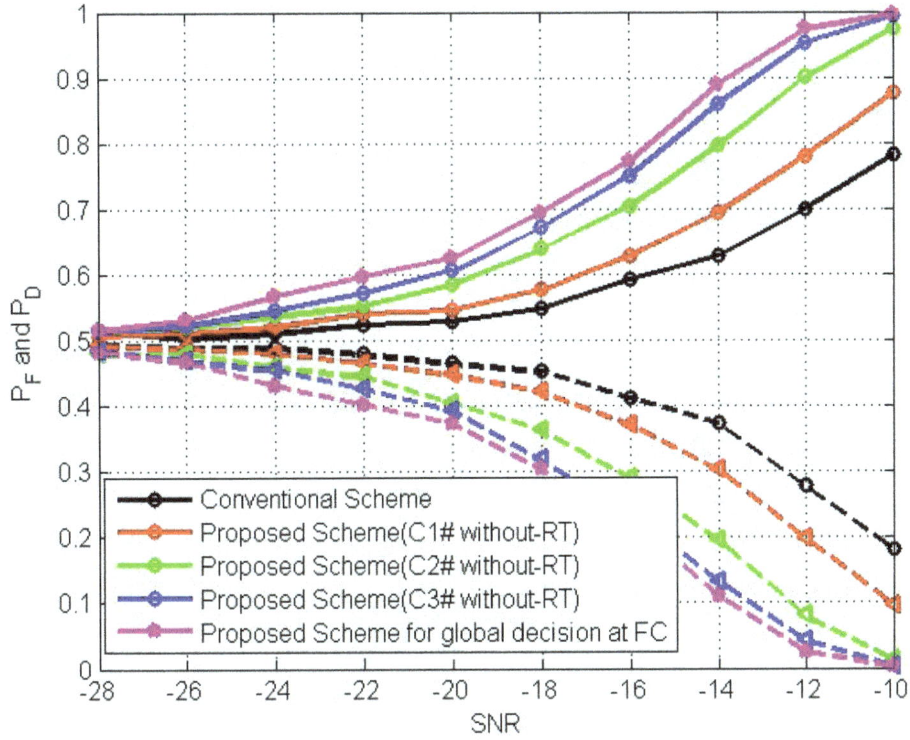


**Figure 9.** ROC curves of the proposed 5G scheme without cluster reporting time and the 4G scheme where SNRs of the PU's signal at the nodes are from –28 to –10 dB.

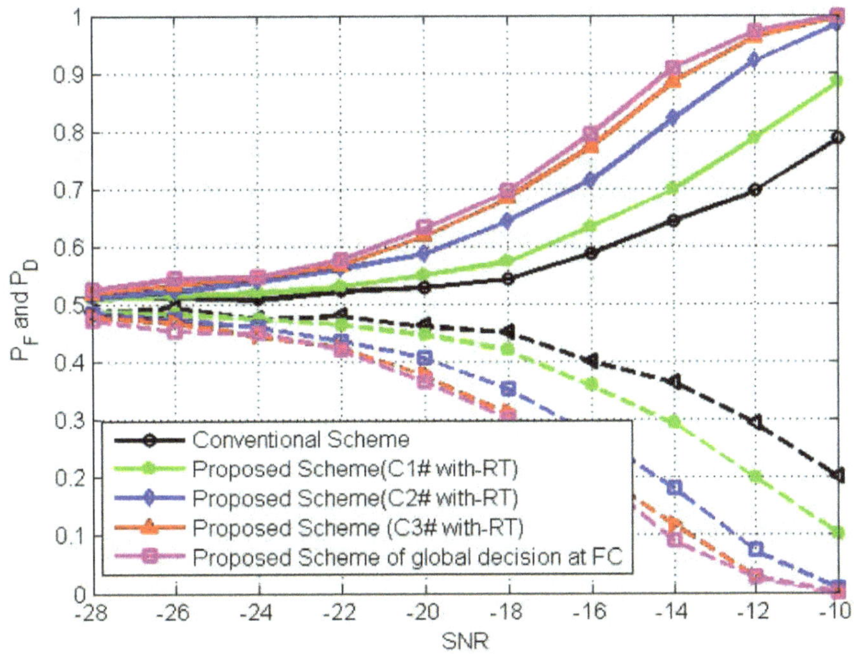


**Figure 10.** ROC curves of the proposed 5G scheme with cluster reporting time and the 4G scheme where SNRs of the PU's signal at the nodes are from –28 dB to –10 dB.

| SNR | | −28 | −26 | −24 | −22 | −20 | −18 | −16 | −14 | −12 | −10 |
|---|---|---|---|---|---|---|---|---|---|---|---|
| 4G scheme | | 0.516 | 0.5042 | 0.5119 | 0.5248 | 0.5295 | 0.5487 | 0.5933 | 0.6286 | 0.6994 | 0.7825 |
| Proposed 5G Scheme | Cluster 1 | 0.5073 | 0.5122 | 0.5209 | 0.5421 | 0.5473 | 0.5775 | 0.6290 | 0.6944 | 0.7810 | 0.8776 |
| | Cluster 2 | 0.5154 | 0.5208 | 0.5378 | 0.5533 | 0.5860 | 0.6408 | 0.7055 | 0.7973 | 0.9006 | 0.9747 |
| | Cluster 3 | 0.5149 | 0.5232 | 0.5453 | 0.5737 | 0.6061 | 0.6727 | 0.7507 | 0.8605 | 0.9528 | 0.9949 |
| | Global | 0.5160 | 0.5324 | 0.5682 | 0.5968 | 0.6264 | 0.6957 | 0.7733 | 0.8896 | 0.9734 | 0.9965 |

**Table 1.** Probability of detection (PD) without cluster reporting time under SNR versus number of clusters.

| SNR | | −28 | −26 | −24 | −22 | −20 | −18 | −16 | −14 | −12 | −10 |
|---|---|---|---|---|---|---|---|---|---|---|---|
| 4G scheme | | 0.516 | 0.5042 | 0.5119 | 0.5248 | 0.5295 | 0.5487 | 0.5933 | 0.6286 | 0.6994 | 0.7825 |
| Proposed 5G scheme | Cluster 1 | 0.5112 | 0.5170 | 0.5207 | 0.5316 | 0.5517 | 0.5743 | 0.6342 | 0.6993 | 0.7883 | 0.8835 |
| | Cluster 2 | 0.5135 | 0.5236 | 0.5407 | 0.5628 | 0.5882 | 0.6445 | 0.7153 | 0.8217 | 0.9206 | 0.9844 |
| | Cluster 3 | 0.5205 | 0.5346 | 0.5474 | 0.5684 | 0.6191 | 0.6845 | 0.7728 | 0.8849 | 0.9625 | 0.9972 |
| | Global | 0.5261 | 0.5460 | 0.5495 | 0.5790 | 0.6327 | 0.6963 | 0.7949 | 0.9093 | 0.9722 | 0.9995 |

**Table 2.** Probability of detection (PD) with cluster reporting time under SNR versus number of clusters.

cluster reporting time and the 4G CRN are illustrated in **Figure 9**. For our proposed 5G CRN scheme, it can be seen that the probability of detection increases as sensing time, $T_s^{nj}$, increases.

The ROC curves of the proposed 5G CRN scheme with cluster reporting time versus the 4G scheme are shown in **Figure 10**. **Figures 9** and **10** show that the probability of detection in the proposed 5G scheme with cluster reporting time is better than the proposed 5G scheme without cluster reporting time.

In **Tables 1** and **2**, the exact values of detection probabilities in the proposed 5G and 4G CRNs are shown. The gain of sensing performance can be verified with the results. For example, the proposed method with a cluster reporting time can detect the spectrum with nearly 100% detection probability whereas the 4G one detects the PU's signal with 78% of detection probability in −10 dB SNR.

## 7. Conclusion

In this chapter, we propose the superallocation and cluster-based cooperative spectrum sensing in a 5G CRN. The proposed 5G scheme can achieve better sensing performance in comparison with the cluster-based cooperative spectrum sensing 4G cognitive radio network. By rescheduling the reporting time slots of SUs and CHs, longer sensing durations are guaranteed for SUs depending on the order of reporting times of SU and CH. With simulations, the gain of performance is verified (**Tables 1** and **2**).

## Author details

Md Sipon Miah[1]*, Md Mahbubur Rahman[1] and Heejung Yu[2]

*Address all correspondence to: sipon@ice.iu.ac.bd

1 Department of Information and Communication Engineering, Islamic University, Kushtia, Bangladesh

2 Department of Information and Communication Engineering, Young Man University, Busan, South Korea

## References

[1] P. Demestichas, A. Georgakopoulos, D. Karvounas, K. Tsagkaris, V. Stavroulaki. J. Lu, C. Xiong, and J. Yao, "5G on the horizon: Key challenges for the radio-access network," *IEEE Vehicular Technology Magazine*, vol. 8, no. 3, pp. 47–53, Sep. 2013.

[2] A. Zakrzewska, S. Rupp, and M. Berger, "Towards converged 5G mobile networks-challenges and current trends," in *Proceedings of the ITU Kaleidoscope Academic Conference*, St. Petersburg, Russian Federation, pp. 39–45, Jun. 2014.

[3] J. Mitola and G. Q. Maguire, "Cognitive radio: Making software radios more personal," *IEEE Personal Communications*, vol. 6, no. 4, pp. 13–18, 1999.

[4] B. Makki and T. Eriksson, "On the ergodic achievable rates of spectrum sharing networks with finite backlogged primary users and an interference indicator signal," *IEEE Transactions on Wireless Communications*, vol. 11, no. 9, pp. 3079–3089, 2012.

[5] S. Haykin, "Cognitive radio: Brain-empowered wireless communications," *IEEE Journal of Selected Areas in Communications*, vol. 23, no. 2, pp. 201–220, 2005.

[6] S. Enserink and D. Cochran, "A cyclostationary feature detector," in *Proceeding of Asilomar Conference on Signals, Systems and Computers*, vol. 2, pp. 806–810, 1994.

[7] F. F. Digham, M. S. Alouini, and M. K. Simon, "On the energy detection of unknown signals over fading channels," *IEEE Transactions on Communications*, vol. 55, no. 1, pp. 21–24, 2007.

[8] Y. Zeng and Y. C. Liang, "Maximum-minimum eigenvalue detection for cognitive radio," in *Proceeding of IEEE 18$^{th}$ International Symposium on Personal, Indoor and Mobile Radio Communications*, pp. 1–5, 2007.

[9] I. F. Akyildiz, W. Y. Lee, M. C. Vuran, and S. Mohanty, "Next generation/dynamic spectrum access/cognitive radio wireless networks: A survey," *Computer Network*, vol. 50, no. 13, pp. 2127–2159, Sep. 2006.

[10] D. Cabric, S. M. Mishra, and R W. Brodersen, "Implementation issues in spectrum sensing for cognitive radios," in *Proceeding of Conference Record of the Thirty-Eighty Asilomar Conference Signals, Systems and Computers*, vol. 1, pp. 772–776, 2004.

[11] A. Ghasemi and E. S. Sousa, "Collaborative spectrum sensing for opportunistic access in fading environments," in *Proceeding of 1st IEEE International Symposium on New Frontiers in Dynamic Spectrum Access Networks*, pp. 131–136, 2005.

[12] M. Mustonen, M. Matinmikko, and A. Mammela, "Cooperative spectrum sensing using quantized soft decision combining," in *Proceeding of 4th International Conference on Cognitive Radio Oriented Wireless Networks and Communications*, pp. 1–5, 2009.

[13] S. Zarrin and T. J. Lim, "Composite hypothesis testing for cooperative spectrum sensing in cognitive radio," in *Proceeding of ICC'09 IEEE International Conference on Communications*, pp. 1–5, 2009.

[14] J. F. Chamberland and V. V. Veercivalli, "The impact of fading on decentralized detection in power constrained wireless sensor networks," in *Proceeding of IEEE International Conference on ICASSP'4*, vol. 3, pp. 837–840, 2004.

[15] T. C. Aysal, S. Kandeepan, and R. Piesiewicz, "Cooperative spectrum sensing over imperfect channels," in *Proceeding of IEEE GLOBECOM Workshops*, pp. 1–5, 2008.

[16] T. C. Aysal, S. Kandeepan, and R. Piesiewicz, "Cooperative spectrum sensing with noisy hard decision transmissions," in *Proceeding of IEEE International Conference on Communications*, pp. 1–5, 2009.

[17] C. Sun, W. Zhang, and K. Ben, "Cluster-based cooperative spectrum sensing in cognitive radio systems," in *Proceeding of IEEE International Conference on Communications, ICC'07*, pp. 2511–2515, Jun. 2007.

[18] W. Xia, S. Wang, W. Liu, and W. Chen, "Cluster-based energy efficient cooperative spectrum sensing in cognitive radios," in *Proceeding of 5th IEEE International Conference on Wireless Communications, Networking and Mobile Computing, WiCom '09*, pp.1–4, 2009.

[19] J. Jin, H. Xu, H. Li, and C. Ren, "Superposition-based cooperative spectrum sensing in cognitive radio networks," in *Proceeding of International Conference on Computer Application and System Modeling* (*ICCASM*), vol. 4, pp. 342–346, 2010.

[20] W. B. Heinzelman, A. P. Chandrakasan, and H. Balakrishnan, "An application-specific protocol architecture for wireless micro sensor networks,", vol. 1, no. 4, pp. 660–670, 2002.

[21] T. Yucek and H. Arslan, "A survey of spectrum sensing algorithms for cognitive radio applications," in *Proceeding of IEEE Communications Surveys & Tutorials*, vol.11, no. 1, pp. 116–130, 2009.

[22] H. Urkowitz, "Energy detection of unknown deterministic signals," in *Proceeding of IEEE*, vol. 55, no. 4, pp. 523–531, Apr. 1967.

[23] V. I. Kostylev, "Characteristics of energy detection of quasideterministic radio signals," *Radiophysics and Quantum Electronics*, vol. 43, no. 10, pp. 833–839, Oct. 2000.

[24] V. I. Kostylev, "Energy detection of a signal with random amplitude," in *Proc. of IEEE International Conference on Communications, ICC*, vol. 3, pp. 1606–1610, 2002.

[25] K. X. Thuc and K. Insoo, "Cooperative spectrum sensing using Kalman filter based adaptive Fuzzy system for cognitive radio networks," *KSII Transactions on Internet and Information Systems*, vol. 6, no. 1, pp. 287–304, Jan. 2012.

[26] S. Boyed and L. Vandenberghe, "Convex Optimization," Cambridge, UK: Cambridge University Press, 2003.

[27] H. Mathis, "On the Kurtosis of digitally modulated signals with timing offsets," in *Proceeding of 3rd IEEE Signal Processing Workshop Signal Processing Advances Wireless Communications*, pp. 86–89, Mar. 2001.

6

# Analysis of Candidate Waveforms for 5G Cellular Systems

Ayesha Ijaz, Lei Zhang, Pei Xiao and Rahim Tafazolli

**Abstract**

Choice of a suitable waveform is a key factor in the design of 5G physical layer. New waveform/s must be capable of supporting a greater density of users, higher data throughput and should provide more efficient utilization of available spectrum to support 5G vision of "everything everywhere and always connected" with "perception of infinite capacity". Although orthogonal frequency division multiplexing (OFDM) has been adopted as the transmission waveform in wired and wireless systems for years, it has several limitations that make it unsuitable for use in future 5G air interface. In this chapter, we investigate and analyse alternative waveforms that are promising candidate solutions to address the challenges of diverse applications and scenarios in 5G.

**Keywords:** waveform modulation, 5G requirements, orthogonal frequency division multiplexing, universal filtered multicarrier, generalized frequency division multiplexing, filterbank multicarrier, windowed orthogonal frequency division multiplexing, filtered orthogonal frequency division multiplexing

## 1. Introduction

Orthogonal frequency division multiplexing (OFDM), which uses a square window in time domain allowing a very efficient implementation, has been adopted as the air interface in several wireless communication standards, including third generation partnership (3GPP) long-term evolution (LTE) and IEEE 802.11 standard families due to the associated advantages such as:

- "Robustness against multipath fading
- Ease of implementation

- Efficient one-tap frequency domain equalization enabled by the use of cyclic prefix (CP)
- Straightforward and simple extension to very large multiple-input multiple-output (MIMO) and high gain beam forming solutions" [1]

Despite its advantages, OFDM suffers from a number of drawbacks including high peak-to-average power ratio (PAPR) and high side lobes in frequency. OFDM requires stringent time synchronization to maintain the orthogonality between different user equipments (UEs). Therefore, signalling overhead increases with the number of UEs in an OFDM-based system. Moreover, it has high sensitivity to carrier frequency offset (CFO) mismatch between different devices. All these drawbacks hinder the adoption of OFDM in the 5G air interface [1] to achieve the following key characteristics currently envisioned for 5G wireless networks:

- 1000 × higher mobile data volume per geographical area
- 10–100 × more connected devices
- 10–100 × higher typical user data rate
- 10 × lower energy consumption
- End-to-end latency of <1 ms
- Ubiquitous 5G access including in low density areas

These fundamental characteristic are envisioned based on following scenarios specified by the 5G research community [2, 3]:

1. **Bitpipe communication:** Broadcasting dense content (such as 3D or 4k video) in small-sized densely deployed cells demands several tens of Mbps to achieve a good quality of experience (QoE). An increased bandwidth and a physical (PHY) layer with high spectrum efficiency is required in this scenario. Therefore, the 5G network must rely on advanced digital communication techniques including MIMO for diversity and multiplexing, massive MIMO to improve the system spectrum efficiency, higher order modulation and efficient coding schemes, adaptive small cell clustering, multicell cooperative transmission, inter-cell interference management and efficient spectrum allocation with cognitive radios (CR).
2. **Internet of things (IoT):** This scenario targets sensory and data collecting use cases such as smart grid, health and environmental measurements and monitoring, transportation, etc. This scenario is mainly characterized by small data packets and massive connections of devices with limited power source. It does not require large channel bandwidth, and duty cycle is generally low while power saving is mandatory. The IoT devices must be able to achieve reliable communication with a loose synchronization or even asynchronous for higher energy efficiency.
3. **Tactile internet:** This scenario focuses on special applications and use cases of IoT and vertical industries with real-time constraints such as internet of vehicles (IoV) and industrial control. These new applications require very low end-to-end latency (ms-level) and high reliability (nearly 100%). The air interface and network forwarding delays need to be

reduced significantly to achieve the sub-millisecond latency requirement. Therefore, shorter frame length with minimal or no overhead, multiple access technologies which can enable grant-free transmission, and solutions for reducing network forwarding delays must be adopted. Technologies such as advanced coding and space/time/frequency diversity must be utilized for reliable data transmission.

4. **Wireless regional area network (WRAN):** This scenario focuses on coverage of low populated remote areas which suffer from low data rates and unreliable solutions. While wired technologies have limited coverage, current wireless networks operating in licensed frequencies have relatively small cell sizes which make them economically unfeasible in sparsely populated areas. The 5G networks must address large coverage areas using dynamic using dynamic channel allocation based on CR with low out of band emission (OBE) and efficiently deal with the multipath effects by reducing the impact of the CP in the overall data rate [2].

The requirements of different scenarios can be impacted by the choice of waveforms. Therefore, to address the drawbacks of OFDM and enable the aforementioned characteristics, different physical-layer waveforms are being investigated for 5G networks. The waveforms currently under consideration include filtered orthogonal frequency division multiplexing (FOFDM) [4], windowed orthogonal frequency division multiplexing (WOFDM) [5], filterbank multicarrier (FBMC) [6], generalized frequency division multiplexing (GFDM) [7] and universal filtered multicarrier (UFMC) [2]. These waveforms are being investigated to analyse their impacts on the following fundamental requirements of 5G [8]:

- Capabilities for supporting massive capacity and massive connectivity
- Support for an increasingly diverse set of services, application and users—all with extremely diverse requirements, e.g. efficient support for short-burst transmissions, IoT and massive machine type communications (mMTC)
- Flexible and efficient use of all available non-contiguous spectrum for wildly different network deployment scenarios

In this chapter, we analyse performance of alternative waveforms in terms of OBE, bit error rate (BER), time and frequency efficiency, PAPR, computational complexity and sensitivity to CFO and time offset (TO). This comparison will help determine the suitability of the candidate waveforms in different scenarios for 5G networks.

## 2. Candidate waveforms

### 2.1. Filtered orthogonal frequency division multiplexing

Large OBE, due to the rectangular shaping of the temporal signal, is one of the main shortcomings of the OFDM used in LTE. **Figure 1** shows the power spectral density (PSD) function of an OFDM waveform with carrier spacing set to 15 kHz, FFT size of 1024 and 72 samples long CP. We can observe loss of spectral efficiency due to the partial use of available bandwidth to fit in an 8 MHz emission spectrum mask (ESM).

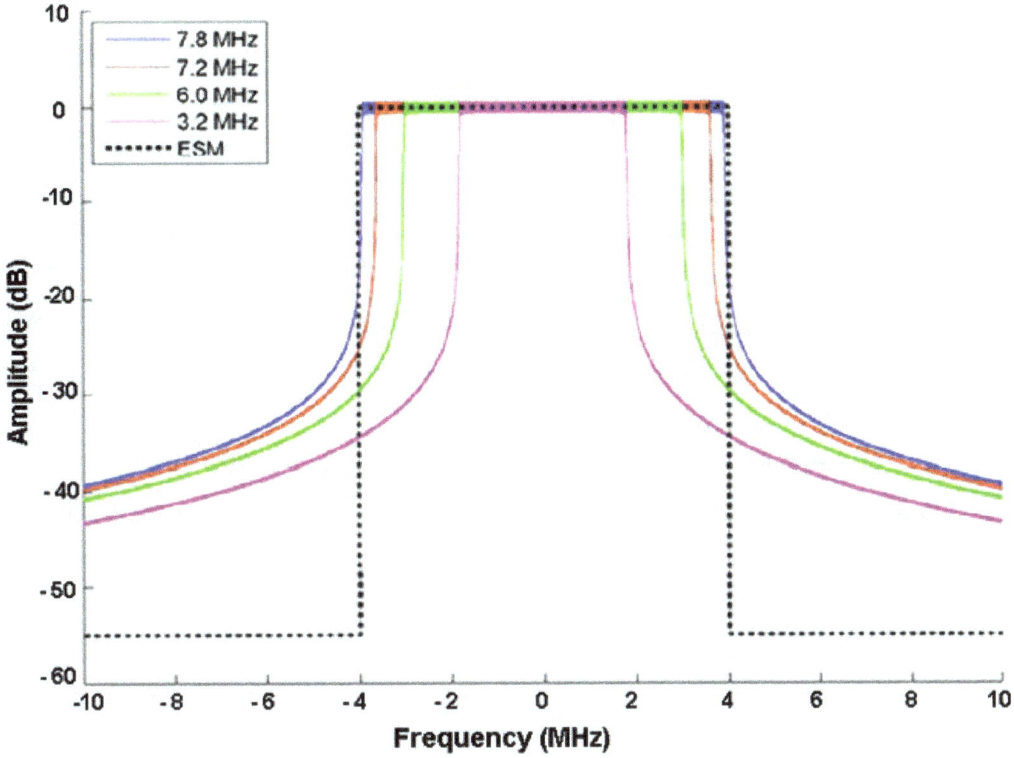


**Figure 1.** Power spectral density of CP-OFDM centred on the active carrier [9].

The problem of large OBE is alleviated in FOFDM using transmit filter cascaded after the modulator as shown in **Figure 2**. At the transmitter, the information bit sequence is encoded into a coded bit sequence which goes through interleaver (Π) and is mapped into QPSK/QAM symbols. Then, serial to parallel (S/P) conversion takes place and a set of N symbols are mapped onto orthogonal subcarriers using inverse fast Fourier transform (IFFT). The output from IFFT block is converted into serial data followed by CP insertion. In order to provide robustness against inter-symbol interference (ISI) and inter-carrier interference (ICI), the length of the CP must be longer than the channel impulse response. The OFDM signal is filtered by a transmit pulse shaping filter (TX filter) before transmission over the multipath fading channel. At the receiver, a receive pulse shaping filter (RX filter) is used and the signal is converted back to the frequency domain using fast Fourier transform (FFT) operation after CP removal. This is followed by one-tap equalization (the equalizer is labelled as equation in **Figure 2**) to mitigate the channel effect. The equalized signal is fed to a soft demapper, and its output is subsequently de-interleaved ($\Pi^{-1}$) and decoded to recover the information bearing signal [4].

Suitably designed filters can suppress the large side lobes of OFDM making FOFDM more bandwidth efficient while preserving the orthogonality among subcarriers. In this document, we have used a square root raised cosine (SRRC) filter, with roll-off factor $\alpha = 0.3$ truncated to 3 symbol interval ($T_r = 3T$ where $T$ is the symbol duration) on each side of the peak at the transmitter, and the receiver filter is matched to the transmit filter. Time and frequency domain

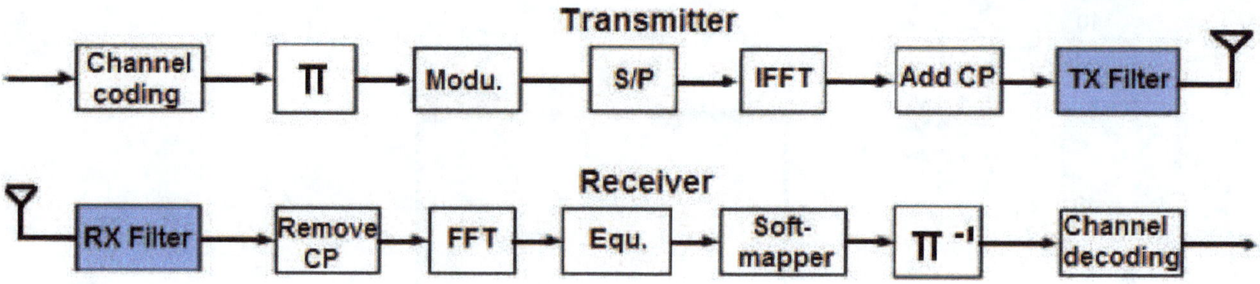


**Figure 2.** Transmitter and receiver structure of FOFDM [4].

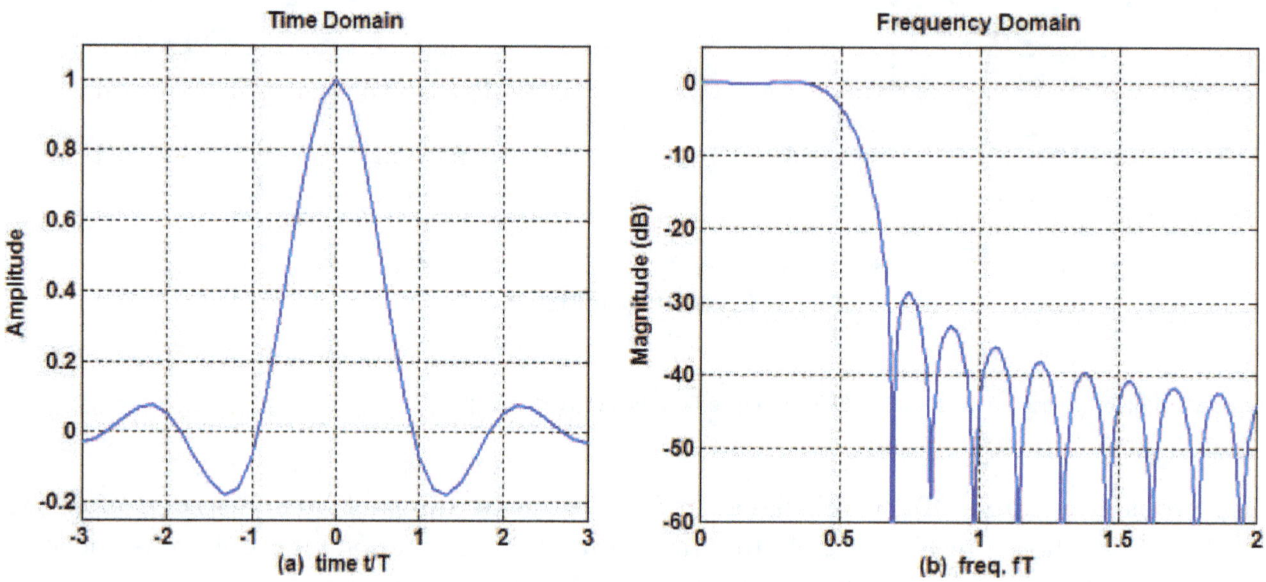


**Figure 3.** SRRC filter characteristics (a) time domain: the $x$-axis is normalized to the symbol interval $T$, the pulse is normalized to a peak value of unity (b) frequency domain: the frequency axis is normalized to the symbol rate $1/T$, the magnitude of the spectra, normalized to peak value of unity, is plotted in dB scale.

characteristics of such a filter are shown in **Figure 3** wherein $x$-axis for time and frequency is normalized to symbol interval $T$ and symbol rate $1/_T$, respectively.

Although FOFDM shows better spectral containment as compared to OFDM, however, when available spectrum fragments are not contiguous, filtering becomes challenging since a separate filter needs to be dynamically designed for each available chunk of spectrum.

### 2.2. Windowed orthogonal frequency division multiplexing

Windowed OFDM is similar to conventional OFDM, however, it uses a non-rectangular transmit window smoothing the edges of the rectangular pulse to provide better spectral containment and reduce ACI. Eq. (**1**) shows such a pulse shape in which roll-off portions are of a raised cosine shape

$$p[n] = \begin{cases} 0.5\left(1 + cos\left\{\{\pi\left(1 + \frac{n}{\beta N_T}\right)\}\right\}\right), & 0 \leq n < \beta N_T \\ 1, \ \beta N_T \leq n < N_T \\ 0.5\left(1 + cos\left\{\pi \frac{n - N_T}{\beta N_T}\right\}\right), & N_T \leq n \leq (\beta + 1) N_T - 1 \end{cases} \quad (1)$$

In Eq. (1), $0 \leq \beta$ 1, is the roll-off factor which controls the length of the roll-off portion of the non-rectangular window, i.e. $\beta(N + N_{CP})$, where $N_{CP}$ is the length of CP in samples. Due to multiplication of CP with a non-unity function, orthogonality will be in general lost in a multipath channel. In order to preserve orthogonality, an extended CP is used in WOFDM and the original samples of the CP are kept outside the roll-off part of the windowing function. Improved PSD side lobe decay in WOFDM can save the guard band overhead of the current OFDM deployments, e.g. 10% overhead in LTE. However, the use of extended CP in WOFDM reduces its spectral efficiency as compared to OFDM. Therefore, both frequency and time domain overheads need to be taken into account to determine overall improvement in spectral efficiency as compared to OFDM. WOFDM also uses a cyclic suffix (CS) after each data block in addition to the CP before each data block. The spectral loss due to additional overhead of CS is partly compensated by overlapping the CP and CS of consecutive symbols.

### 2.3. Filter bank multicarrier

Filter bank multicarrier applies filtering on a per-subcarrier basis and is considered as an attractive alternative to OFDM to provide improved out-of-band spectrum characteristics. Since subcarrier filters are narrow in frequency and thus require long filter lengths (normally at least $4T$ to preserve an acceptable ISI and ICI), the symbols are overlapping in time. To comply with the real orthogonality principle, offset-QAM (OQAM) can be applied and, therefore, FBMC is not orthogonal in the complex domain. The most common FBMC technique is the FBMC/OQAM, which is also known as OFDM with offset quadrature amplitude modulation (OFDM/OQAM ) [10].

In FBMC, the prototype filter needs to be carefully designed to minimize or eliminate ISI and ICI while keeping the side lobes small. These prototype filters are implemented using an efficient technique called polyphase implementation, which uses multi-rate signal processing techniques to reduce the complexity by joint implementation of all synthesis or analysis filters in the filter bank. The transmitted signal in FBMC is the sum of the outputs of a bank of $N$ filters, whose length is given by $L = N \times p$, where $N$ is the FFT size and $p$ is the length of each polyphase filter. We have used an isotropic orthogonal transform algorithm (IOTA) prototype function with $p = 6$, for use in FBMC system, which is well-localized in time and frequency domain as shown in **Figure 4**.

Since subcarriers can be better localized in FBMC due to more advanced prototype filter design, therefore the CP can be removed resulting in improved spectral efficiency as compared to OFDM. This is in addition to the spectral efficiency gain due to reduced guard band in FBMC. However, FBMC/OQAM incurs an overhead due to transition times (tails) at both ends

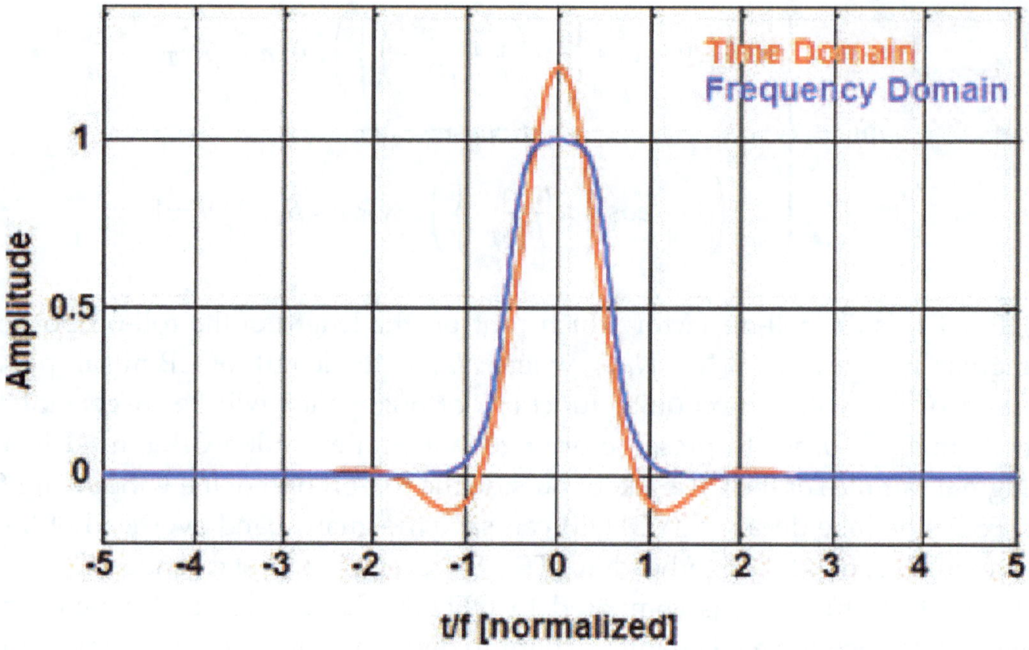


**Figure 4.** Time and frequency response of IOTA prototype function. Time domain pulse is normalized to average power of unity. The $x$-axis is normalized to the symbol interval $T$, the frequency axis for spectra is normalized to the symbol rate $1/T$ and the frequency domain spectrum is normalized to peak value of unity.

of a transmission burst and an overhead due to the $T/2$ time offset between the OQAM symbols [11] (total tail duration is equal to length of the prototype filter). Although solutions have been proposed to remove signal tails of OFDM/OQAM signals [11], however, the overhead cannot be removed totally, without increasing its sensitivity to time and frequency misalignments, and it increases the latency of communication.

### 2.4. Universal filtered multicarrier

As the name implies, UFMC is also a filtered multicarrier modulation scheme using suitably designed filters to reduce OBE like FOFDM and FBMC and combines the benefits of the two schemes. UFMC applies filtering to chunks of contiguous subcarriers instead of single subcarriers (as in FBMC) or the complete band (as in FOFDM). **Figure 5** shows the block diagram of a UFMC transmitter with total bandwidth divided into $B$ sub-bands where the time-domain transmit vector $x$ for a particular multicarrier symbol is the superposition of the sub-band-wise filtered components, with filter length $L$ and FFT length $N$. The transmit signal can be mathematically described as follows:

$$x = \sum_{i=1}^{B} F_i V_i s_i \tag{2}$$

where $S_i$ is the transmit vector containing $n_i$ complex QAM symbols for transmission in $i$th sub-band. For each of $B$ sub-band, indexed $i$, $S_i$ is transformed to time-domain by the IDFT-matrix $V_i$ with dimensions $[N \times n_i]$. $N$ is the required number of samples per symbol to

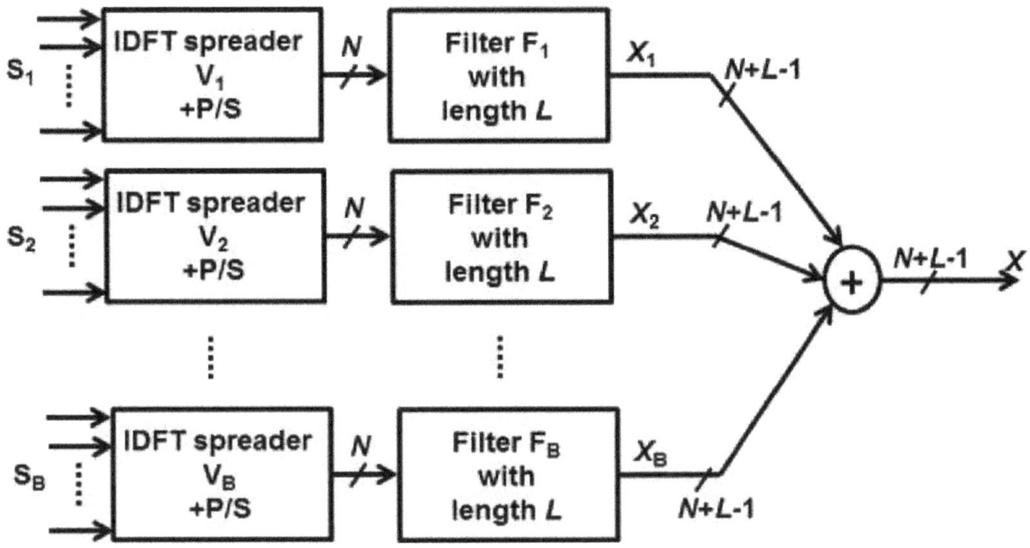


**Figure 5.** UFMC transmitter.

represent all sub-bands without introducing aliasing (i.e. $N$ depends on the overall covered bandwidth). "$V_i$ includes the relevant columns of the inverse Fourier matrix according to the respective sub-band position within the overall available frequency range. $F_i$ is a Toeplitz matrix with dimensions $[(N + L - 1) \times N]$, composed of the filter impulse response, performing linear convolution" [2]. Unlike OFDM, CP can be dropped in UFMC and its additional symbol duration overhead is used to introduce sub-band filters. Since filtering is applied to a sub-band, these filters can be shorter [2] (UFMC filters are in the order of an OFDM CP) than the per-subcarrier filters of an FBMC system improving the suitability of UFMC for communicating in short bursts, compared to FBMC. Moreover, orthogonality is still maintained between subcarriers. Since the same filter can be used for each sub-band, spectral holes can be dynamically utilized without posing a challenge in implementation as compared to FOFDM.

We have used Dolph-Chebyshev filters with side-lobe-attenuation equal to 40 dB and filter length $L$ equal to one sample larger than the CP length in an LTE system. **Figure 6** depicts the impulse and frequency response for an exemplary setting with $L = 73$ and $N = 1024$.

Since UFMC modulates each data symbol at the same time and the same frequency as in OFDM, its receiver [2] can demodulate legacy OFDM signals and UFMC modulated signal can be directly demodulated by the legacy OFDM receiver. This feature makes UFMC-based system backwards compatible with the legacy OFDM systems [12]; a feature missing in FBMC.

### 2.5. Generalized frequency division multiplexing

GFDM is a block-based, non-orthogonal multicarrier transmission scheme capable to spread data across a two-dimensional (time and frequency) block structure (multi-symbols per multicarriers). The block-based transmission in GFDM is enabled by circular pulse shaping of the individual subcarriers. "The main difference between OFDM and GFDM is that the latter

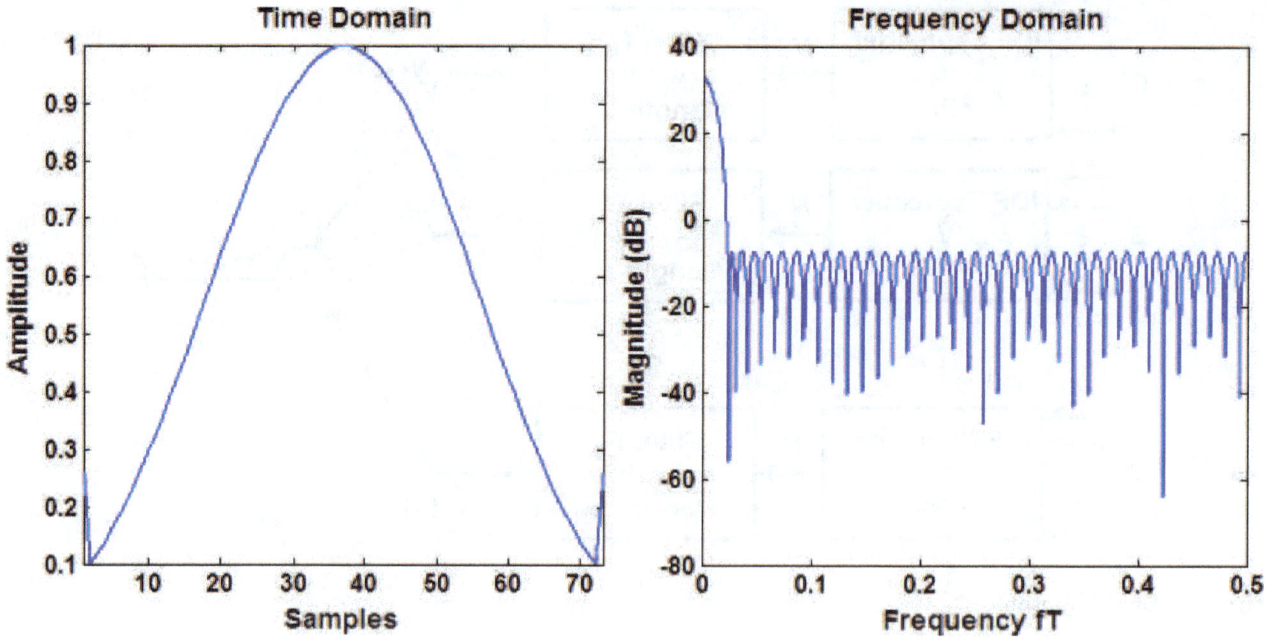


**Figure 6.** Chebyshev filter characteristics in time and frequency domain. The time domain pulse is normalized to a peak value of unity. The frequency axis is normalized to the symbol rate $1/T$.

transmits $MN$ data symbols per frame using $M$time slots with $N$subcarriers where each data symbol is represented by a pulse shape $g(t)$, whereas OFDM transmits $N$ data symbols using one time slot with $N$ subcarriers, where each symbol is filtered by a rectangular pulse shape" [2]. GFDM cannot only model the spectrum shape by choosing an appropriate pulse shape to provide a very low OBE, frequency spacing between subcarriers is also more flexible in GFDM than in OFDM which allows for a higher flexibility for spectrum fragmentation. GFDM can achieve higher spectral efficiency since it does not need guard band to avoid adjacent channel interference (ACI).

The baseband block diagram of a GFDM transceiver system is given in **Figure 7**. The data symbols to be transmitted on $i$th subcarrier, $d_i = d_i(0), \ldots, d_i(M-1)]^T$, are first up-sampled by the factor of $N$ to form an impulse train $s_i(n) = \Sigma_{m=0}^{M-1} d_i(m)\delta(n-mN), n = 0, \ldots, NM-1$. This signal is then circularly convolved with the prototype filter and up-converted to its corresponding subcarrier frequency. The resulting signals for all subcarriers are summed up to form the GFDM symbol $x(n)$ given below:

$$x(n) = \sum_{i=0}^{N-1} \sum_{m=0}^{M-1} d_i(m) g_{\{(n-mN) \bmod MN\}} e^{j\frac{2\pi in}{N}}, \; n = 0, \ldots, NM-1 \tag{3}$$

where $g_l$ is the $l$th coefficient of the prototype filter. Circular filtering helps to remove the latency associated with the prototype filter transient intervals when conventional linear convolution is used like in the FBMC schemes. We have used an SRRC filter with roll-off factor $\alpha = 0.3$ in the GFDM-based link level simulator. The impulse response and frequency domain characteristics for the prototype filter are given in **Figure 8** for $N = 128$ and $M = 7$.

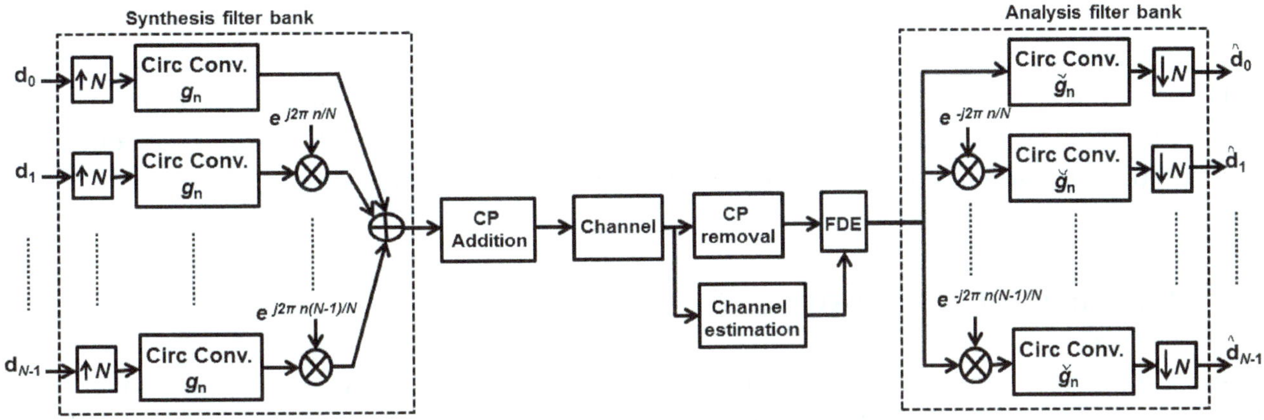


**Figure 7.** Block diagram of a GFDM transceiver system [7].

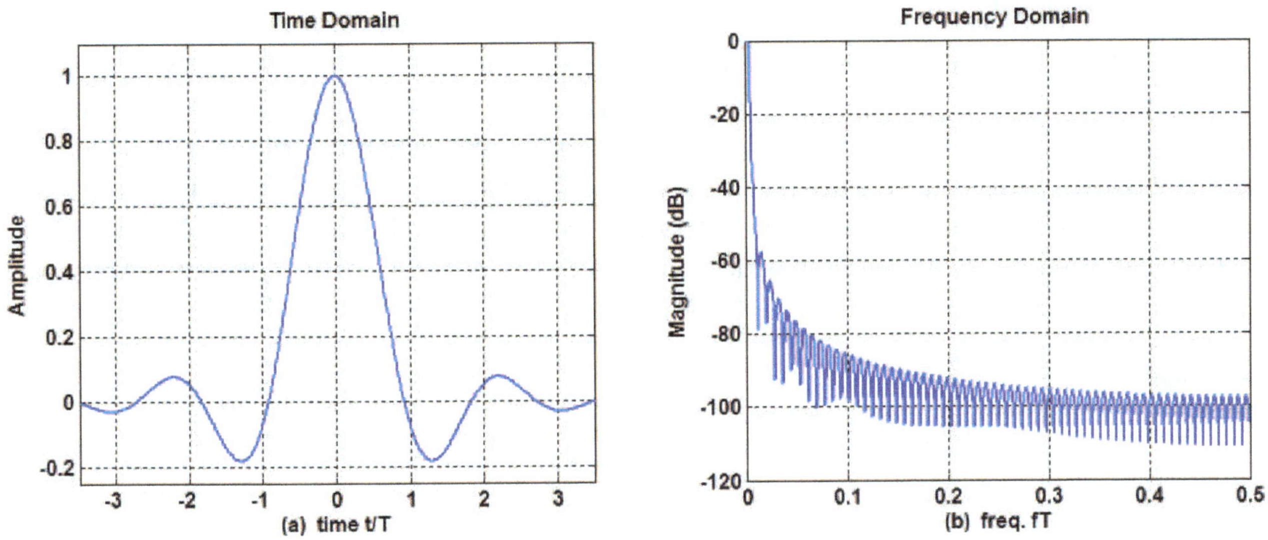


**Figure 8.** Time and frequency domain characteristics of an SRRC filter in GFDM transmitter (a) time domain: the $x$-axis is normalized to the symbol interval $T$, the pulse is normalized to a peak value of unity (b) frequency domain: the frequency axis is normalized to the symbol rate $1/T$, the magnitude of the spectra, normalized to peak value of unity, is plotted in dB scale.

Based on Eq. (3), GFDM signal $\boldsymbol{x} = [x(0), \ldots, x(MN-1)]^T$ can also be formulated as $\boldsymbol{x} = \boldsymbol{Ad}$ where $\boldsymbol{A}$ is an $MN \times MN$ modulation matrix whose elements can be represented as:

$$[A]_{nm} = g_{\{(n-mN)\text{mod } MN\}} e^{j\frac{2\pi nm}{N\ M}} \tag{4}$$

Lastly, on the transmitter side, a cyclic prefix of $N_{CP}$ samples is added to the GFDM data block to produce $\tilde{x}$. Since it uses only one CP for $M$ time slots (i.e. one block) rather than a CP for each slot (i.e. multicarrier symbol) as is the case in OFDM, it has higher spectral efficiency than the latter. GFDM turns into OFDM when $M = 1$ and $\boldsymbol{A}$ is an $N \times N$ inverse Fourier matrix. In CP-based GFDM systems, frequency domain equalization (FDE) can be performed after CP removal to compensate for the multipath channel impairments. The received signal, after channel equalization, can be demodulated after using linear receivers such as zero forcing

(ZF), matched filter (MF) and minimum mean square error (MMSE) receivers. While MF receiver maximizes signal-to-noise ratio (SNR) per subcarrier, it cannot completely remove ICI. Self-interference due to non-orthogonality of the neighbouring subcarriers and time slots can be removed using ZF receiver at the expense of noise enhancement. MMSE receiver can be used to make a trade-off between self-interference and noise enhancement [2].

## 3. Comparison of waveforms

Now, we present simulation results and discuss performance of the candidate waveforms. Based on the characteristics of these waveforms, we discuss their suitability for the scenarios which are being foreseen for 5G networks. The simulation parameters are given in **Table 1**.

### 3.1. Power spectrum

**Figure 9** shows power spectral density of different waveforms assuming non-contiguous fragments of spectrum are available for transmission. In **Figure 9**, two available spectrum fragments are separated by an unavailable band while the spectrum at the two edges is also not used for transmission. It is observed that UFMC and FBMC reduce the OBE by reducing spectral leakage from the transmission subcarriers to the unused neighbouring band. Hence, these waveforms are more suitable candidates, as compared to OFDM, for applications that

| Parameter | Settings | | | |
|---|---|---|---|---|
| MCM schemes | OFDM, WOFDM, FOFDM, FBMC, UFMC, GFDM | | | |
| Subcarrier spacing ($\Delta f$) | 15 KHz | | | |
| Resource block size | 12 subcarriers | | | |
| *Sub-band size for UFMC* ($D$) | 12 subcarriers | | | |
| *No. of MC symbols per subframe* ($M$) | 7 | | | |
| Bandwidth | 5 MHz | | | |
| FFT size ($N$) | 512 | | | |
| Encoder | Turbo coding, rate 1/3, 1 | | | |
| CP length (samples) ($N_{CP}$) | 32 for OFDM, FOFDM and GFDM. 0.25 × FFT size for WOFDM. 0 for FBMC and UFDM | | | |
| Channel model | Extended pedestrian A (EPA) [13], AWGN | | | |
| Channel estimation | Ideal | | | |
| Equalizer | 1-tap MMSE FDE | | | |
| Sub-frames | 10,000 | | | |
| Filters | FOFDM | FBMC | UFMC | GFDM |
| | RRC filter $\alpha = 0.3$ $L = 13$ | IOTA pulse $p = 6$ | Dolph-Chebyshevside lobe attenuation = 40 dB $L = 33$ | RRC filter $\alpha = 0.1$ |

**Table 1.** Simulation settings.

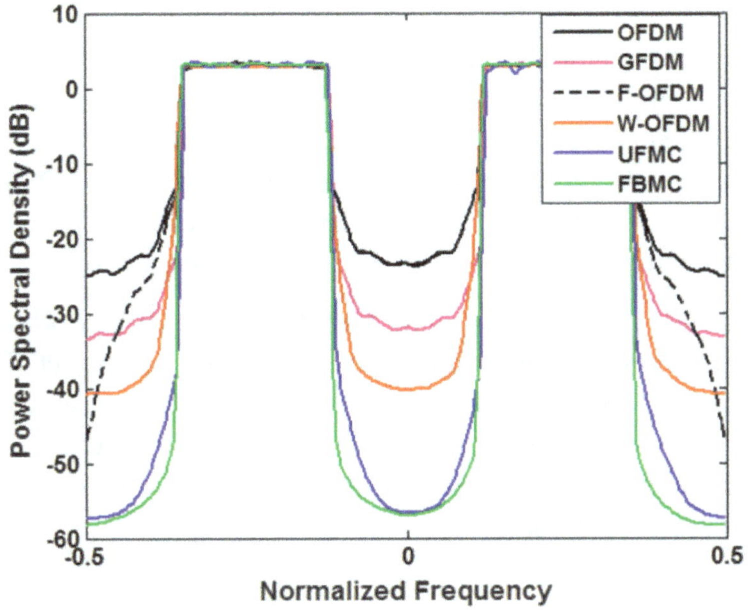


**Figure 9.** Power spectral density of waveforms with fragmented spectrum around the centre frequency.

have strict ACI requirements such as in cognitive radio (CR). It also implies that these waveforms will not need large guard bands to avoid ACI, thereby, improving spectral efficiency and facilitating carrier aggregation. WOFDM also shows considerably lower OBE as compared to OFDM. However, OBE of GFDM is not significantly lower than OFDM due to the abrupt changes of the signal value between GFDM blocks.

Although FOFDM has lower side lobes as compared to OFDM in the two unused bands at the edges, its OBE to the unavailable band between the available fragments is the same as that of OFDM. This is due to the use of filter over the whole band in FOFDM using OFDM as the underlying technology. Therefore, FOFDM cannot efficiently utilize non-contiguous chunks of spectrum.

### 3.2. Bit error rate performance

Having analysed the PSD properties of transmitted signal using different MCM schemes, we now analyse the BER performance of different waveforms assuming only one transmitter and receiver using the entire bandwidth for data transmission and no interferer in adjacent frequency bands. We first simulate the BER performance in an AWGN only channel using the QPSK (OQPSK for FBMC) modulation without error correction coding. Then BER performance was simulated using a rate 1/3 turbo code in the extended pedestrian A (EPA) channel [13] assuming perfect channel knowledge to analyse the performance of different waveforms in frequency selective channel. The results were obtained by averaging BER over 10,000 subframes transmitting 7 MC symbols per subframe. For FBMC, we used hard truncation by discarding two FFT blocks on both sides of the transmit matrix to reduce the overhead caused by filter tails. Similarly, hard truncation was employed to completely remove filter tails in FOFDM. Although CP is not needed for OFDM, GFDM, FOFDM in the AWGN channel, it is still used here to comply with the standard system configuration. Simulation results presented in **Figure 10** show that all schemes have comparable BER performance in the AWGN channel in the absence of ACI. Slight discrepancy in the performances of different waveforms as compared to the theoretical performance is due to the overhead imposed by the CP or filter tails. WOFDM shows 1 dB degradation due to the largest overhead, i.e. 25% of FFT size. FBMC shows 0.5 dB degradation as compared to the theoretical performance of QPSK in an AWGN channel while other waveforms are very close to the theoretical curve.

**Figure 10** also shows BER performance using QPSK/OQPSK with code rate = 1/3 in an EPA channel using parameters specified in **Table 1** and assuming perfect knowledge of noise variance is available for MMSE equalizer. It is observed that all waveforms, except WOFDM and FBMC, show similar performance as that of OFDM. While loss in WOFDM is due to greater CP overhead, FBMC also shows similar performance as that of WOFDM in multipath fading channel under consideration.

### 3.3. Time-frequency efficiency

Time-frequency efficiency $r_{TF}$ which depends on the characteristics of the underlying waveform of an air interface is an important parameter to compare the performance of different waveforms. It is defined as follows [14]:

$$r_{TF} = r_T.r_F = \frac{L_D}{L_D + L_T} \times \frac{N_u}{N'} \tag{5}$$

where $r_T$ is "the efficiency in time domain relating the information carrying body ($L_D$) of the burst/subframe to its overall length including the tails ($L_T$)" [14]. Hence, length of the cyclic prefix and the filters are of relevance for $r_T$. $r_F$ is the efficiency in frequency domain, and it is the ratio of number of usable subcarriers $N_u$ (i.e. excluding guard carriers) to the overall number of subcarriers $N'$ within the usable band.

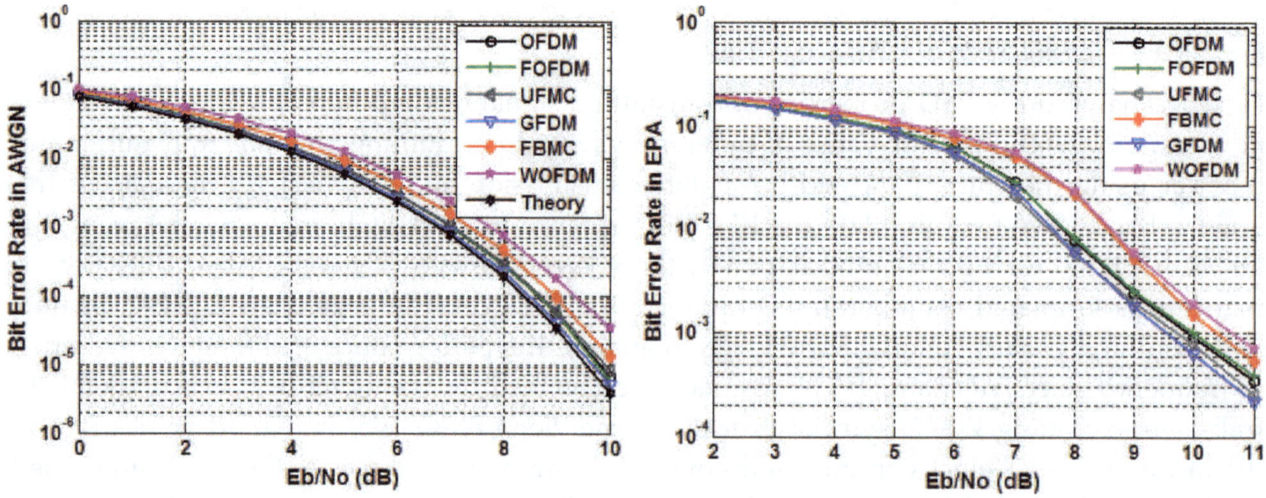


**Figure 10.** BER for QPSK/OQAM in AWGN (code rate = 1) and EPA (code rate = 1/3) channel.

Here, we present time domain efficiency taking into account basic signal characteristics,i.e. how many data symbols may be included into a given time-frequency block for a certain CP and filter length without reflecting on other overheads such as pilot symbols.

### 3.3.1. *Time domain efficiency*

As shown in Eq. (**5**), time domain efficiency is given by $r_T = L_D/(L_D + L_T)$. If we assume the burst to contain $M$ multicarrier symbols (each comprising of $N$ samples), the length of the information carrying body of the transmitted signal is $L_D = MN$. The tails of different waveforms, with design specifications given in Section 2, are given below:

$$L_{T,OFDM} = MN_{CP} \tag{6}$$

$$L_{T,F\text{-}OFDM} = MN_{CP} \tag{7}$$

$$L_{T,W\text{-}OFDM} = MN_{CP} = 0.25 \times MN \tag{8}$$

$$L_{T,FBMC} = N \tag{9}$$

$$L_{T,UFMC} = M(L-1) \tag{10}$$

$$L_{T,GFDM} = N_{CP} \tag{11}$$

**Figure 11** shows time domain efficiency of candidate waveforms versus the frame/burst size ranging from 1 to 20 MC symbols per frame/burst with FFT size ($N$) equal to 1024 and CP length equal to 72 samples. The length of UFMC filter, i.e. $L = 73$. It can be observed from these results that FOFDM using hard truncation has similar time domain efficiency as OFDM as is also evident from Eqs. (**6**) and (**7**). Time-domain overhead for both schemes is proportional to the frame size ($M$) and CP length. Therefore, their time-efficiency is constant for a fixed size of CP. This is also true for WOFDM, however, it has lower efficiency than OFDM due to longer CP. GFDM has the highest efficiency due to its block-based nature using one CP per frame. FBMC, on the other hand, has significantly lower efficiency than OFDM particularly for very short burst sizes. Its performance approaches that of OFDM for the design used by LTE (indicated by black vertical line), i.e.14 MC symbols per frame outperforms OFDM for longer bursts.

### 3.3.2. *Frequency domain efficiency*

As shown in Eq. (**5**), frequency domain efficiency is given by $r_F = N_u/N'$. Using LTE as reference and assuming a transmission bandwidth of 10 MHz with subcarrier spacing 15 kHz, the number of subcarriers $N'$ fitting into the given bandwidth is:

$$N' = \frac{10\text{ MHz}}{15\text{ kHz}} = 666 \tag{12}$$

According to the LTE standard, number of subcarriers actually carrying data is $N_{U,\ OFDM} = 600$. For FBMC, with very low out-of band radiation as shown in **Figure 9**, one guard subcarrier at each side of the band is sufficient and thus $N_{U,\ FBMC} = 664 - (N_g - 1)$ where $N_g$ reflects the number of users sharing the band. Since FBMC is not orthogonal with

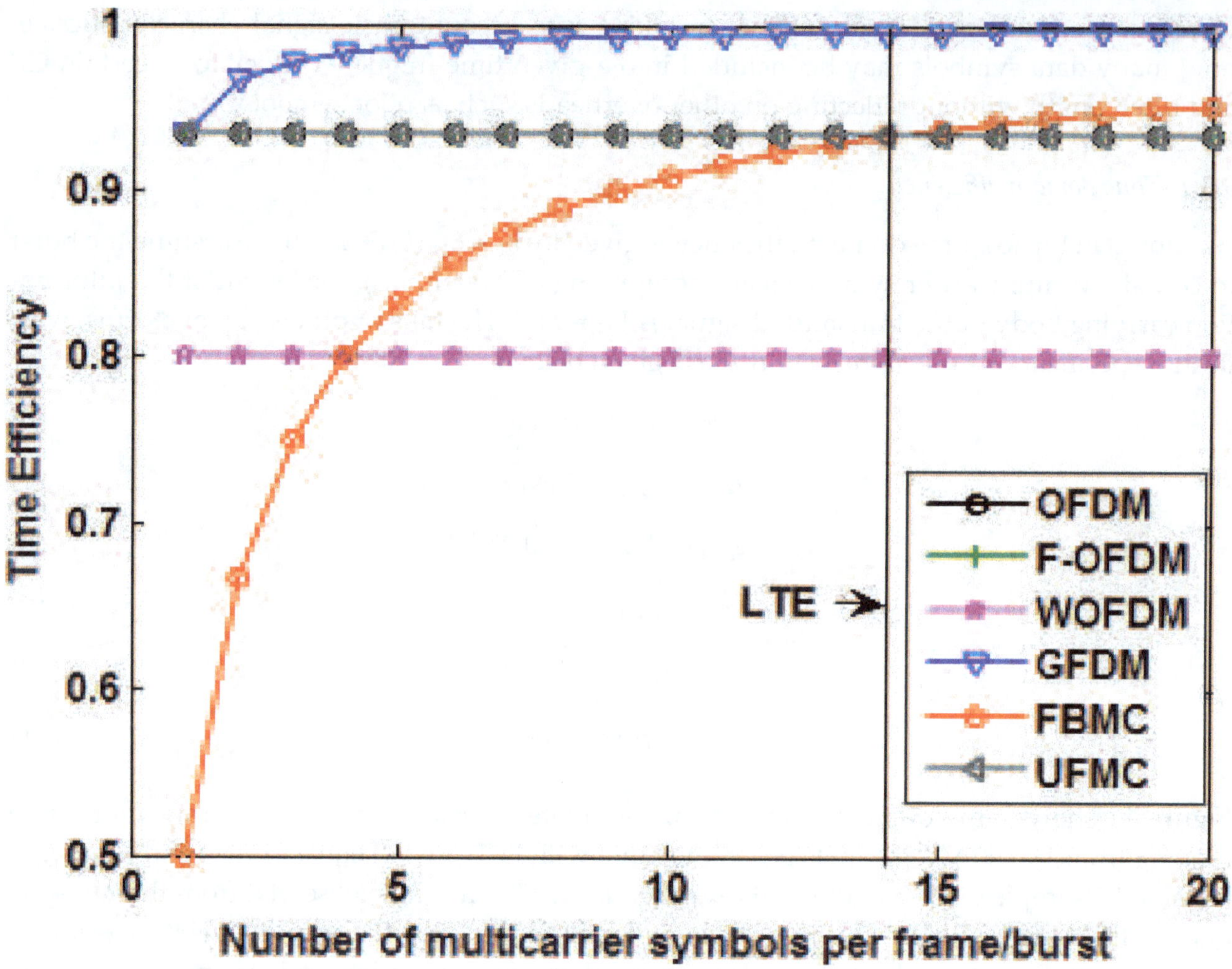


**Figure 11.** Time domain efficiency versus burst size.

respect to the complex plane, an additional guard subcarrier is needed to separate UL transmissions [if complex precoding is applied (the same holds for DL transmissions)] of users being allocated adjacent in frequency. "This is necessary as the transmissions of different users are experiencing different channel gains introducing multi-user interference at the allocation edges. Hence, $N_g$ is equal to the number of users sharing the transmission time interval (assuming continuous user allocations)" [14]. Assuming a scenario where whole bandwidth is available for single user transmission, $N_{U,\ FBMC}$ = 664. GFDM, UFMC and WOFDM designed for very low OBE, as shown in **Figure 9**, also need one subcarrier guard at each side of the band. Therefore, $N_{U,\ UFMC} = N_{U,\ WOFDM} = N_{U,\ GFDM} = 664$.

Since FOFDM, with an SRRC filter design as given in Section 2.1, does not exhibit very low OBE as compared to OFDM, $N_{U,\ FOFDM}$ is expected to be quite similar to $N_{U,\ OFDM}$ and this value needs to be decided after further careful investigation of the OBE characteristics and spectral emission mask requirements in different scenarios. For the sake of analysis, we choose it arbitrarily to be equal to $N_{U,\ OFDM}$.

### 3.3.3. *Overall time-frequency efficiency*

Assuming a single user occupying the whole bandwidth, i.e. $N_g = 1$, **Figure 12** shows the comparison of time-frequency efficiency of different waveforms versus the number of multicarrier symbols per burst. Since frequency domain efficiency of all the waveforms except OFDM and FOFDM is nearly unity, their overall efficiencies remain unchanged. However, overall time-frequency efficiency of OFDM and FOFDM reduces by 10%. Therefore, we observe that while time-domain efficiency of UFMC design under consideration is similar to that of OFDM, its overall efficiency is better due to lower guard band required for UFMC. It can also be observed that the overall time-frequency efficiency of FBMC approaches the efficiency of OFDM when burst size approaches 5, and it exhibits greater efficiency for burst sizes exceeding 5 multicarrier symbols. Based on these analytical results, we can conclude that both UFMC and GFDM are more suitable for short burst transmissions as compared to other MCM schemes. FBMC is more suitable for long burst transmission and is inefficient for short burst communication.

## 3.4. Peak-to-average power ratio performance

Peak-to-average power ratio (PAPR) measures the envelope variation of a waveform and is defined as the peak amplitude of the waveform divided by its root-mean-square value. Large

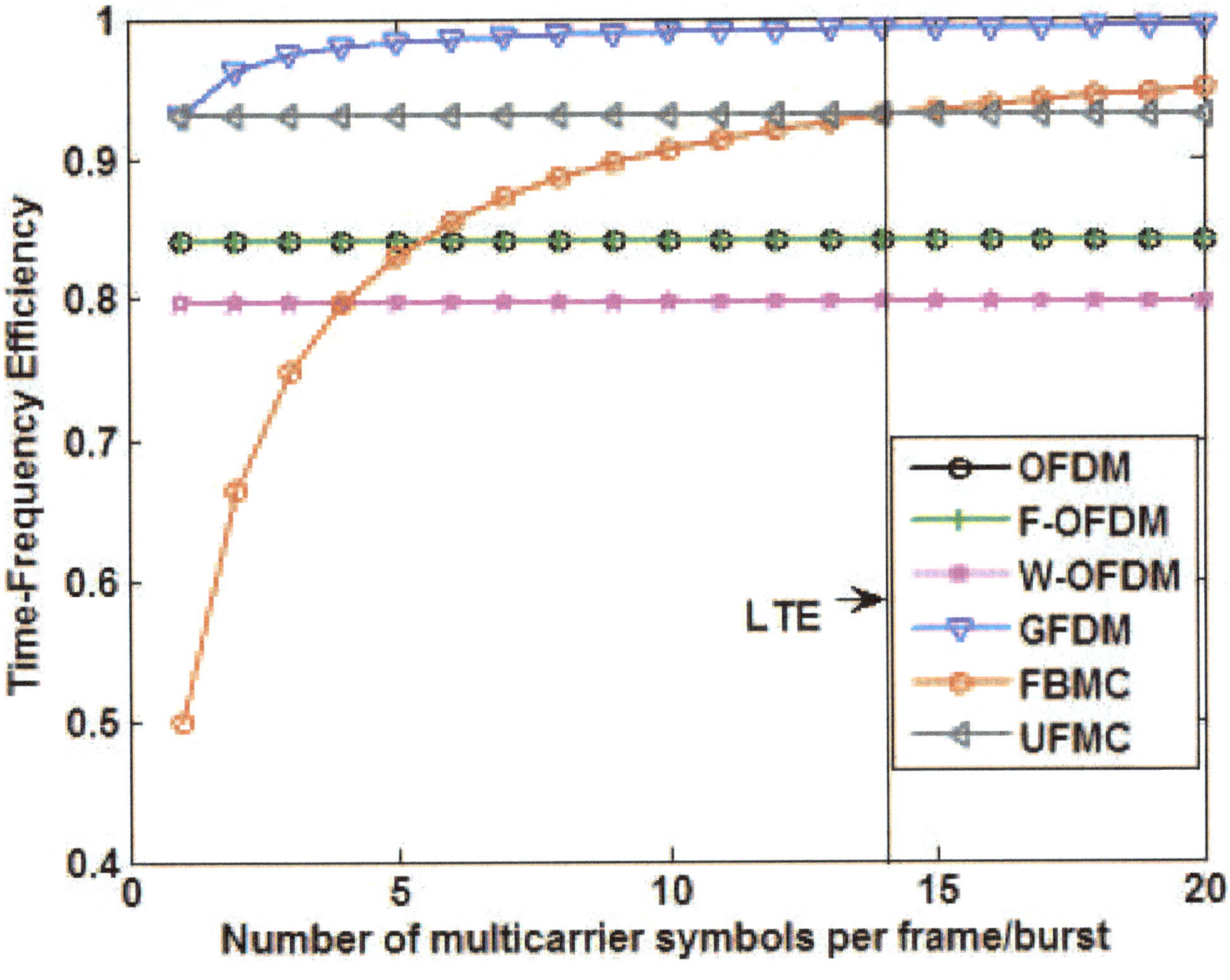


**Figure 12.** Time-frequency efficiency versus burst size.

PAPR requires power amplifiers to have a very large linear range. Otherwise, the nonlinearity leads to signal distortion, which causes spectral regrowth and higher BER. It was gathered from the literature survey [15] that all multicarrier candidate waveforms suffer from large PAPR. **Figure 13** presents the PAPR performance comparison of different waveforms and confirms the findings from the literature as it is seen that all the candidate waveforms exhibit large PAPR. Comparing the relative performance, we observe that OFDM and WOFDM have the lowest PAPR while FOFDM shows the highest PAPR. Other MCM schemes using filter to limit OBE also show higher PAPR as compared to OFDM. A general observation from these results is that use of filters in MCM schemes to limit OBE, increases the PAPR due to interference/overlapping among the time domain samples of filtered signals.

### 3.5. Impact of CFO

In this section, we present results of simulations carried out to analyse the impact of carrier frequency offset on the BER performance of different waveforms. Simulations were performed

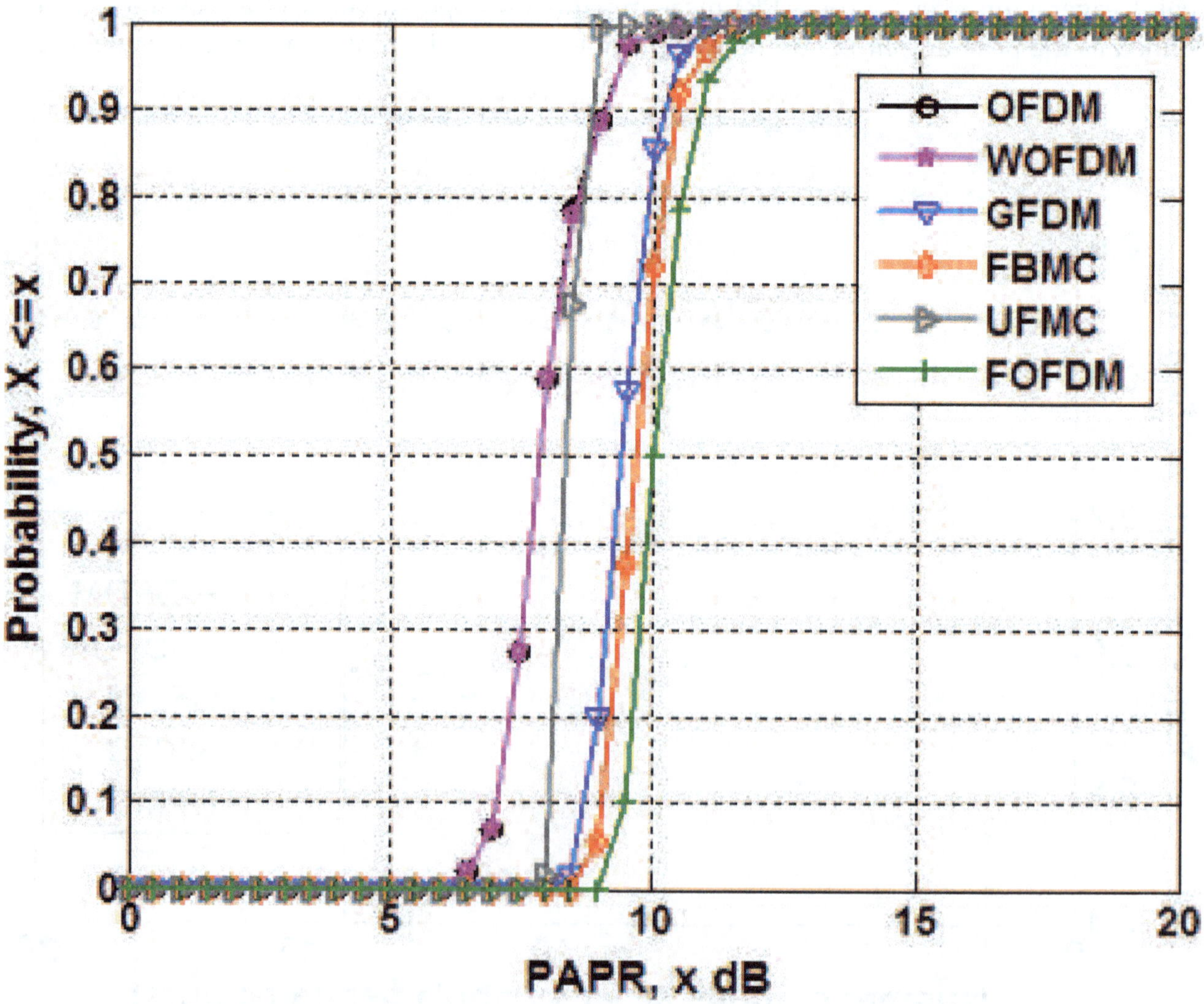


**Figure 13.** PAPR performance of candidate waveforms.

using parameters as given in **Table 1** for QPSK in an AWGN channel only, hence, the channel does not introduce any impairment.

**Figure 14** shows the raw BER of QPSK assuming $\in$ = 0.05, 0.1, where $\in = f'T$ is the normalized CFO, i.e. the frequency offset $f'$ normalized by the subcarrier spacing $^1/_T$. Note that this is the residual CFO and is not compensated for in the channel equalization block. It is observed from simulation results that all the waveforms show similar level of degradation, approximately 2 dB, in BER performance for $\in$ = 0.05 as compared to the BER performance shown in **Figure 10** for a perfectly synchronized receiver in an AWGN channel. However, the degradation in FBMC is comparatively larger, approximately 2.5 dB, as compared to other waveforms. This is due to the intrinsic interference in the FBMC scheme and the degradation becomes worse when normalized CFO increases to 0.1 due to increased level of intrinsic interference in FBMC. Comparing the results of $\in$ = 0.05 and $\in$ = 0.1, it can be seen that for larger value of CFO, all waveforms except FBMC show approximately 10.5 dB degradation and also tend to exhibit an error floor for higher values of Eb/No where inter-carrier interference becomes dominant due to larger CFO. Large degradation in the BER performance of FBMC indicates the need for intrinsic interference cancellation techniques or re-designing filters with even better localized pulse shapes to make FBMC more robust to CFO.

### 3.6. Impact of time offset

In this section, we present BER performance of different waveforms to analyse their sensitivity to timing offset (TO). We simulated BER performance for two different arbitrary values of TO, i.e. 80 and 150 samples in AWGN channel only. Hence, it is ensured that the channel itself does not introduce any time spreading. Simulation results given in this section were obtained by estimating channel using noise-free samples of received signal. We know from the literature survey that due to intrinsic interference in FBMC, it requires special pilot design, e.g. auxiliary

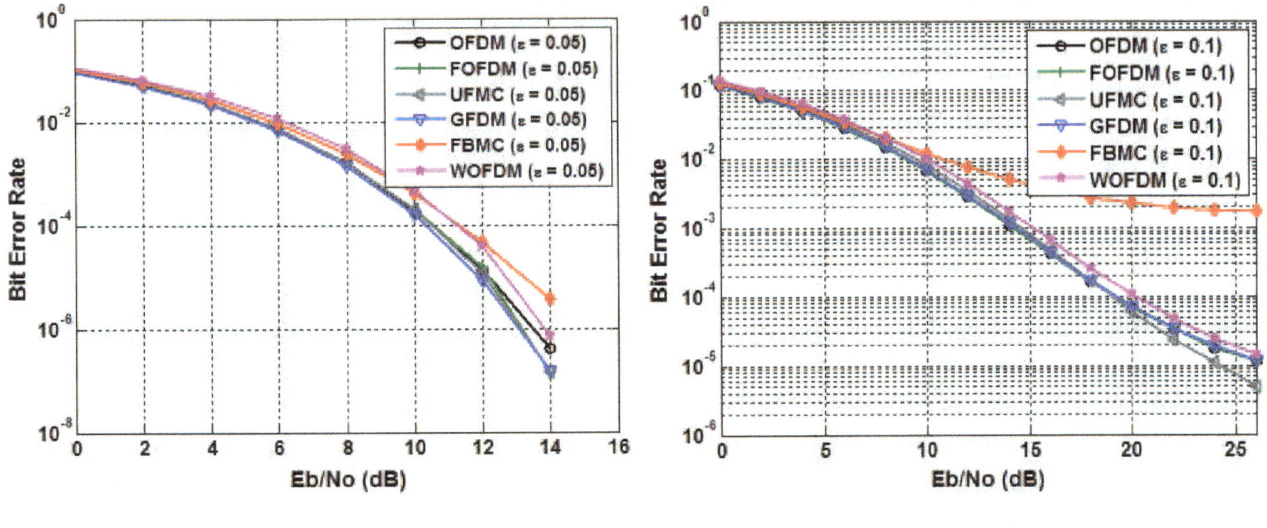


**Figure 14.** BER of QPSK/OQPSK in AWGN for $\epsilon$ = 0.05, 0.1.

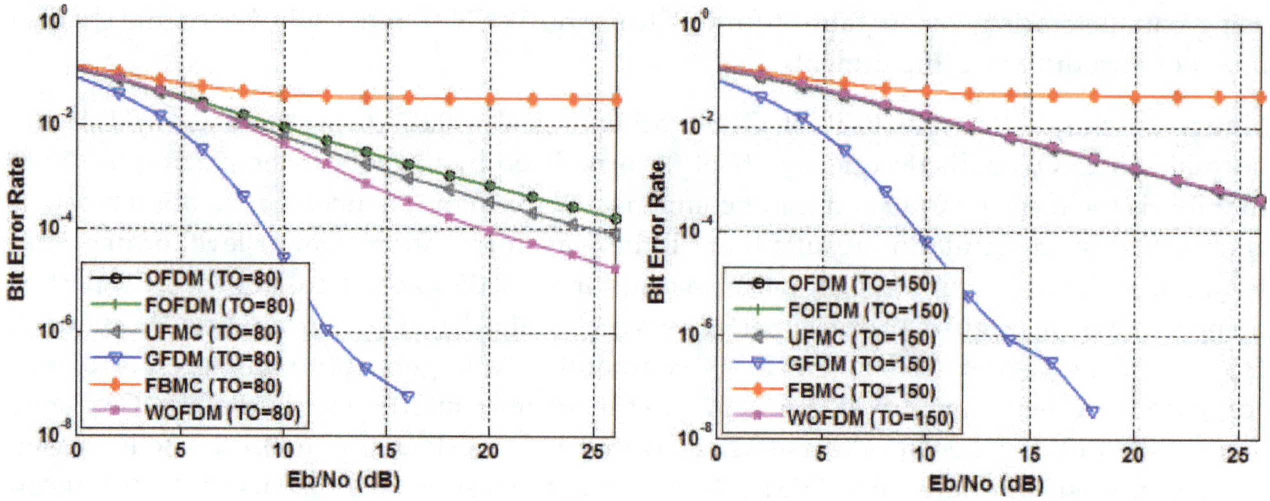


**Figure 15.** BER of QPSK/OQPSK in AWGN for TO = 80, 150 samples.

pilots [16], for channel estimation. Otherwise, the performance is severely degraded as can be seen in simulation results presented in **Figure 15** for TO = 80 and TO = 150 samples.

### 3.7. Computational complexity

The final figure of merit to be considered in this chapter is the computational complexity of different waveforms. In this section, computational complexity is evaluated in terms of number of real multiplications for each MCM Scheme. It is assumed that $N_u(N_u \leq N)$ subcarriers are loaded with transmitted symbols. A pair of $N$-point FFT and IFFT (via Split Radix FFT) with complexity $\mu_{FFT\&IFFT} = 2(N\log_2 N{-}3N + 4)$ is used as the component in the efficient implementations of relevant MCM schemes.

**Table 2** shows the computational complexity of the 5G candidate waveforms in terms of total number of required real multiplications per burst comprising of $M$ multicarrier symbols (each MC symbol comprising of $N$ subcarriers). While calculating complexity of UFMC and GFDM, it is assumed that each complex multiplication can be performed using three real multiplications. Complexity of OFDM comprises of IFFT and FFT complexity at the transmitter and receiver. FOFDM includes the added complexity due to transmit and receive filters. In FOFDM, it is assumed that the transmit filtering and adding CP could be combined such that the filtering is only performed once for the CP samples [12]. WOFDM has added complexity as compared to OFDM due to windowing that is a point wise multiplication operation. Complexity of UFMC transmitter is calculated based on number of real multiplication required for direct implementation of the operations given in **Figure 5**. Receiver complexity is derived based on the complexity of 2$N$ point FFT operation performed at the UFMC receiver [2]. Complexity of FBMC is based on real multiplications required for filter, frequency shifting and FFT and IFFT operations in FBMC transceiver [10]. Complexity of GFDM is based on the low complexity transceiver architecture given in [7] in addition to the MN point FFT and IFFT operations required at the GFDM receiver to enable 1-tap FDE.

| MCM | Number of real multiplications per burst | Normalized complexity |
|---|---|---|
| OFDM | $M\left(2(N\log_2 N-3N+4)\right)$ | 1 |
| FOFDM | $M\left(2(N\log_2 N-3N+4)+2NL+2(N+N_{CP})L\right)$ | 4.8427 |
| WOFDM | $M(2(N\log_2 N-3N+4)+2(N+0.25N))$ | 1.1785 |
| UFMC | $M\left((2N\log_2 2N-6N+4)+\frac{N'}{D}(N\log_2 N-3+4+2LN)\right)$ | 601.89 |
| FBMC | $M\left(4(N\log_2 N-3N+4)+4N+8Np\right)$ | 5.7122 |
| GFDM | $6MN\left(M+\log_2 N)+2(MN\log_2 MN-3MN+4\right)$ | 11.8231 |

**Table 2.** Complexity of MCM schemes.

The last column of **Table 2** shows the complexity of each MCM scheme normalized to the OFDM complexity for $M = 14$, $N = 1024$, $D = 12$, $p = 6$, $N_{CP} = 72$, $N' = 664$, $L = 72$ (*UFMC*), and $L = 13$ (*FOFDM*). It is observed that as compared to OFDM, WOFDM has the lowest complexity. FOFDM and FBMC are approximately five and six times more complex than OFDM, while GFDM is nearly 12 times more complex as compared to OFDM. The highest complexity is shown by UFMC. The complexity of UFMC is directly proportional to the number of sub-bands which in turn depends on the sub-band size. It must be noted that more efficient ways of implementation, e.g. polyphase implementation given in [9], can reduce the complexity of UFMC by nearly 4.5 times. Using a smaller FFT size per sub-band in UFMC can also attain significant reduction in complexity

## 4. Summary

The waveforms for 5G networks should address certain challenges to meet the diverse set of requirements for future wireless communications. This chapter has described different candidate waveforms and some preliminary simulation results are presented to compare their performance with OFDM and verify the comparisons given in the literature summarized in **Table 3**. Based on the simulation results given in this chapter, performance of different waveforms as compared to OFDM is summarized in **Table 4**.

It is observed that while most of the results match the comparison found in the literature, TO and CFO resiliency of FBMC does not match the results in **Table 3** [15]. This is due to the fact that we have not taken into account any intrinsic interference cancellation techniques or FBMC-specific pilot design for improved channel estimation.

While 5G candidate waveforms show better spectral containment than OFDM making them suitable for carrier aggregation, other factors such as spectral efficiency, synchronization requirements and computational complexity need to be taken into account in order to find the most suitable techniques and corresponding tradeoffs for different 5G scenarios. However, this needs further simulations and analysis particularly in multi-user scenarios according to

| Figure of merit | OFDM | FOFDM | WOFDM | FBMC | GFDM | UFMC |
|---|---|---|---|---|---|---|
| PAPR | High | High | High | High | Moderate (for SC-FDE) | High |
| OBE | High | Low | Low | Low | Low | Low |
| SE | Low | Low | Low | High | High | High |
| Computational complexity | Low | Moderate | Moderate | High | High | High |
| Short-burst traffic | No | No | No | No | Yes | Yes |
| Fragmented spectrum | No | No | Yes | Yes | Yes | Yes |
| TO resiliency | Poor | Poor | Moderate | Good | Good | Good |
| CFO resiliency | Poor | Poor | Moderate | Good | Good | Good |

**Table 3.** Comparison of different MCM schemes [15].

| Figure of merit | FOFDM | WOFDM | FBMC | GFDM | UFMC |
|---|---|---|---|---|---|
| PAPR | High | Similar | High | High | High |
| OBE | Low (in sidebands only) Similar (in fragments between available bands) | Low | Low | Slightly lowerLow (using guard symbols or windowing [2]) | Low |
| Time-frequency efficiency | Similar | Low | High for longer bursts | High | High |
| Computational complexity | Moderate | Similar | Moderate | Moderate | High |
| Short-burst traffic | No | No | No | Yes | Yes |
| Fragmented spectrum | No | Yes | Yes | Yes | Yes |
| TO resiliency | Similar | Better | Poor | Better | Better |
| CFO resiliency | Similar | Lower | Poor | Better | Similar |

**Table 4.** Summary of performance of different MCM schemes as compared to OFDM.

the propagation conditions of different 5G use cases and scenarios to understand the suitability of each candidate waveform in that specific environment.

## Author details

Ayesha Ijaz*, Lei Zhang, Pei Xiao and Rahim Tafazolli

*Address all correspondence to: a.ijaz@surrey.ac.uk

Institute for Communication System (ICS), Home of 5G Innovation Centre (5GIC), University of Surrey, Guildford, UK

## References

[1] Ijaz, A., et al. Enabling massive IoT in 5G and beyond systems: PHY radio frame design considerations. IEEE Access. 2016;**4**:3322–3339.

[2] 5GNOW deliverable D3.2_v1.3. 5G waveform candidate selection. 2014. Available at: http://www.5gnow.eu

[3] NGMN. 5G white paper. Available at: https://www.ngmn.org/uploads/media/NGMN_5G_White_Paper_V1_0.tif

[4] Xiao, P., Toal, C., Burns, D., Fusco, V., Cowan, C. Transmit and receive filter design for OFDM based WLAN systems. In: International Conference Wireless Communications and Signal Processing (WCSP); October 2010; IEEE; pp. 1–4.

[5] Bala, E., Li, J., Yang, R. Shaping spectral leakage: a novel low-complexity transceiver architecture for cognitive radio. IEEE Vehicular Technology Magazine. 2013;**8**(3):38–46.

[6] Siohan, P., Siclet, C., Lacaille, N. Analysis and design of OFDM/OQAM systems based on filterbank theory. IEEE Transactions on Signal Processing. 2002;**50**(5):1170–1183.

[7] Farhang, A., Marchetti, N., Doyle, L.E. Low complexity transceiver design for GFDM (forthcoming). arXiv:1501.02940.

[8] 5G: a technology vision. Huawei Technologies Co., Ltd.; 2013. Available at: https://www.scribd.com/document/251024709/5G-a-Technology-Vision

[9] Noguet, D., Gautier, M., Berg, V. Advances in opportunistic radio technologies for TVWS. EURASIP Journal on Wireless Communications and Networking. 2011;**2011**(1). DOI: 10.1186/1687-1499-2011-170.

[10] Du, J., Xiao, P., Wu, J., Chen, Q. Design of isotropic orthogonal transform algorithm-based multicarrier systems with blind channel estimation. IET Communications. 2012;**6**(16):2695–2704.

[11] Abdoli, M.J., Jia, M., Ma, J. Weighted circularly convolved filtering in OFDM/OQAM. In: IEEE 24th Annual International Symposium on Personal, Indoor, and Mobile Radio Communications (PIMRC); 8 September; 2013. pp. 657–661.

[12] Li, J., Bala, E., Yang, R. Resource block filtered-OFDM for future spectrally agile and power efficient systems. Physical Communication. 2014;**11**:36–55.

[13] 3GPP TS 36.104, Technical Specification Group Radio Access Network; Evolved Universal Terrestrial Radio Access (E-UTRA).Base Station (BS) radio transmission and reception, v8.2.0; May 2008.

[14] Schaich, F., Wild, T., Chen, Y. Waveform contenders for 5G-suitability for short packet and low latency transmissions. In: IEEE 79th Vehicular Technology Conference (VTC Spring); May 2014; pp. 1–5.

[15] Farhang, A., Marchetti, N., Figueiredo, F., Miranda, J.P. Massive MIMO and waveform design for 5th generation wireless communication systems. In: 1st International Conference on 5G for Ubiquitous Connectivity (5GU); November 2014; IEEE; pp. 70–75.

[16] Stitz, T., Ihalainen, T., Viholainen, A., Renfors, M. Pilot-based synchronization and equalization in filter bank multicarrier communications. EURASIP Journal on Advances in Signal Processing. 2010;(1). DOI: 10.1155/2010/741429.

7

# Selective Control Information Detection in 5G Frame Transmissions

Saheed A. Adegbite and Brian G. Stewart

**Abstract**

Control signalling information within wireless communication systems facilitates efficient management of limited wireless resources, plays a key role in improving system performance of 5G systems. This chapter focuses detection of one particular form of control information, namely, selective control information (SCI). Maximum-likelihood (ML) is one of the conventional SCI detection techniques. Unfortunately, it requires channel estimation, which introduces some implementation constraints and practical challenges. This chapter uses generalized frequency division multiplexing (GFDM) to evaluate and demonstrate the detection performance of a new form of SCI detection that uses a time-domain correlation (TDC) technique. Unlike the ML scheme, the TDC technique is a form of blind detection that has the capability to improve detection performance with no need for channel estimation. In comparison with the ML based receiver, results show that the TDC technique achieves improved detection performance. In addition, the detection performance of the TDC technique is improved with GFDM receivers that use the minimum mean square error (MMSE) scheme compared with the zero-forcing (ZF) technique. It is also shown that the use of a raised cosine (RC) shaped GFDM transmit filter improves detection performance comparison with filters that employ root raised cosine (RRC) pulse shape.

**Keywords:** 5G frame, blind detection, generalised frequency division multiplexing (GFDM), minimum mean square error (MMSE), physical control channel

## 1. Introduction

New physical layer architecture developments are under consideration for future 5G wireless systems to meet growing demands for even higher data rates and increasing data traffic. In comparison with the classical orthogonal frequency division multiplexing (OFDM) used in 4G, 5G physical layer architectures adopt a new type of frequency division multiplexing based on filtered OFDM in an attempt to improve spectral efficiency, increase data throughput and reduce latency [1]. In addition, 5G systems are designed to enable flexible resource allocation and configurable system architecture based on various communication scenarios, varied traffic and user needs [2]. To meet these challenges, various forms of system-critical control information are required to be transmitted through the use of both shared and dedicated physical control channels to facilitate efficient management of 5G system resources and to achieve optimum system performance. This chapter discusses and describes the use of a time-domain blind detection technique that uses time-domain correlation (TDC) between the transmitted control information and the received control information as a means of detection.

Control signals are important in wireless systems as they carry essential signalling information between the user equipment (UE) and the base station to facilitate successful detection of payload user data. Hence, successful detection of these control signals is a key to achieving the required system performance in 5G systems. In general, 5G control signals carry both user-specific and network-level information such as scheduling grant, user allocation, adaptive modulation and coding schemes (AMC), 5G frame configuration and power control. In practical wireless systems, an erroneous detection of control signals triggers re-transmission and causes transmission delays, which will ultimately degrade system performance. As a consequence, control signals are normally encoded using a large number of subcarriers to ensure robust and error-free detection [3].

The focus of this chapter is to address detection challenges of a specific category of wireless control signals called selective control information (SCI). An example of SCI encountered in 4G and implemented in 5G is the control format indicator (CFI) carried by the physical control format indicator channel (PCFICH). The CFI is used to inform the receiver about the signal format of the physical downlink control channel (PDCCH) and is a form of SCI because the actual CFI value ranges between 1 and 4 [3]. Hence, the encoded CFI information can be chosen (i.e. selected) from a small number of candidate CFI information values, which are known at both transmitting and receiving ends of the system [4]. The PDCCH carries major downlink control information (DCI) that represents various types of network configuration and system variables including power control, resource allocations and scheduling grants. A more detailed discussion on CFI can be found in [3]. Another example of SCI is the control information used to encode the type of modulation scheme of payload user data. In summary, SCI is a type of control information that is selective from a deterministic set of candidate information sequences known at both the transmitter and the receiver [5].

In the literature, the maximum likelihood (ML) detection scheme is considered as the standard detection technique for decoding SCI because it is more computationally efficient solution, in terms of hardware implementation, compared with methods such as the K-best list sphere detector (K-LSD) and successive interference cancellation (SIC) [6]. An example of a practical hardware implementation of the ML estimation method for the decoding of the PCFICH is described in [7]. Unfortunately, the ML detection scheme imposes a practical constraint in that it requires channel estimation at the receiver. In theory, the detection performance of the ML estimation technique can be enhanced through the use of an advanced channel estimation technique such as linear minimum mean square error (LMMSE). However, the need for channel estimation requires additional transmission overhead in the form of pilot signals to facilitate pilot-assisted channel estimation and also increases computational complexity at the receiver. Therefore, the need for channel estimation makes the ML scheme an unattractive and unsuitable solution in practical systems and in the occurrence of severe fading channel [6].

Unlike the ML detection method, the TDC solution discussed in this chapter is a form of blind detection technique in that it requires neither channel estimation nor channel equalisation at the receiver. The TDC technique is designed to address the practical challenges of the ML estimation method and to improve detection performance of essential control signalling information adopted in 5G systems. To demonstrate the potential use of the TDC detection technique in 5G systems and its advantage over the ML detection method, this chapter investigates, through MATLAB simulations, the detection performance of the TDC detection technique using the well-known generalised frequency division multiplexing (GFDM) architecture being considered for 5G. In this study, the detection performance is evaluated using the block error rate (BLER) metric. The effects of GFDM demodulation techniques and transmit filter pulse shapes are studied and investigated to further understand and demonstrate the potential use of the TDC technique in a practical GFDM system. In comparison with the classical OFDM technique, GFDM performs subcarrier-level filtering to minimise or manage out-of-band (OOB) radiation and improve spectral efficiency in 5G. The roll-off factor $\alpha$ of the transmit filter plays a key role towards controlling the OOB. Therefore, the pulse shape and the roll-off factor of the transmit filter will impact detection performance. Using filters with root-raised-cosine (RRC) and raised-cosine (RC) responses, one aspect of this chapter will investigate the dependency between shape of the transmit filter and detection performance of the TDC technique. Another aspect of this chapter will also investigate the influence of the roll-off factor of each chosen filter type on the detection performance.

In practice, GFDM demodulation can be implemented using techniques such as zero-forcing (ZF), minimum mean square error (MMSE) and matched filtering (MF) [2]. In this chapter, only the ZF and MMSE are considered because of the self interference caused by the use of the MF technique. The impact of these two GFDM demodulation methods on the detection performance of the TDC technique is studied so as to further understand and highlight the limitations and/or robustness of the TDC detection technique for 5G systems.

## 2. SCI Transmission and Reception

This section briefly describes the basic transmitter/receiver architecture used to encode and decode the SCI.

### 2.1. SCI Transmission

A detailed description of the GFDM transmitter is presented in [2]. **Figure 1** describes a block diagram representation of the considered GFDM transmitter architecture.

Let $\boldsymbol{d}$ be the transmitted source data of length $N$, which may consist of control signalling formation, payload user data and some preambles. In GFDM, modulated subcarrier symbols in $\boldsymbol{d}$ are formatted into a 2D time-frequency GFDM block of dimension $K$ by $M$ where $K$ and $M$,

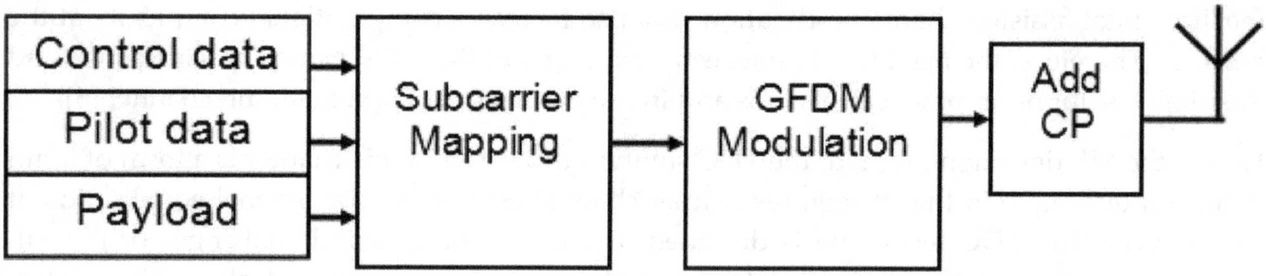


**Figure 1.** GFDM transmitter.

respectively, represent the number of subcarriers (in the frequency-domain) and the number of subsymbols (in the time-domain) [8]. For $0 \leq k \leq K - 1$ and $0 \leq m \leq M - 1$, where $k$ and $m$ are arbitrary indices of the subcarrier and subsymbol, respectively, each subcarrier symbol in $\boldsymbol{d}$ can be denoted by $\boldsymbol{d}_{k,m}$, and $\boldsymbol{d}$ can be represented as

$$\boldsymbol{d} = [\boldsymbol{d}_{0,0}\ \boldsymbol{d}_{1,0}\ \ldots\ \boldsymbol{d}_{K-1,0}\ \boldsymbol{d}_{0,1}\ \ldots\ \boldsymbol{d}_{K-1,1}\ \boldsymbol{d}_{k,m}\ \ldots\ \boldsymbol{d}_{K-1,M-1}]. \tag{1}$$

*2.1.1. Subcarrier mapping*

**Figure 2** shows a diagrammatic representation of the considered subcarrier mapping scheme. For simplicity, in this chapter, it is assumed that $d$ consists of (1) a pilot sequence, $d_p$ of size $N_p$; (2) an SCI sequence vector, $d_c$ of size $N_c$; and (3) other forms of control/payload information, $d_r$ of size $N_d$. Thus, $N = N_p + N_c + N_d$.

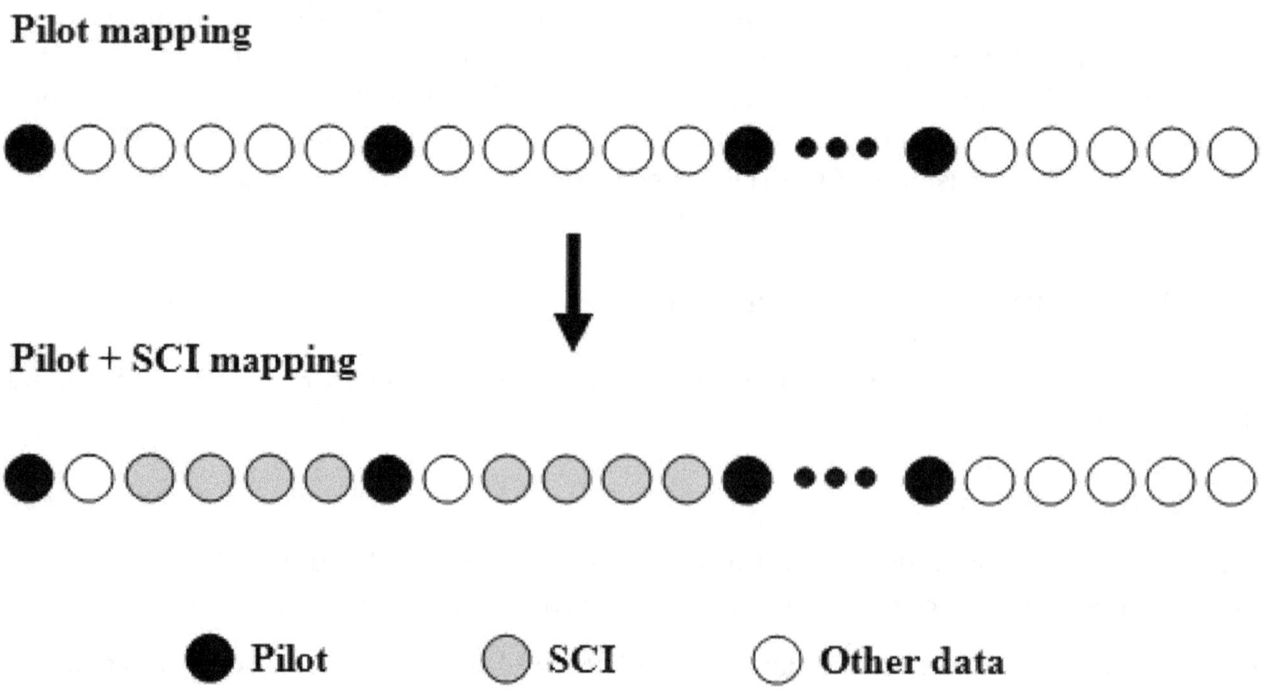


**Figure 2.** Subcarrier mapping.

### 2.1.1.1. Subcarrier mapping: pilots

In an attempt to mimic practical 5G frame structures, some preambles in the form of reference signals or pilots are embedded with the transmitted signal. In practical systems, reference signals are often adopted to facilitate channel estimation and synchronisation so as to improve data recovery performance of payload user data. Within the considered subcarrier mapping, some pilots are embedded within $d$ at regular intervals. As an example, a pilot spacing of six is considered in this study because currently, there is no standard specification for pilot spacing in 5G.

### 2.1.1.2. Subcarrier mapping: SCI

After pilot subcarrier allocation, SCI subcarriers are allocated as indicated in **Figure 2**. In the considered mapping, it is assumed that the size of the SCI sequence $d_c$ is a multiple of 4 so that elements of $d_c$ are mapped in groups of 4 in a similar manner to a form of resource element mapping in 4G. The four subcarriers in each group are mapped to un-allocated subcarriers in-between two consecutive pilot positions.

Let $\mathcal{C}$ represent a set of candidate information, which consists of $U$ different SCI sequences, that is,

$$\mathcal{C} = \{C_1, C_2, C_u \ldots C_U\} \tag{2}$$

where each $\mathbf{C}_u$ is of the same size as $\boldsymbol{d}_c$ and each element of $\mathbf{C}_u$ is a complex-valued QPSK-modulated symbol of unity magnitude. For $0 \le c \le N_c - 1$, where $c$ is an arbitrary index, each element of $\mathbf{C}_u$ is denoted by $\mathbf{C}_u[c]$. The complex conjugate $\mathbf{C}_u[c]^*$ is mathematically equivalent to $1/\mathbf{C}_u[c]$.

As an example, assume that the encoded SCI is used to carry information about the modulation scheme of payload user data. In the case of 4G and also 5G, there is a finite number of known modulation types and each type can be encoded into an SCI sequence $\mathbf{C}_u$ where $1 \le u \le U$. **Table 1** shows an example of the mapping of $C_u$ to a modulation type. Thus, the transmitted SCI sequence is uniquely identified by the index $u$ given that $\mathcal{C}$ is deterministic and known. Hence, a block-level detection is performed at the receiver in order to determine an estimate of $\overline{u}$, from which the type of modulation or any other form of control information is automatically determined [9]. It is important to note that a block-level detection procedure

| SCI index, $u/\overline{u}$ | SCI, $d_c$ | Modulation |
|---|---|---|
| 1 | $C_1$ | 4–QAM |
| 2 | $C_2$ | 16–QAM |
| 3 | $C_3$ | 64–QAM |
| 4 | $C_4$ | 256–QAM |

**Table 1.** An example of SCI encoding scheme.

used for the recovery of control information is entirely different from the usual subcarrier-level or one-tap equalisation associated with the recovery of payload user data [5].

Given that the transmitted SCI sequence $d_c$ is chosen from a finite set $\mathcal{C}$, let $\overline{u}$ define the index of the selected and transmitted SCI sequence vector, such that:

$$d_c = C_{\overline{u}} \text{ where } C_{\overline{u}} \in \mathcal{C}. \tag{3}$$

After SCI mapping, all other remaining un-allocated subcarriers are assigned to other forms of data $\boldsymbol{d}_r$. It is important to note that the main focus of this chapter is on the detection of the SCI index $u$ that corresponds to $\boldsymbol{d}_c$.

### 2.1.2. *Transmitted signal*

Let $\boldsymbol{x}$ be the time-domain GFDM signal of length $N$. For $0 \leq n \leq N-1$, each element $x[n]$ is derived from Ref. [2]

$$\boldsymbol{x}[n] = \sum_{k=0}^{K-1} \sum_{m=0}^{M-1} g_{k,m}[n] \, \boldsymbol{d}_{k,m} \tag{4}$$

where $g_{k,m}[n]$ represents a time and frequency shifted form of a transmit filter $g[n]$. Each $g_{k,m}[n]$ is given as [2]

$$g_{k,m}[n] = g[(n-mK)\,\mathbf{mod}\,N] \exp\left(-j2\pi \frac{k}{K} n\right) \tag{5}$$

where **mod** is the modulo function.

Let $\boldsymbol{A}$ be the transmit filter matrix where

$$\boldsymbol{A} = [g_{0,0}\, g_{1,0} \cdots g_{K-1,0}\, g_{0,1} \cdots g_{K-1,1}\, g_{k,m} \cdots g_{K-1,M-1}]. \tag{6}$$

Then, the GFDM signal can also be expressed by Michailow et al. [2]

$$\boldsymbol{x} = \boldsymbol{A}\boldsymbol{d}. \tag{7}$$

Finally, the GFDM signal $\mathbf{x}$ is further extended by a cyclic prefix (CP) to mitigate channel fading and reduce inter-symbol interference (ISI).

## 2.2. SCI Detection

In this chapter, SCI decoding is implemented using the ML and the TDC detection techniques. **Figure 3** shows the block diagram representation of the conventional ML-based SCI detection scheme. It is important to note that the considered receiver architecture for decoding SCI is slightly different from typical GFDM receiver for decoding payload user data. For instance, in a typical GFDM receiver, QAM demodulation is required to determine an estimate of

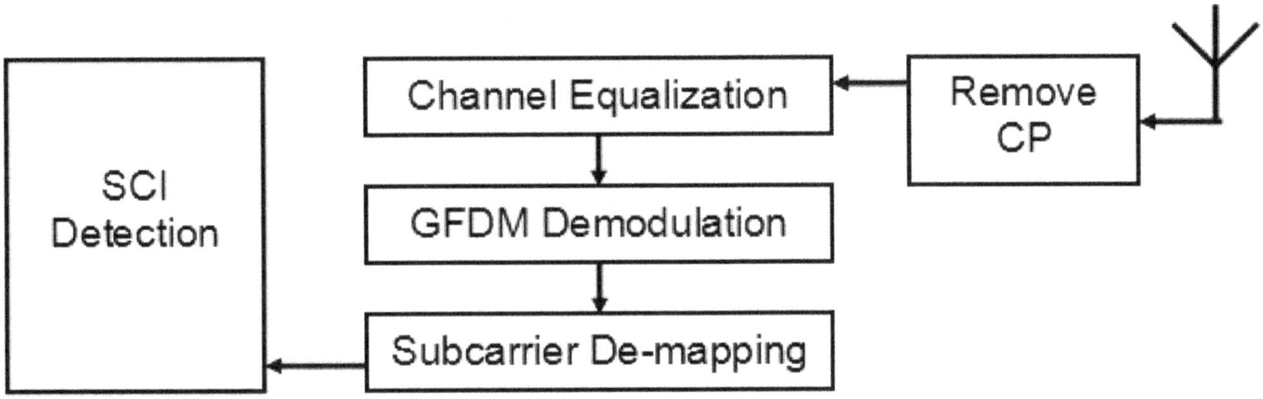


**Figure 3.** ML-based receiver architecture.

transmitted bitstream. However, in the considered receiver, a form of SCI decoding is implemented instead of QAM demodulation. Unlike QAM demodulation, SCI decoding produces a scalar value that represents an estimate of the SCI index $\overline{u}$.

After CP removal at the receiver, let $y$ be the received signal after the transmission over a transmission channel medium with channel matrix $\boldsymbol{H}$, corrupted with additive white Gaussian noise $v$ with variance $\sigma_v^2$, as expressed by Michailow et al. [2], thus

$$y = \boldsymbol{H}\,\boldsymbol{A}\boldsymbol{d} + v. \tag{8}$$

The next stage involves GFDM demodulation, which serves to mitigate the inter-carrier interference (ICI) cause by the filtering process at the transmitter. Let $\hat{\boldsymbol{B}}$ be a $N \times N$ receiver matrix, which is used for GFDM demodulation.

In the ZF-based GFDM receiver, $\hat{\boldsymbol{B}}$ is computed using

$$\hat{\boldsymbol{B}} = (\boldsymbol{A}^H\boldsymbol{A})\boldsymbol{A}^H \tag{9}$$

where $\boldsymbol{A}^H$ denotes an Hermitian or conjugate transpose of $\boldsymbol{A}$. In the MMSE-based receiver, the receiver matrix $\hat{\boldsymbol{B}}$ is, however, determined from

$$\hat{\boldsymbol{B}} = \left(\frac{\sigma_v^2}{\sigma_d^2}\boldsymbol{I} + \boldsymbol{A}^H\boldsymbol{A}\right)^{-1}\boldsymbol{A}^H. \tag{10}$$

From the expression in Eq. (10), $\boldsymbol{I}$ is the identity matrix, and $\sigma_d^2$ is the variance of $\boldsymbol{d}$. Using $\hat{\boldsymbol{B}}$, the output of the GFDM demodulation block is computed from

$$\hat{\boldsymbol{d}} = \hat{\boldsymbol{B}}y. \tag{11}$$

Hence, the received SCI subcarriers $\hat{\boldsymbol{d}}_c$ are represented as a subset of $\hat{\boldsymbol{d}}$. The next stage involves the SCI decoding where an estimate of the index $\overline{u}$ is determined given that the set $\mathcal{C}$ is also known at the receiver. In this case, the decoded SCI can be directly determined through an estimate of the SCI index $\hat{u}$.

#### 2.2.1. ML scheme

The ML detection technique uses a form of Euclidean distance minimisation function. Let $\boldsymbol{H}_c$ represent the frequency-domain representation of sub-channel coefficients that correspond to the SCI subcarrier locations.

Let $\hat{u}$ denote an estimate of $\overline{u}$. Then, using the ML decision criterion, $\hat{u}$ is determined through

$$\hat{u} = \underset{u;\, C_u \in \mathcal{C}}{\arg\min} \, |\hat{\boldsymbol{d}}_c - \boldsymbol{H}_c \boldsymbol{C}_u|^2. \tag{12}$$

The expression in Eq. (12) suggests that detection performance of the ML estimation method depends on the channel coefficients $\boldsymbol{H}_c$. In this chapter, the ML decision is implemented using perfect channel estimation. However, in practical systems, channel estimation is implemented as described in [10]. Unfortunately, the need for channel estimation increases both design and computational complexities, and erroneous channel estimation is expected to produce erroneous estimation of $\hat{u}$. This is the main practical challenge associated with the use of the ML estimation method in 5G wireless systems.

## 3. TDC Detection Technique

The TDC technique uses a form of signal correlation as a means of detection. A time-domain detection approach is considered because studies from, for example, [11] and [12] have shown that it offers robust decoding even in the presence of ISI [13]. With regard to SCI specifically, the TDC technique uses a correlation that exists between $\hat{d}_c$ and each possible candidate SCI $\mathbf{C}_u$ within $\mathcal{C}$ is used to determine an estimate of the transmitted SCI [5]. **Figure 4** shows the block diagram representation of the GFDM receiver that uses the TDC-based SCI detection scheme.

### 3.1. Discrete Correlation Theorem

The applied correlation within the TDC detection technique can be explained using the well-known discrete correlation theorem (DCT). Based on the DCT, a correlation of two arbitrary time-domain signals $q_1$ and $q_2$ (of the same size) is obtained from [14]

$$\mathrm{CORR}\{q_1, q_2\} = \mathrm{IFFT}\{Q_1 \times Q_2^*\} \tag{13}$$

where * represents the complex conjugation, and $Q_1$ and $Q_2$ are, respectively, the frequency-domain representations of $q_1$ and $q_2$, that is,

$$Q_1 = \mathrm{FFT}\{q_1\} \quad \text{and} \quad Q_2 = \mathrm{FFT}\{q_2\} \tag{14}$$

where FFT $\{\cdot\}$ denotes the fast Fourier transform (FFT) function.

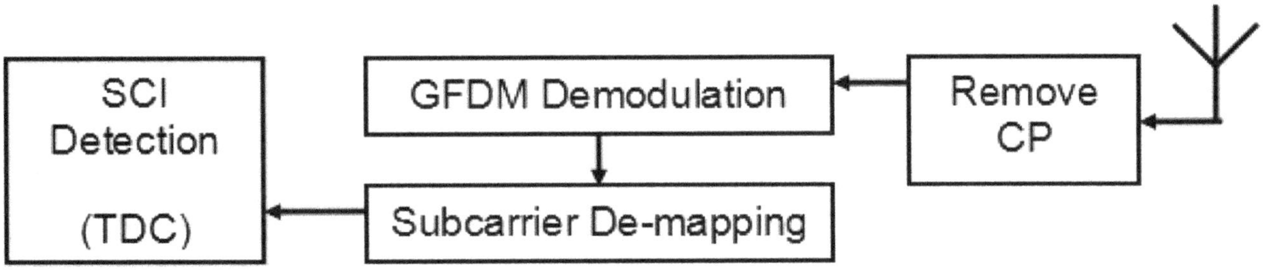


**Figure 4.** TDC-based receiver architecture.

In a TDC-based receiver, a complex-valued term $\boldsymbol{Z}_u$ is first computed in a similar manner to the DCT definition in Eq. (13). Thus, $\boldsymbol{Z}_u$ is given by

$$\begin{aligned}\boldsymbol{Z}_u &= \hat{\boldsymbol{d}}_{\mathbf{c}} \times \boldsymbol{C}_u^* \\ &= (\boldsymbol{H}_c \hat{\boldsymbol{d}}_c + \boldsymbol{V}_{\mathbf{c}})) \times \boldsymbol{C}_u^* \\ &= (\boldsymbol{H}_c \hat{\boldsymbol{d}}_c \boldsymbol{C}_u^*) + (\boldsymbol{V}_c \boldsymbol{C}_u^*) \\ &= (\boldsymbol{H}_c \hat{\boldsymbol{d}}_c \boldsymbol{C}^*{}_u) + \left(\boldsymbol{V}'_{\mathbf{c}}(u)\right)\end{aligned} \tag{15}$$

where $\boldsymbol{V}_c$ is the frequency-domain representation of AWGN components of the SCI subcarriers and $\boldsymbol{V}'_c(u) = \boldsymbol{V}_c \boldsymbol{C}_u^*$. For $0 \le c \le N_c - 1$, $Z_u$ is a vector of size $N_c$ and may be represented as

$$Z_u = [Z_u[0],\ Z_u[1],\ Z_u[c] \ldots Z_u[N_c{-}1]]. \tag{16}$$

When there is a strong correlation between $\hat{\boldsymbol{d}}_c$ and $\mathbf{C}_u$, then the expression in Eq. (15) can be approximated to

$$\boldsymbol{Z}_u \approx \begin{cases} \boldsymbol{H}_c + \boldsymbol{V}'_{\mathbf{c}}(u), & u = \overline{u} \\ \boldsymbol{H}_c \hat{\boldsymbol{d}}_c \boldsymbol{C}_u^* + \boldsymbol{V}'_{\mathbf{c}}(u), & \text{otherwise.} \end{cases} \tag{17}$$

By omitting the noise terms in Eq. (17) for simplicity, the expression in Eq. (17) is reduced to

$$\boldsymbol{Z}_u \approx \begin{cases} \boldsymbol{H}_c, & u = \overline{u} \\ \boldsymbol{H}_c \hat{\boldsymbol{d}}_c \boldsymbol{C}_u^*, & \text{otherwise.} \end{cases} \tag{18}$$

From the expression in Eq. (17), it can be seen that the same channel term $H_c$ and identical noise term $V'_{\mathbf{c}}(u)$ are present in both $Z_{u=\overline{u}}$ and $Z_{u\neq\overline{u}}$ terms when $u = \overline{u}$ and $u \neq \overline{u}$, respectively. Thus, without loss of generality, a simplified representation of the main difference between each value of $Z_{u=\overline{u}}$ and $Z_{u\neq\overline{u}}$ is further reduced to

$$\boldsymbol{Z}_u[c] \approx \begin{cases} \mathbf{1}, & u = \overline{u} \\ \hat{\boldsymbol{d}}_c \boldsymbol{C}_u^*, & \text{otherwise.} \end{cases} \tag{19}$$

In a similar manner to the expression in Eq. (13), let $z_u$ be the time-domain equivalent $Z_u$ obtained from

$$z_u = \underset{N_c\text{-point}}{\text{IFFT}} \{Z_u\}. \tag{20}$$

In this chapter, $z_u$ will be referred to as the TDC function. For $0 \le w \le N_c - 1$, $z_u$ is written as

$$z_u = [z_u[0],\ z_u[1],\ z_u[w] \ldots z_u[N_c-1]]. \tag{21}$$

It should be noted that the IFFT operation in Eq. (20) requires no zero padding if the value of $N_c$ is a power of 2. However, in cases (not shown in this chapter) where $N_c$ is not a power of 2, zero padding can be applied as required in FFT algorithms with no degradation in performance.

### 3.2. TDC Decision Criterion

From the approximation of $\mathbf{Z}_{\overline{u}}$ in Eq. (19), the magnitudes of $\mathbf{Z}_{\overline{u}}$ result in an impulse function in a similar manner to an auto-correlation function. Hence, the magnitude $z_{\overline{u}}[w]$ (derived from $\mathbf{Z}_{\overline{u}}$) can be approximated to

$$|z_{\overline{u}}[w]| = \begin{cases} 1, & w = 0 \\ 0, & 1 \le w \le N_c - 1 \end{cases} \tag{22}$$

where $|\cdot|$ is the magnitude of a complex-valued variable. Otherwise, $|z_u[w]| > 0$ when $u \ne \overline{u}$.

Using the approximation in Eq. (22), the mean value of $|z_{\overline{u}}|$ is

$$\begin{aligned} \mathbf{E}\{|z_{\overline{u}}|\} &= \frac{1}{N_c} \sum_{w=0}^{N_c-1} |z_{\overline{u}}[w]| \\ &\approx 1/N_c \end{aligned} \tag{23}$$

where $\mathbf{E}$ is the expectation function. Similarly, from the definition in Eq. (22), $E\{|z_{u \ne \overline{u}}|\}$ is expected to be larger than $E\{|z_{\overline{u}}|\}$ because the corresponding magnitudes of $z_{u \ne \overline{u}}[w]$ are non-zero. Therefore,

$$\mathbf{E}\{|z_{\overline{u}}|\} \ll E\{|z_{u \ne \overline{u}}|\}. \tag{24}$$

Thus, in the presence of the channel fading term $H_c$, the expression in Eq. (24) is still valid since the resulting time-domain functions $z_{\overline{u}}$ and $z_{u \ne \overline{u}}$ are both affected by the same channel component.

The expression in Eq. (24) therefore implies that an estimate of $\overline{u}$ corresponds to the $u$-index of the time-domain function with the minimum mean value amongst all $U$ time-domain functions. Therefore, the TDC detection criterion is defined by Saheed et al. [5]

$$\hat{u} = \arg \min_u E\{|z_u|\}. \tag{25}$$

From the expressions in Eqs. (15)–(25), it can be noted that the TDC detection technique requires no channel estimation. The main potential drawback of the TDC technique is the need for multiple IFFTs, which may increase computational complexity of the TDC-based receiver, particularly in limited practical cases, where the number of candidate SCI $U$ is large. However, this may not be a critical problem due to increased use of high-speed digital signal processors (DSPs) with efficient implementation of FFT.

### 3.3. Rayleigh Distribution

The hypothesis in Eq. (25) suggests that the distribution of $|z_u|$ may follow a Rayleigh distribution. Let $x$ be a continuous random variable. By letting $x = |z_u[w]|$, the Rayleigh probability distribution function (PDF) of $x$ can be described by Walck [15]

$$\boldsymbol{P}(x) = \frac{x}{\lambda^2} \exp(-x^2/2\lambda^2), x > 0 \tag{26}$$

where $\lambda$ is the Rayleigh scale parameter, which indicates the point (the value of $x$) at which the PDF $P(x)$ is maximum [15]. As a function of $\lambda$, the mean of $x$, $E(x)$, is expressed by Walck [15]

$$\boldsymbol{E}(x) = \lambda\sqrt{\frac{\pi}{2}}. \tag{27}$$

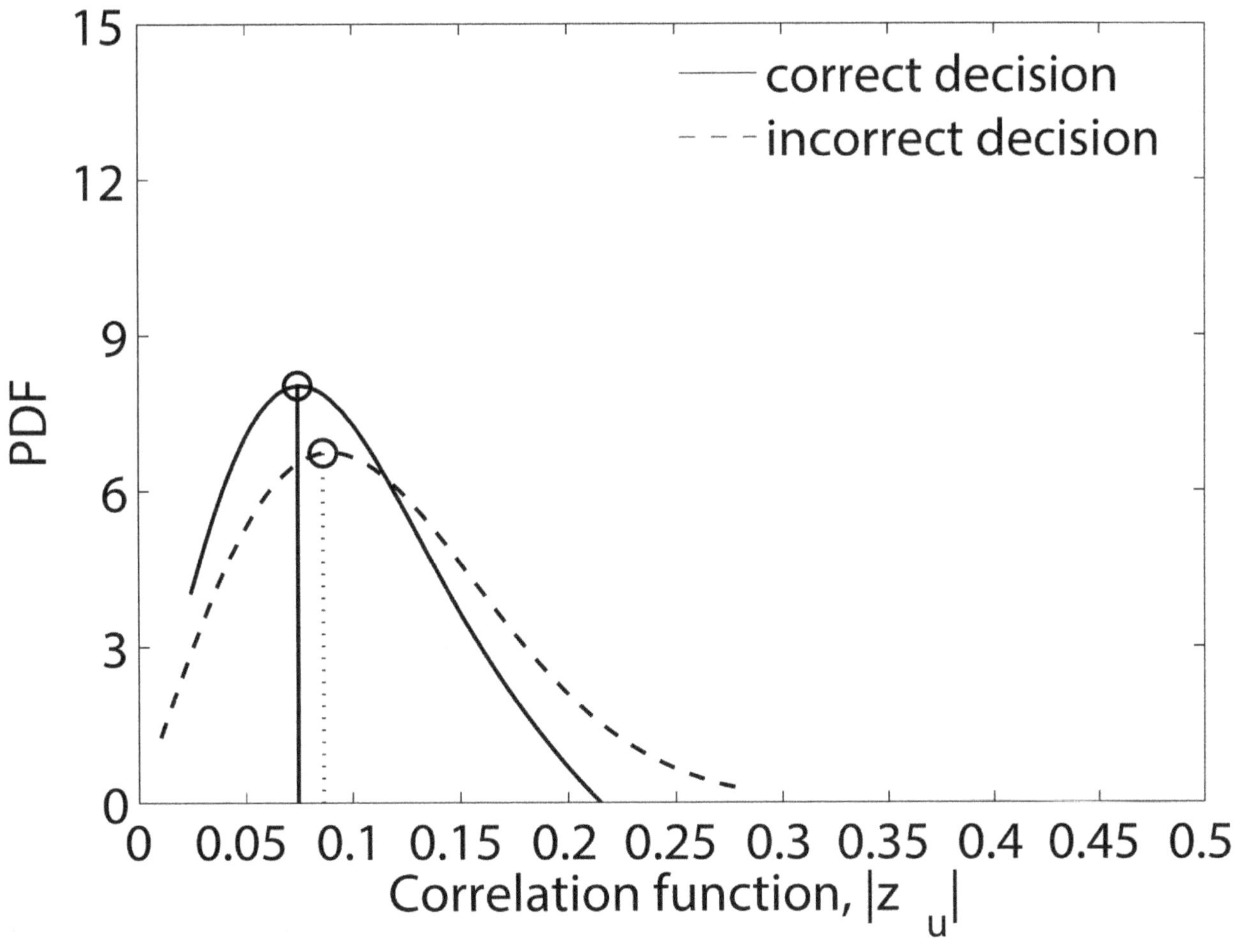


**Figure 5.** Distribution of $|z_u|$.

The expression for $E(x)$ indicates a linear relationship between $\lambda$ and $E(x)$. Hence, in relation to the TDC decision criterion in Eq. (25), the value of $\lambda$ is expected to be smaller for a correct decision compared with the case of an incorrect decision. As an example, **Figure 5** shows the Rayleigh PDF of $|z_u|$ in the presence of a multipath channel fading and transmit signal-to-noise ratio (SNR) of 6 dB. Results in **Figure 5** indicate that the value of $\lambda$ is smaller when $u = \overline{u}$ compared with when $u \neq \overline{u}$. Therefore, amongst all the $U$ correlation functions, the TDC-based decision criterion minimises the mean of $|z_u|$.

## 4. Detection Performance

This section presents the numerical detection performance of the TDC detection technique in comparison with the ML scheme. MATLAB simulations demonstrate the effect of the GFDM demodulation technique and the filter pulse shape characteristics on the detection performance of the TDC technique.

### 4.1. Simulation Set-up

Simulations consider that a GFDM system with $K = 64$, $M = 9$, $N_c = 32$, $U = 4$ and the size of CP is set to 16. Simulation is based on transmission over a frequency-selective Rayleigh fading channel known as the extended pedestrian type A (EPA), with a root mean square (RMS) delay spread, $\tau_{\text{rms}}$ of 45ns [16]. **Table 2** shows the power-delay profile of the EPA channel [17].

| **Channel parameters** | **1** | **2** | **3** | **4** | **5** | **6** | **7** |
|---|---|---|---|---|---|---|---|
| Path delay, ns | 0 | 30 | 70 | 90 | 110 | 190 | 410 |
| Power, dB | 0.0 | −1.0 | −2.0 | −3.0 | −8.0 | −17.2 | −20.8 |

**Table 2.** EPA fading channel.

#### *4.1.1. Block error rate*

For user data, bit error rate (BER) is often used as the detection performance metric. However, in the case of control information, the BLER is the customary detection performance metric [5]. To compute the BLER, an error count between the actual value $\overline{u}$ that corresponds to the selected sequence $\boldsymbol{d}_c$ and its estimate $\hat{u}$ obtained at the receiver is evaluated. An erroneous block exists when $\overline{u} \neq \hat{u}$. Otherwise, the detection is considered error-free.

For each SNR level, the BLER is computed as [5]

$$\text{BLER} = \frac{1}{N_{\text{BLK}}} \sum_{i=1}^{N_{\text{BLK}}} \boldsymbol{F}_i \tag{28}$$

where $N_{\text{BLK}}$ is the number of OFDM symbol blocks (for a given SNR level). For $1 \leq i \leq N_{\text{BLK}}$, $\boldsymbol{F}_i$ is computed from

$$F_i = \begin{cases} \mathbf{1} & \mathbf{if} \quad \overline{u} \neq \hat{u} \\ 0 & \text{otherwise.} \end{cases} \tag{29}$$

The BLER produced by each SCI decoding technique is evaluated as a function of the GFDM demodulation technique, filter type and filter roll-off factor parameter. The frequency-domain response of the considered RC filter with a roll-off factor $\alpha$ is given by Michailow et al. [2]

$$G_{\mathrm{RC}}[f] = \frac{1}{2}\left[1-\cos\left(\pi \mathrm{lin}_{\alpha}\left(\frac{f}{M}\right)\right)\right]. \tag{30}$$

Thus, the frequency-domain response of the RRC filter response is derived as

$$G_{\mathrm{RRC}}[f] = \sqrt{G_{\mathrm{RC}}[f]}. \tag{31}$$

## 4.2. Numerical Results

### *4.2.1. Detection performance with ZF-based GFDM receiver*

Using the ZF-based GFDM demodulation technique, **Figure 6** shows the BLER performance of the TDC technique based on an RRC shaped filter with roll-off factor of 0.1, 0.5 and 0.9.

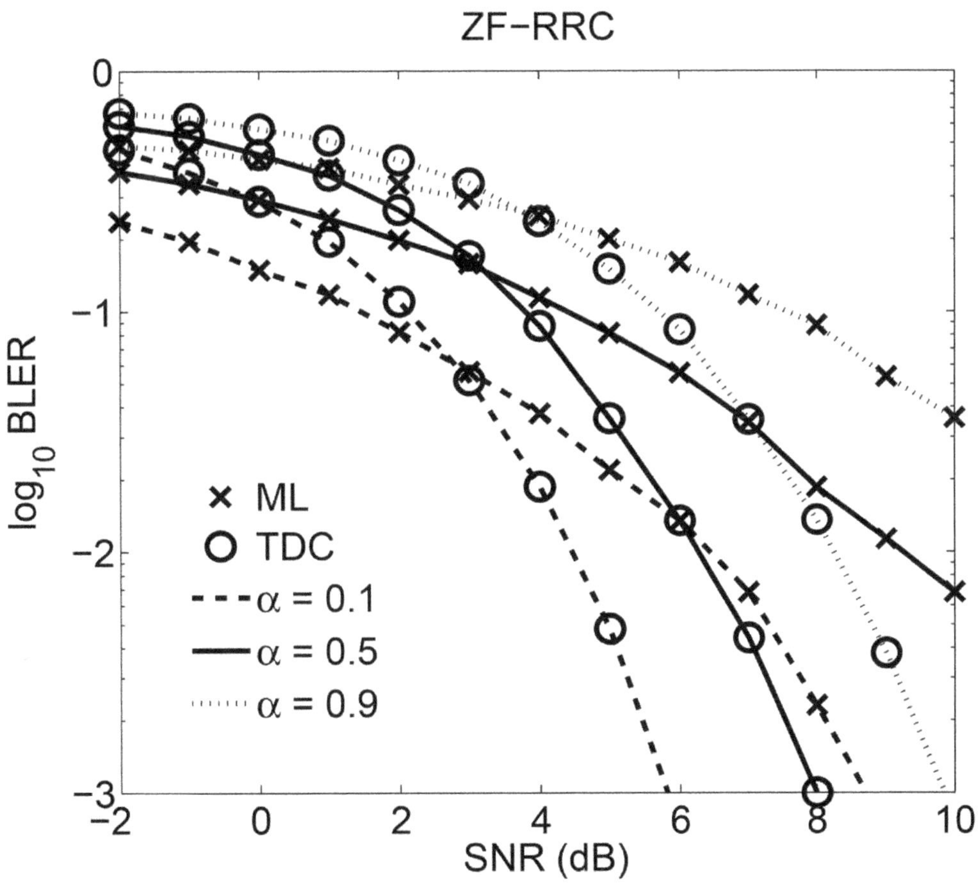


**Figure 6.** BLER comparison of the ML/TDC techniques with ZF, RRC shaped filter and roll-off factor $\alpha$ =[0.1, 0.5 and 0.9].

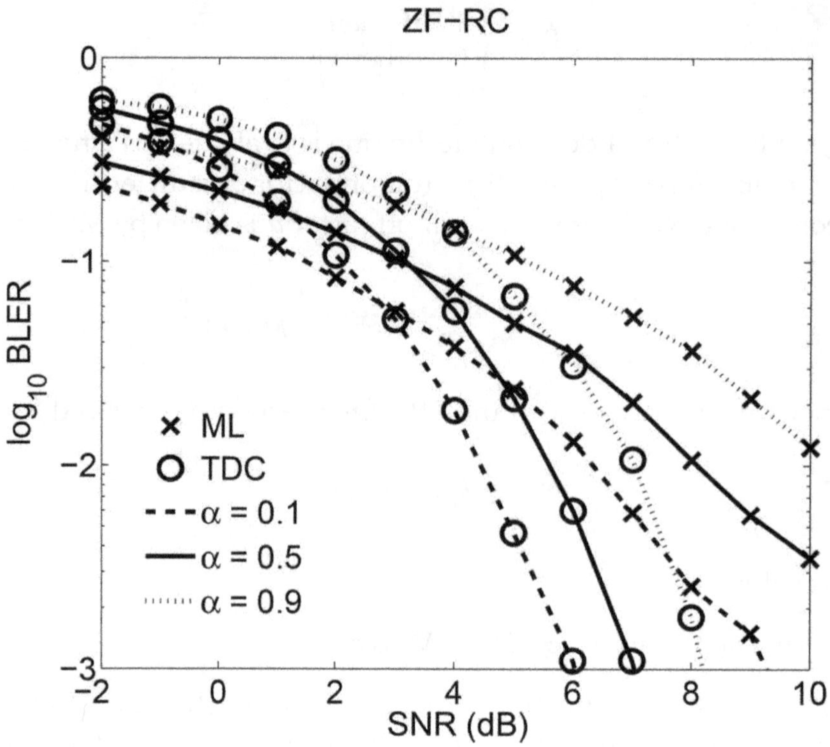


**Figure 7.** BLER comparison of the ML/TDC techniques with ZF, RC shaped filter and roll-off factor $\alpha$ =[0.1, 0.5 and 0.9].

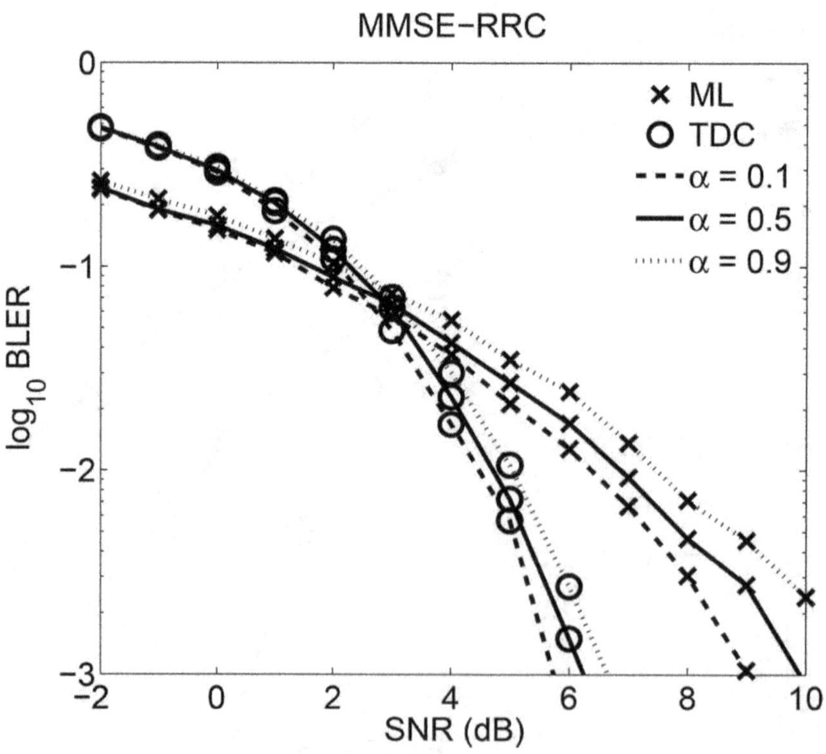


**Figure 8.** BLER comparison of the ML/TDC techniques with MMSE, RRC shaped filter and roll-off factor $\alpha$ =[0.1, 0.5 and 0.9].

**Figure 7** shows similar results for an RC shaped filter. Results in **Figures 6** and **7** show that the detection performance of both the ML and TD techniques is greatly influenced by the choice of the roll-off factor of each form of transmit filter. Results in **Figures 6** and **7** also show that the TDC techniques improve detection performance compared with the ML method.

In **Figures 6** and **7**, it can be seen that detection performance is degraded as the roll-off value is increased from 0.1 to 0.9. This can be attributed to the increasing level of the inherent noise enhancement factor of the ZF scheme as the roll-off factor is increased, as suggested within a major 5G research study highlighted in [2]. The RC shaped filter produces a slightly improved detection performance compared with the RRC shaped filter due to less inherent ICI in the RC shaped filter compared with the RRC filter, as suggested in [18].

*4.2.2. Detection performance with MMSE-based GFDM receiver*

Similarly, using the MMSE-based GFDM demodulation, **Figure 8** shows the BLER comparison with the use of an RRC shaped filter. **Figure 9** shows the same results using the RC shaped filter. Results in **Figures 8** and **9** show that the TDC technique improves detection performance in comparison with the ML method. Results in **Figures 8** and **9** also show that the detection performance is not significantly influenced by the value of the roll-off parameter of RC/RRC

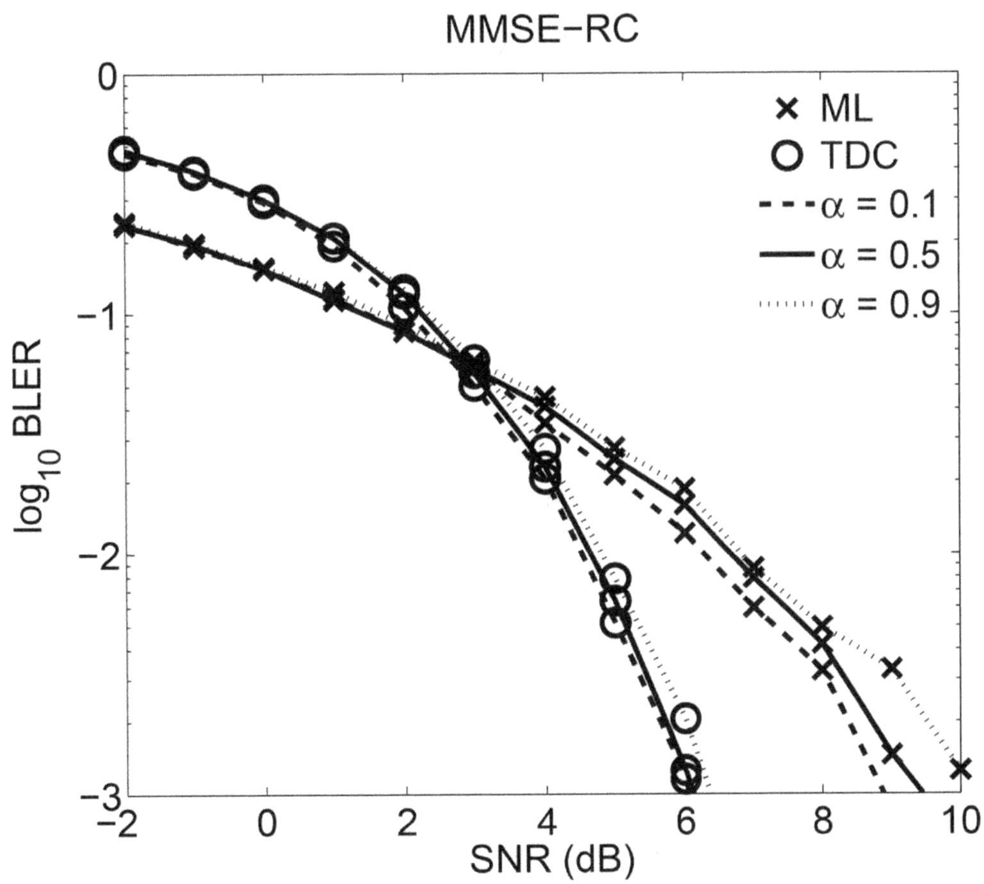


**Figure 9.** BLER comparison of the ML/TDC techniques with MMSE, RC shaped filter and roll-off factor $\alpha$ =[0.1, 0.5 and 0.9].

| Target BLER | GFDM receiver | Roll-off, $\alpha$ | Estimated SNR (dB) | | | |
|---|---|---|---|---|---|---|
| | | | RC | | RRC | |
| | | | ML | TDC | ML | TDC |
| 1% | ZF | 0.1 | 6.4 | 4.6 | 6.5 | 4.6 |
| | | 0.5 | 8.1 | 5.8 | 9.3 | 6.5 |
| | | 0.9 | >10.0 | 6.9 | ≫10.0 | 8.3 |
| | MMSE | 0.1 | 6.3 | 4.7 | 6.5 | 4.7 |
| | | 0.5 | 6.7 | 4.8 | 6.9 | 4.9 |
| | | 0.9 | 6.9 | 5.0 | 7.6 | 5.2 |
| 0.1% | ZF | 0.1 | 9.4 | 6.2 | 8.8 | 6.0 |
| | | 0.5 | ≫10.0 | 7.4 | ≫10.0 | 7.9 |
| | | 0.9 | ≫10.0 | 8.6 | ≫10.0 | 10.0 |
| | MMSE | 0.1 | 8.9 | 5.9 | 9.1 | 6.3 |
| | | 0.5 | 9.6 | 6.4 | 9.9 | 6.7 |
| | | 0.9 | >10.0 | 6.7 | >10.0 | 6.9 |

**Table 3.** Estimated SNR (dB) at BLER levels of 1 and 0.1%.

shaped filter types. This is because the MMSE scheme produces no inherent noise enhancement. It is important to note that similar observations were also highlighted within a recent study found in [19].

#### *4.2.3. Estimated SNR at target BLER of 1 and 0.1%*

**Table 3** shows the approximate SNR (in dB) required to achieve, for example, target BLER levels of 1 and 0.1%.

In summary, presented results in **Figures 6–9** support existing observations on the effects of filter shapes and the type of GFDM demodulation technique on the detection performance of the GFDM system. These results also show that the TDC technique has a robust detection performance capability and is potentially applicable in 5G systems.

## 5. Conclusions

This chapter introduced a TDC detection technique for SCI decoding and presented its detection performance using the GFDM architecture for 5G systems. Unlike the ML method of SCI detection, the TDC scheme requires no channel estimation and has no extra system overhead associated with channel estimation. It is shown that the TDC technique improves detection performance when compared with the conventional ML method.

With the ZF-based receiver, the BLER performance of the TDC technique is degraded as the roll-off value of the RC and RRC shaped filter is increased from 0.1 to 0.9. However, with the MMSE receiver, the detection performance of the TDC technique is relatively similar for different filter roll-off values. Hence, with the ZF-based receiver, the detection performance of the TDC technique is largely influenced by the choice of the roll-off value of the transmit filter. Furthermore, with the ZF-based receiver, the RC shaped transmit filter improved the BLER performance of the TDC technique compared with the RRC shaped filter of the same roll-off factor. Therefore, the TDC technique is a viable and an attractive SCI decoding solution for 5G systems.

## Acknowledgements

We would like to thank Dr. Funmilayo Ogunkoya of the Electrical and Electronics Engineering department of Obafemi Awolowo University (OAU), Nigeria, for her support and insight towards producing some of the presented simulation results.

## Author details

Saheed A. Adegbite[1]* and Brian G. Stewart[2]

*Address all correspondence to: sa.adegbite@gmail.com

1 School of Engineering and Built Environment, Glasgow Caledonian University, Glasgow, UK

2 Department of Electronics and Electrical Engineering, University of Strathclyde, Glasgow, UK

## References

[1] 3GPP TSG-RAN R1-162248. Waveform for the next generation radio interface, 2016.

[2] N. Michailow, M. Matthé, I. S. Gaspar, A. N. Caldevilla, L. L. Mendes, A. Festag, and G. Fettweis. Generalized frequency division multiplexing for 5th generation cellular networks. *IEEE Transactions on Communications*, 62(9):3045–3061, 2014.

[3] 3GPP Technical Specification (TS) 36.211 v12.0.0. Evolved universal terrestrial radio access (E-UTRA); Physical Channels and Modulation, 2013.

[4] S. A. Adegbite, S. G. McMeekin, and B. G. Stewart. Improved PCFICH decoding in LTE systems. In *The 21st IEEE International Workshop on Local and Metropolitan Area Networks*, IEEE, Beijing, China, pages 1–6, 2015.

[5] S. A. Adegbite, S. G. McMeekin, and B. G. Stewart. A time-domain control signal detection technique for OFDM. *EURASIP Journal on Wireless Communications and Networking*, 2016(1):1–10, 2016.

[6] S. J. Thiruvengadam and L. M. A. Jalloul. Performance analysis of the 3GPP-LTE physical control channels. *EURASIP Journal on Wireless Communications and Networking*, 2010(1):1–10, 2010.

[7] S. Abbas, S. J. Thiruvengadam, and S. Susithra. Novel receiver architecture for LTE-A downlink physical control format indicator channel with diversity. *VLSI Design*, 2014:1–7, 2014.

[8] I. S. Gaspar, L. L. Mendes, N. Michailow, and G. Fettweis. A synchronization technique for generalized frequency division multiplexing. *EURASIP Journal on Advances in Signal Processing*, 2014(1):1–10, 2014.

[9] S. A. Adegbite, S. G. McMeekin, and B. G. Stewart. A selective control information detection scheme for OFDM receivers. *Telecommunication Systems*, 1–11, 2016. DOI: 10.1007/s11235-016-0154-6

[10] S. Ehsanfar, M. Matthe, D. Zhang and G. Fettweis, "A Study of Pilot-Aided Channel Estimation in MIMO-GFDM Systems," WSA 2016; 20th International ITG Workshop on Smart Antennas, Munich, Germany, 2016, pp. 1–8.

[11] H. Zamiri-Jafarian, H. Khoshbin, and S. Pasupathy. Time-domain equalizer for OFDM systems based on SINR maximization. *Communications, IEEE Transactions on*, 53(6):924–929, 2005.

[12] J. Balakrishnan, R. K. Martin, and C. R. Johnson. Blind, adaptive channel shortening by sum-squared auto-correlation minimization (SAM). *IEEE Transactions on Signal Processing*, 51(12):3086–3093, 2003.

[13] H. Minn and V. K. Bhargava. An investigation into time-domain approach for OFDM channel estimation. *IEEE Transactions on Broadcasting*, 46(4):240–248, 2000.

[14] S. D. Stearns and D. R. Hush. *Digital Signal Processing with Examples in MATLAB®*, Second Edition. Taylor & Francis, Florida, USA, 2002.

[15] Christian Walck. Handbook on statistical distributions for experimentalists. *Internal Report (SUF-PFY/96-01), University of Stockholm, Sweden*, pages 138–139, 2007.

[16] 3GPP Technical Specification (TS) 36.101 v12.0.0. Evolved universal terrestrial radio access (E-UTRA); User Equipment (UE) Radio Transmission and Reception, 2013.

[17] S. Adegbite, B. G. Stewart, and S. G. McMeekin. Least squares interpolation methods for LTE system channel estimation over extended ITU channels. *International Journal of Information and Electronics Engineering*, 3(4):414–418, 2013.

[18] B. M. Alves, L. L. Mendes, D. A. Guimaraes, and I. S. Gaspar. Performance of GFDM over frequency-selective channels-invited paper. In *Proceedings of International Workshop on Telecommunications*, Santa Rita do Sapucai, Brazil, 2013.

[19] N. Michailow, S. Krone, M. Lentmaier, and G. Fettweis. Bit error rate performance of generalized frequency division multiplexing. In *76th Vehicular Technology Conference (VTC Fall), 2012 IEEE*, IEEE, Quebec City, Canada, pages 1–5, 2012.

# 8

# Physical-Layer Transmission Cooperative Strategies for Heterogeneous Networks

Syed Saqlain Ali, Daniel Castanheira, Adão Silva and Atílio Gameiro

**Abstract**

The deployment of small cells within the boundaries of a macro-cell is considered to be an effective solution to cope with the current trend of higher data rates and improved system capacity. In the current heterogeneous configuration with the mass deployment of small cells, it is preferred that these two cell types coexist over the same spectrum, because acquiring additional spectrum licenses for small cells is difficult and expensive. However, the coexistence leads to cross-tier/inter-system interference. In this context, this contribution investigates interference alignment (IA) methods in order to mitigate the interference of macro-cell base station towards the small cell user terminals. More specifically, we design a diversity-oriented interference alignment scheme with space-frequency block codes (SFBC). The main motivation for joint interference alignment with SFBC is to allow the coexistence of two systems under minor inter-system information exchange. The small cells just need to know what space-frequency block code is used by the macro-cell system and no inter-system channels need to be exchanged, contrarily to other schemes recently proposed. Numerical results show that the proposed method achieves a performance close to the case where full-cooperation between the tiers is allowed.

**Keywords:** interference alignment (IA), space-frequency block codes (SFBC), downlink (DL), heterogeneous networks (HetNets), small-cell system, macro-cell system

## 1. Introduction

Due to new generation of wireless user equipment and the proliferation of bandwidth-intensive applications (such as video, mobile broadband modems, tablets and mobile data applications) and the corresponding network load are increasing in exponential manner, where most of this new data traffic is generated indoors. To improve the coverage and provide boost in network capacity, cellular operators are urged to explore different methods, where massive multiple input multiple output (MIMO) [1] and heterogeneous network [2] concepts are two promising technologies to cope with the increased demand for higher data rates as demanded by 5G [3]. Massive MIMO is a large-scale multiuser MIMO strategy that has the capability of communicating with dozens of users at the same time and frequency band. Moreover, the concept of massive MIMO-aided HetNets recently attracted the attention of research community [4]. In this chapter, we focus on the heterogeneous network scenario, where the small cells (SCs) coexist with macro-cells which allow more users to be served. Apart from the capability to provide higher data rates, SCs offer other advantages, such as they are low-power wireless access points (APs) and have low deployment cost, they operate inside the coverage area of a macro-cell, creating a heterogeneous network [5, 6] and they offer great benefits for both operators and users, who get higher data rates, get better coverage and avail new services [7].

Inspired by the features and potential advantages of the small-cell networks, their development and deployment have gained considerable interest in the wireless industry and research communities. On the other hand, these networks also come up with their own challenges. There are significant technical issues related to self-organization, backhauling and interference management that still need to be addressed for their successful rollout and operation [8]. Furthermore, due to huge deployment of SCs within the boundaries of a macro-cell and the cost involved in acquiring additional frequency licenses for small-cells, it is preferred that the macro- and small cells coexist over the same spectrum. However, the coexistence of two systems will result in a number of challenges, namely related to interference management [9], i.e. the cross-tier/inter-system interference. In a coexistence scenario, being the owner of the spectrum, the macro-cell system has the access priority to the available radio spectrum and in the literature of cognitive radio (CR) [10, 11], the macro-cell terminals are denominated as primary users/system; however, the small-cell terminals can only opportunistically access the free space resources of the macro-cell system without generating any interference to it and are denominated secondary. In this context, heterogeneous networks require more dynamic planning and if the system is not carefully designed then it will cause significant interference that affects the performance of both macro-cell and small-cell systems.

In order to cancel interference in heterogeneous networks, different interference mitigation techniques have been proposed [12, 13]. One of the recent and effective approaches to deal with interference issues in heterogeneous networks is the interference alignment (IA) technique [14]. The concept of IA has emerged as an essential approach to align an arbitrary large number of interferers and achieve the maximum degree of freedom (DoF) in interference channels [15, 16]. The problem of limited inter-system information exchange in heterogeneous-based systems using IA has been addressed in some publications [17, 18]. In Ref. [19], it was shown that only 1 bit of information exchange is required between the macro- and small cells to achieve full diversity order at the macro-cell. This work assumed the knowledge of the cross-tier channel at the small cells. Furthermore, the concept of IA has been jointly used with CR in order to mitigate interference in heterogeneous networks. In Ref. [20], authors proposed a practical joint IA and cognitive communication technique in order to deal with the interference of small-cell user terminals (UTs) towards the macro-base station. In this work, three IA methods with different levels of inter-system information exchange were proposed, namely: the coordinated, static and uncoordinated approaches. The first method achieves the best performance with very high feedback requirements while the uncoordinated and static methods require no feedback but at the expense of performance degradation. Therefore, to overcome the limitations of coordinated and uncoordinated-static methods, the authors in Ref. [21] investigated a coordinated one-bit method for the uplink of heterogeneous networks.

One of the key aspects in coordinated-based systems is the amount of feedback that needs to be exchanged between the cooperating identities [22], in order to define the overhead requirements needed by the network to avail the benefits from cooperation. When full-coordination is allowed between the two systems, it achieves the best performance and maximum diversity order. On the other hand, when no information is exchanged, the diversity is reduced to minimum as demonstrated in Refs. [20, 21]. In this context, the design of practical schemes that can provide close to optimal performance with limited information exchange is of paramount importance. Therefore, in Ref. [23] we proposed IA-based schemes for the downlink of heterogeneous systems under limited inter-system information exchange. In Ref. [23], we design a new IA-based scheme for the considered heterogeneous systems. Namely, the coordinated $2n$-bit approach, which is an extension of the 2-bit method proposed in Ref. [24]. Moreover, to demonstrate the further reduction of information exchange between the two systems, we proposed a joint IA and space-frequency block code (SFBC) approach [25]. In this chapter, we present the schemes mentioned in Refs. [23, 25] for a general number of antennas at each terminals and for the case where OFDM modulation is considered. Furthermore, for our SFBC-based schemes, we consider a general formulation of the diversity-oriented joint IA and SFBC method that can be applied for any SFBC. For this new method, the small cells just need to sense what SFBC is used by the macro-cell system and no inter-system channels need to be exchanged, contrarily to the previously proposed approaches.

The rest of the chapter is structured as follows: Section 2 introduces the system and signal models for macro-cell and small-cell systems with and without SFBC. In Section 3, we start by summarizing the related work and then the joint IA and SFBC schemes are derived in detail. In Section 4, we discuss the performance ad information exchange requirements for all the methods. In Section 5, we present the numerical results and performance comparison of the proposed methods with others from the literature. Finally, conclusions are provided in Section 6.

## 2. System model

Let us consider the downlink of a heterogeneous network, where a set of $K$ small-cells are overlaid within the boundaries of a macro-cell, both sharing the same spectrum as depicted in **Figure 1**. The $K$ small-cell base stations (SBSs) are able to cooperate through a backhaul network (e.g. radio over fibre) to a central unit (CU) that allows joint processing of transmitted signals. In this work, we consider the downlink case, i.e. the base stations (BSs) are sending information to the corresponding user equipment (UE). We consider OFDM-based terminals with $N_c$ available subcarriers, but the proposed methods also work with generalized frequency division multiplexing (GFDM), since similarly to OFDM the transmit signals are a linear

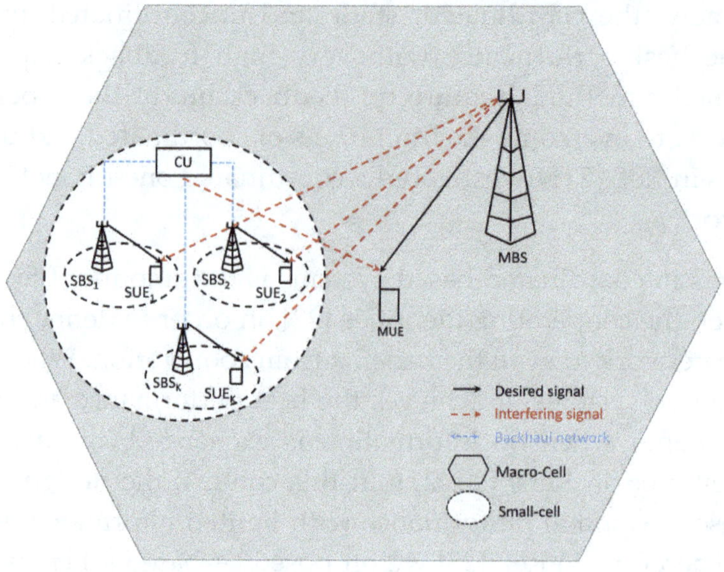


**Figure 1.** System model: $N$ small cells within the coverage area of macro-cell.

combination of the data symbols [26]. The transmit power per subcarrier for macro-base station (MBS) and SBSs is constraint to $P_m$ and $P_s$, respectively. We consider that the MBS serves only one user equipment, macro UE (MUE), per subcarrier,[1] and the SBS $k$ serves only the small-cell user equipment $k$ ($SUE_k$) $k = \{1,\ldots K\}$.

## 2.1. Signal model without SFBC

First, we describe the signal model for the macro- and small-cell systems for the case where no SFBC is employed at the MBS [23]. The block diagram of the considered systems is presented in **Figure 2**. At the macro-cell system, we assume that the MBS and MUE have $M_m$ and $N_m$ antennas, respectively. The transmitted signal ($\mathbf{x}_m^{f_n}$) at the MBS on subcarrier $f_n$ is given by

$$\mathbf{x}_m^{f_n} = \gamma_m(\mathbf{V}_m^{f_n}\mathbf{d}_m^{f_n}), \tag{1}$$

where $\gamma_m^2 = P_m/tr(\mathbf{V}_m^{f_n H}\mathbf{V}_m^{f_n})$, $\mathbf{V}_m^{f_n}\in\mathbb{C}^{M_mN_m}$ and $\mathbf{d}_m^{f_n}\in\mathbb{C}^{N_m}$ denote a normalizing constant, the precoder and the transmitted symbols at the MBS, respectively. The received signal in the frequency domain at the MUE ($\mathbf{y}_m^{f_n}\in\mathbb{C}^{N_m}$) can be mathematically expressed as

$$\mathbf{y}_m^{f_n} = \underbrace{\mathbf{G}_1^{f_n}\mathbf{x}_m^{f_n}}_{\text{Desiredsignal}} + \underbrace{\mathbf{G}_2^{f_n}\mathbf{x}_s^{f_n}}_{\text{Interference}} + \mathbf{n}_m^{f_n}. \tag{2}$$

where $\mathbf{x}_s^{f_n}\in\mathbb{C}^{M_sK}$ , $\mathbf{G}_1^{f_n}\in\mathbb{C}^{N_mM_m}$ ,$\mathbf{G}_2^{f_n}\in\mathbb{C}^{N_mM_sK}$ and $\mathbf{n}_m^{f_n}\in\mathbb{C}^{N_m}$ denote the overall transmitted signal

[1]Considering an OFDM/A-based system, the total number of macro-cell users can be significantly larger than one, since different set of resources can be allocated to different users.

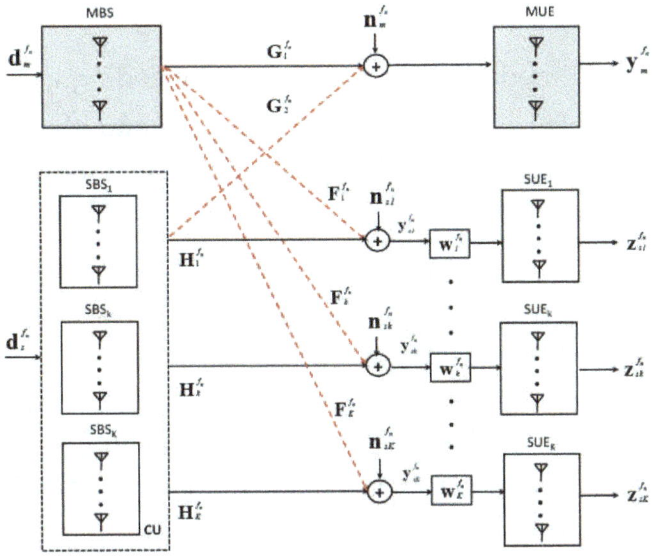


**Figure 2.** Block diagram of the considered system.

at the small-cell system, the channel between MBS and MUE, the overall channel between CU and MUE (i.e. the channels between the SBSs and the MUE) and the zero-mean white Gaussian noise with variance $\sigma^2$, respectively [23]. We assume that at the MBS only $\mathbf{G}_1^{f_n}$ is known and it has no knowledge about the existence of a small-cell system. Furthermore, we assume that the MUE is a high mobility equipment and then $\mathbf{G}_1^{f_n}$ and the precoder $\mathbf{V}_m^{f_n}$ (function of macro-cell channel $\mathbf{G}_1^{f_n}$) change on every transmission time interval (TTI).

In the small-cell system, each SBS has $M_s$ transmit and the $SUE_k$ $k = \{1,\ldots K\}$ has $N_s$ receive antennas. The transmitted signal ($\mathbf{x}_s^{f_n}$) at the CU on subcarrier $f_n$ is expressed as

$$\mathbf{x}_s^{f_n} = \gamma_s(\mathbf{V}_s^{f_n}\mathbf{d}_s^{f_n}), \tag{3}$$

where $\mathbf{V}_s^{f_n} \in \mathbb{C}^{M_sK(N_s-N_m)K}$, $\mathbf{d}_s^{f_n} = [\mathbf{d}_{sk}^{f_n}]_{1\le k\le K} \in \mathbb{C}^{(N_s-N_m)K}$, $\mathbf{d}_{sk}^{f_n} \in \mathbb{C}^{N_s-N_m}$ and $\gamma_s^2 = P_s/tr(\mathbf{V}_s^{f_n H}\mathbf{V}_s^{f_n})$denote the overall precoder computed at the CU, the concatenation of the $K$ SBSs transmit symbols, the SBS $k$ transmit symbols and a normalizing constant. The received signal after the filter matrix ($W_k^{f_n}$) at the $SUE_k$ is

$$\mathbf{z}_{sk}^{f_n} = W_k^{f_n}(\underbrace{\mathbf{F}_k^{f_n}\mathbf{x}_m^{f_n}}_{\text{Interference}} + \underbrace{\mathbf{H}_k^{f_n}\mathbf{x}_s^{f_n}}_{\text{Desiredsignal}} + \mathbf{n}_{sk}^{f_n}), \tag{4}$$

where $\mathbf{F}_k^{f_n} \in \mathbb{C}^{N_sM_m}$, $\mathbf{H}_k^{f_n} \in \mathbb{C}^{N_sM_sK}$ and $\mathbf{n}_{sk}^{f_n} \in \mathbb{C}^{N_s}$ denote the channel between the MBS and $SUE_k$, the overall channel between the SBSs and $SUE_k$ and the zero-mean white Gaussian noise with variance $\sigma^2$ at $SUE_k$, respectively. We consider that the SUEs are low mobility terminals[2] and then the channel $\mathbf{F}_k^{f_n}$ can be considered as quasi-static which reduces the overhead required for their estimation [23].

[2]Since the terminals associated with the small cells are mainly indoor/pedestrian users.

## 2.2. Signal model with SFBC

Now, we consider the signal model with space-frequency coding at the MBS. We consider a block fading MIMO channel, i.e. $\mathbf{G}_1^{f_n} = \mathbf{G}_1 \text{for} f_n = 1, \ldots, F$ and the channel is independent between different blocks of $F$ subcarriers. Thus, the system equation mentioned in Eq. (2), over one block is [27]

$$\mathbf{Y}_m = \mathbf{G}_1 \mathbf{X}_m + \mathbf{I}_s + \mathbf{N}_m, \tag{5}$$

where$\mathbf{Y}_m = [\mathbf{y}_m^1, \ldots, \mathbf{y}_m^F]$ is the received signal matrix, $\mathbf{X}_m = [\mathbf{x}_m^1, \ldots, \mathbf{x}_m^F]$ is the transmitted signal, $\mathbf{I}_s = [\mathbf{G}_2^1 \mathbf{x}_s^1, \ldots, \mathbf{G}_2^F \mathbf{x}_s^F]$ is the inter-tier interference and $\mathbf{N}_m = [\mathbf{n}_m^1, \ldots, \mathbf{n}_m^F]$ is the zero-mean white Gaussian noise with variance $\sigma^2$. The macro-cell system employs an SFBC to encode $S_m$ complex symbols $d_m^1, \ldots, d_m^{S_m}$ chosen from an $r$-QAM constellation [25]. We consider linear dispersion codes (LD) of the form Ref. [28]

$$\mathbf{X}_m = \sum_{s=1}^{S_m} (\mathbf{A}_m^s \Re\{d_m^s\} + \mathbf{B}_m^s \Im\{d_m^s\}), \tag{6}$$

where $d_m^s = \Re\{d_m^s\} + j\Im\{d_m^s\}$,$m = 1, \ldots, S_m$, $\mathbf{A}_m^s$ and $\mathbf{B}_m^s$ are the codeword matrices. The rate of the LD code is

$$R = \frac{S_m}{F} \log_2(r), \text{bits/subcarrier} \tag{7}$$

Therefore, by rewriting Eq. (5) in column-stacked form we obtain [25]

$$\mathbf{y}_m = (\mathbf{I}_F \otimes \mathbf{G}_1)\mathbf{x}_m + \mathbf{i}_s + \mathbf{n}_m = \mathcal{G}_1 \mathbf{V}_m \mathbf{d}_m + \mathbf{i}_s + \mathbf{n}_m. \tag{8}$$

where $\mathcal{G}_1 = \mathbf{I}_F \otimes \mathbf{G}_1$, $\mathbf{x} = \text{vec}(\mathbf{X})$ is $N_m F$ dimensional, $\mathbf{i}_s = \text{vec}(\mathbf{I}_s)$ is $M_m F$ dimensional, $\mathbf{x}_m = \text{vec}(\mathbf{X}_m) = \mathbf{V}_m \mathbf{d}_m$ is $M_m F$ dimensional, $\mathbf{d}_m = [\Re\{d_m^1\{, \ldots, \Re\{d_m^{S_m}\}, \Im\{d_m^1\}, \ldots, \Im\{d_m^{S_m}\}]^T$, $\mathbf{V}_m = [\text{vec}(\mathbf{A}_1), \ldots, \text{vec}(\mathbf{A}_{S_m}), \text{vec}(\mathbf{B}_1), \ldots, \text{vec}(\mathbf{B}_{S_m})]$ is an $N_m F 2 S_m$ code generator matrix that is an equivalent representation of the LD code.

At the small-cell system, the signal model for the methods with SFBC is similar to one presented previously. Using a similar procedure as in the previous section for the received signal at SUEs, we obtain [27]

$$\mathbf{y}_{sk} = \mathcal{F}_k \mathbf{V}_m \mathbf{d}_m + \mathcal{H}_k \mathbf{x}_s + \mathbf{n}_m, \tag{9}$$

where $\mathbf{y}_{sk} = [(\mathbf{y}_{sk}^1)^T, \ldots, (\mathbf{y}_{sk}^F)^T]^T$, $\mathcal{F}_k = \text{diag}(\mathbf{F}_k^1, \ldots, \mathbf{F}_k^F)$, $\mathcal{H}_k = \text{diag}(\mathbf{H}_k^1, \ldots, \mathbf{H}_k^F)$, $\mathbf{x}_s = [(\mathbf{x}_s^1)^T, \ldots, (\mathbf{x}_s^F)^T]^T$ and $\mathbf{n}_{sk} = [(\mathbf{n}_{sk}^1)^T, \ldots, (\mathbf{n}_{sk}^F)^T]^T$. To compute the CU transmit signal, a linear precoder is considered, that is the CU transmits

$$\mathbf{x}_s = \mathbf{V}_s \mathbf{d}_s, \tag{10}$$

where $\mathbf{V}_s \in \mathbb{C}^{M_s KFS_s KF}$, $\mathbf{d}_s = [\mathbf{d}_{sk}^{f_n}]_{1 \le k \le K, 1 \le f_n \le F} \in \mathbb{C}^{S_s KF}$ and $\mathbf{d}_{sk}^{f_n} \in \mathbb{C}^{S_s}$ denote the overall precoder computed at the CU, the concatenation of the $K$ SBSs transmit symbols, $\mathbf{d}_{sk}^{f_n}$ is the SBS $k$

transmit symbols, respectively. The transmit power at the CU is constrained to $P_s$, per subcarrier

$$tr(\mathbf{V}_s^{f_n^H}\mathbf{V}_s^{f_n})\leq P_s, \tag{11}$$

The received signal after the filter matrix ($\mathbf{W}_k$) at the $SUE_k$ by taking into account Eqs. (9) and (10) is

$$\mathbf{z}_{sk} = W_k(\mathcal{F}_k\mathbf{V}_m\mathbf{d}_m + \mathcal{H}_k\mathbf{V}_s\mathbf{d}_s + \mathbf{n}_{sk}). \tag{12}$$

## 3. Proposed approaches for precoder and filter matrix design

In this section, we present the design of precoder and filter matrices of the macro-cell and small-cell systems, in order to allow efficient coexistence of the two systems over the same radio spectrum. To design our proposed methods, we consider different levels of cooperation between the two systems. All the methods presented in this chapter are derived for a generic antenna configuration and therefore they are applicable for massive MIMO systems. On the other hand, the complexity will scale depending on the number of transmit antennas. Since the proposed methods involve matrix multiplications and inversions, thus the complexity will be similar to ZF-based precoding in massive MIMO. Moreover, for the sake of simplicity, we just consider one user per MBS but adding more macro-cell user will not impact the performance of both the systems, since interference can be completely removed. First, we summarize the methods presented in Ref. [23] for the case without SFBC. Then, we present in detail the proposed methods in Ref. [25], for the case where IA and SFBC are jointly used.

### 3.1. Methods without SFBC

In this section, we summarized the schemes presented in Ref. [23] for a general number of antennas at each terminal and for the case where OFDM modulation is considered. In Ref. [23], we design a new IA-based scheme for the considered heterogeneous systems. Namely, the coordinated $2n$-bit approach, which is an extension of the 2-bit method proposed in Ref. [24].

#### *3.1.1. Full-coordinated scheme*

For the full-coordinated method, we assume the knowledge of the $\mathbf{G}_1^{f_n}$ channel at the MBS. For the case where the MUE is equipped with single antenna, a maximal ratio transmission (MRT)-based precoder can be employed as in Ref. [24]. When an antenna array is used at the MUE, a ZF or MMSE-based precoders can be used. In this work, we consider the MRT-based precoder at the MBS given by

$$\mathbf{V}_m^{f_n} = \gamma_m \mathbf{G}_1^{f_n^H}, \tag{13}$$

Furthermore, we assumed that the macro-cell system is not aware of the existence of small-cell system within its coverage area and the MBS precoder $\mathbf{V}_m^{f_n}$ is fixed and it will not change due to the presence of SUEs. However, the SUEs can be severely affected by the macro-cell transmission. From Eqs. (1) and (4), we can see that to enforce the zero-interference condition and mitigate the interference coming from MBS, the filter matrix at $SUE_k$ must satisfy

$$\mathbf{W}_k^{f_n}\mathbf{F}_k^{f_n}\mathbf{V}_m^{f_n} = 0, \tag{14}$$

From Eq. (14) it follows that to satisfy the zero-interference condition the filter matrix ($\mathbf{W}_k^{f_n}$) at SUEs is

$$\mathbf{W}_k^{f_n} = \text{null}(\mathbf{F}_k^{f_n}\mathbf{V}_m^{f_n}), \tag{15}$$

$$\mathbf{A}^{f_n} = \text{null}(\mathbf{V}_m^{f_n}). \tag{16}$$

Where $\mathbf{A}^{f_n}$ is the alignment direction that specifies completely the received macro-cell interfering signal towards the SUEs. Using this information, the small cells can align their transmission accordingly without experiencing any interference from the macro-cell system. It can be verified from the zero-interference condition mentioned in Eq. (14) that the DoF available for the small-cell system is $(N_s - N_m)K$.

### 3.1.2. *Uncoordinated-static scheme*

Once again for this scheme, we follow the same procedure (as for the previous method) to remove the interference from MBS at SUEs, but the precoder at MBS is static at the beginning of interaction between the two systems and it will remain constant, i.e. its value do not change every TTI and its value is also known at the small-cell terminals. Therefore, this method requires no inter-system cooperation. For example, we assume the precoder at MBS is the all-ones matrix, i.e. $\mathbf{V}_m^{f_n} = \mathbf{1}$ [23].

### 3.1.3. *Coordinated 2n-bit scheme*

To achieve a trade-off between performance and feedback requirements of the full-coordinated and uncoordinated-static methods, we propose a coordinated $2n$-bit method. To design the alignment direction, we consider the same precoder used for the full-coordinated scheme. Only a quantized version of the alignment vector is exchanged between the two systems [23]. Therefore, we quantize the alignment direction with $2n$ bits ($n$ bits for the real and $n$ bits for the complex part, where $n = 1, 2, 3, ..$). The quantized alignment direction is

$$\mathbf{A}_q^{f_n} = f_Q(Re\{(\mathbf{A}^{f_n})\}) + jf_Q(Im\{(\mathbf{A}^{f_n})\}) \tag{17}$$

where $f_Q(.)$ denotes a quantization function, the $Re\{.\}$ and $Im\{.\}$ are the real and imaginary parts of alignment direction $\mathbf{A}^{f_n}$. In this chapter, for the sake of simplicity, we consider only uniform quantizers. Notice that for this case, the MBS precoder is also quantized, by taking into account the zero-interference condition ($\mathbf{A}_q^{f_n} = \text{null}(\mathbf{V}_{m,q}^{f_n})$), $\mathbf{V}_{m,q}^{f_n}$ is a quantized version of $\mathbf{V}_m^{f_n}$ [23].

### 3.2. Methods with SFBC

In this section, we design new joint IA and SFBC schemes without any information exchange between two systems as compared to the full-coordinated and coordinated $2n$-bit methods, where we need the channel information $\mathbf{G}_1^f$ in order to design the precoder at the MBS and filter matrix at the SUEs. The main motivation behind the use of SFBC at the macro-cell system is that it allows the design of filter matrix at SUEs without having any coordination between the two systems. More specifically, the small-cells just need to sense that the macro-cell system is using an SFBC scheme [23].

### 3.2.1. IA-filter matrix design for methods with SFBC

Now, we present the design of IA-filter matrix at the SUEs for the proposed joint IA and SFBC scheme. We consider that the macro-cell system has no information about the existence of small-cells within its coverage area. In the coexistence scenario, the MBS interferes with the SUEs. From Eq. (12) we can find that to enforce the zero-interference condition and mitigate the interference coming from MBS, the IA-filter matrix at $SUE_k$ must satisfy

$$\mathbf{W}_k\mathcal{F}_k\mathbf{V}_m = \mathbf{0}, \tag{18}$$

In order to cancel the interference coming from MBS towards the $SUE_k$, we need to compute an appropriate filter matrix at the $SUE_k$. From Eq. (18) it follows that to satisfy the zero-interference condition the IA-filter matrix at $SUE_k$ is

$$\mathbf{W}_k = \mathrm{null}(\mathcal{F}_k\mathbf{V}_m), \tag{19}$$

As mentioned in Section 2.2, the precoder $\mathbf{V}_m$ for SFBCs does not depend on the macro-channel and thus there is no need to exchange any information from the macro-cell to the small-cell system to design the IA-filter matrix, contrarily to the full-coordinated and coordinated $2n$-bit methods [23]. For these two cases, the precoder is computed for each channel instance and as the macro-cell terminal is a mobile terminal the equalizer matrix $\mathbf{W}_k$ must be computed on every TTI. This means that the IA-filter matrix must be exchanged between the two systems every TTI. Another possible strategy consists of estimating the equivalent channel $\mathbf{F}_k^{f_n}\mathbf{V}_m^{f_n}$, by listening to the pilot signals, but it will also require a high pilot density [29].

After applying the IA-filter matrix mentioned in Eq. (19) to Eq. (12), we obtain

$$\mathbf{z}_{sk} = \mathbf{W}_k(\mathcal{F}_k\mathbf{V}_m\mathbf{d}_m + \mathcal{H}_k\mathbf{V}_s\mathbf{d}_s + \mathbf{n}_{sk}) = \mathbf{W}_k\mathcal{H}_k\mathbf{V}_s\mathbf{d}_s + \mathbf{W}_k\mathbf{n}_{sk}. \tag{20}$$

From Eqs. (18) and (20) we verify that the interference from MBS is completely removed at SUEs. This is made possible due to the redundancy present in the MBS transmitted data symbols. Once again, for the joint IA and SFBC case due to the zero-interference condition mentioned in Eq. (18), the DoF available at the small-cells is $(N_s - N_m)K$.

#### 3.2.1.1. Interference from small cells to macro-cell

In the previous section, we described how to tackle the interference from the macro- to the small cells. In this section, we describe how to cancel the interference from the small cells to the macro-cells (for all the methods presented in this chapter). Being a small-cell system it should not interfere with the macro-cell system (i.e. the macro-cell has priority to access the available resources). On the other hand, the SUEs should not interfere with each other. We consider that the SBSs are connected via the backhaul network (optical fibre) to a CU in order to perform joint processing of transmitted signals [25]. The CU has enough DoF (i.e. $KM_s$) to cancel both the interference that the SBSs cause in the MUE and the interference between SUEs. The precoding matrix at the CU is based on the ZF criteria, in order to zero force the macro-cell and small-cell channels together. In this context, the ZF precoder $\mathbf{V}_s^{f_n}$, computed at the CU, is given by Ref. [25]

$$\mathbf{V}_s^{f_n} = \mathbf{A}^{f_n^H}(\mathbf{A}^{f_n}\mathbf{A}^{f_n^H})^{-1}, f_n = 1, \ldots, F \tag{21}$$

where $\mathbf{A}^{f_n} = \mathbf{W}^{f_n}\mathbf{H}_{eq}^{f_n}$, $\mathbf{H}_{eq}^{f_n} = [(\mathbf{G}_2^{f_n})^H, (\mathbf{H}_1^{f_n})^H, \ldots, (\mathbf{H}_K^{f_n})^H]^H$ and $\mathbf{W}^{f_n} = diag(\mathbf{I}, \mathbf{W}_1^{f_n}, \ldots \mathbf{W}_k^{f_n}, \ldots \mathbf{W}_K^{f_n})$. The filter matrix $\mathbf{W}_k^{f_n}$ is known at the CU since the channels $\mathbf{F}_k^{f_n}$ are quasi-static, the SUEs may feedback them to the CU without much overhead requirements.

### 3.2.2. Examples for specific SFBC codes

In the following, we consider few examples of diversity-oriented SFBC schemes used at the macro-cell system in order to design the IA-filter matrix of our joint schemes. We considered three SFBC schemes: Alamouti codes [30], quasi-orthogonal codes [31] and Tarokh codes [32] with the data symbols coded in space and frequency as shown in **Figure 3**. Furthermore, from the context of space-time/space-frequency coding literature, the channel between adjacent carriers is assumed to be approximately constant,[3] i.e. $\mathbf{G}_1^{f_m} \approx \mathbf{G}_1^{f_n}, m \neq n \in \mathbb{N}$ [25].

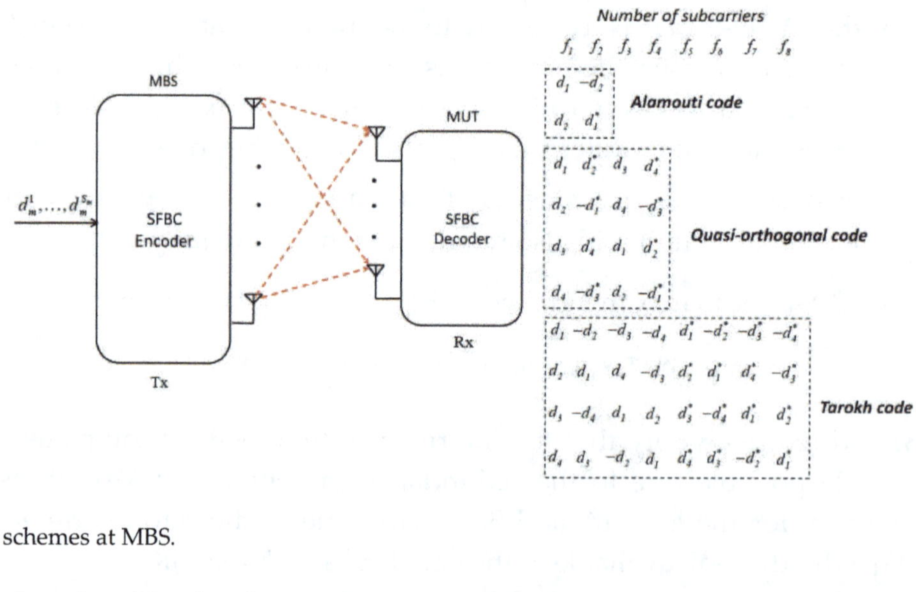


**Figure 3.** SFBC schemes at MBS.

- **Alamouti codes:** For the first case, we employ the standard Alamouti SFBC [30] based scheme at the MBS, with two ($M_m = 2$) antennas at the transmitter and single antenna ($N_m = 1$) at the receiver. For this well-known method, the encoder takes a block of two data symbols, i.e. $d_1$ and $d_2$. For a given subcarrier, two symbols are simultaneously

[3] OFDM-based systems are usually designed so that channels between some adjacent carriers are approximately flat.

transmitted from the two antennas, as shown in **Figure 3**. For the first subcarrier $f_1$, the symbol transmitted from the first antenna is denoted by $d_1$ and from the second one by $d_2$ and over subcarrier $f_2$, $(-d_2)^*$ and $(d_1)^*$ are transmitted from the first and second antennas, respectively [23]. The transmitted signal at the MBS on subcarriers $f_1$ ($\mathbf{x}_m^{f_1}$) and $f_2$ ($\mathbf{x}_m^{f_2}$) is given by

$$\mathbf{x}_m^{f_1} = \begin{bmatrix} d_1 \\ d_2 \end{bmatrix}, \mathbf{x}_m^{(f_2)^*} = \begin{bmatrix} -d_2 \\ d_1 \end{bmatrix} \tag{22}$$

For this case, as mentioned previously, the MBS precoder is applied jointly for $F = 2$ consecutive subcarriers as,

$$\mathbf{V}_m{}^T = \begin{bmatrix} 1 & 0 & 0 & -1 \\ 0 & 1 & 1 & 0 \\ j & 0 & 0 & j \\ 0 & j & -j & 0 \end{bmatrix} \tag{23}$$

As it can be verified from Eq. (23) the macro-cell precoder does not depend on the macro-channel, this means there is no need to exchange any channel information from the macro-cell to the small-cell system to design the IA-filter matrix.

- **Quasi-orthogonal codes:** As verified in Ref. [30], the Alamouti-based scheme is restricted to two antennas at the transmitter side. Therefore, we consider the quasi-orthogonal-based scheme that can be able to use more than two antennas at the transmitter and increase the multiplexing gain. For this case, the transmitter has four ($M_m = 4$) and the receiver has a single antenna ($N_m = 1$), as shown in **Figure 3**. In this method, four pairs of four data symbols are transmitted in parallel. The four data symbols are transmitted over four antennas on four subcarriers, $F = 4$ according to the following encoding [25]

$$\mathbf{x}_m^{f_1} = \begin{bmatrix} d_1 \\ d_2 \\ d_3 \\ d_4 \end{bmatrix}, \mathbf{x}_m^{(f_2)^*} = \begin{bmatrix} d_2 \\ -d_1 \\ d_4 \\ -d_3 \end{bmatrix}, \mathbf{x}_m^{f_3} = \begin{bmatrix} d_3 \\ d_4 \\ d_1 \\ d_2 \end{bmatrix}, \mathbf{x}_m^{(f_4)^*} = \begin{bmatrix} d_4 \\ -d_3 \\ d_2 \\ -d_1 \end{bmatrix} \tag{24}$$

For this case, as mentioned previously, the MBS precoder is applied jointly for $F = 4$ consecutive subcarriers.

$$\mathbf{V}_m{}^T = \begin{bmatrix} 1 & 0 & 0 & 0 & 0 & -1 & 0 & 0 & 0 & 0 & 1 & 0 & 0 & 0 & 0 & -1 \\ 0 & 1 & 0 & 0 & 1 & 0 & 0 & 0 & 0 & 0 & 0 & 1 & 0 & 0 & 1 & 0 \\ 0 & 0 & 1 & 0 & 0 & 0 & 0 & -1 & 1 & 0 & 0 & 0 & 0 & -1 & 0 & 0 \\ 0 & 0 & 0 & 1 & 0 & 0 & 1 & 0 & 0 & 1 & 0 & 0 & 1 & 0 & 0 & 0 \\ j & 0 & 0 & 0 & 0 & j & 0 & 0 & 0 & 0 & j & 0 & 0 & 0 & 0 & j \\ 0 & j & 0 & 0 & -j & 0 & 0 & 0 & 0 & 0 & 0 & j & 0 & 0 & -j & 0 \\ 0 & 0 & j & 0 & 0 & 0 & 0 & j & j & 0 & 0 & 0 & 0 & j & 0 & 0 \\ 0 & 0 & 0 & j & 0 & 0 & -j & 0 & 0 & j & 0 & 0 & -j & 0 & 0 & 0 \end{bmatrix} \tag{25}$$

As seen in the Alamouti code, the macro-cell precoder for this case also does not depend on the macro-channel as verified from Eq. (25); this means there is no need to exchange any channel information from the macro-cell to the small-cell system to design the IA-filter matrix.

- **Tarokh codes:** Once again, for Tarokh codes we assume four antennas ($M_m = 4$) at the transmitter and a single antenna ($N_m = 1$) at the receiver side, as presented in **Figure 3**. The only difference is the number of subcarriers used to transmit the data symbols, for this case eight subcarriers are used, i.e. the Tarokh code that provides the code rate of 1/2. The four data symbols are transmitted over four antennas on eight subcarriers $F = 8$ according to the following encoding [25]

$$\mathbf{x}_m^{f_1} = \begin{bmatrix} d_1 \\ d_2 \\ d_3 \\ d_4 \end{bmatrix}, \mathbf{x}_m^{f_2} = \begin{bmatrix} -d_2 \\ d_1 \\ -d_4 \\ d_3 \end{bmatrix}, \mathbf{x}_m^{f_3} = \begin{bmatrix} -d_3 \\ d_4 \\ d_1 \\ -d_2 \end{bmatrix}, \mathbf{x}_m^{f_4} = \begin{bmatrix} -d_4 \\ -d_3 \\ d_2 \\ d_1 \end{bmatrix}, \mathbf{x}_m^{(f_5)^*} = \begin{bmatrix} d_1 \\ d_2 \\ d_3 \\ d_4 \end{bmatrix}, \mathbf{x}_m^{(f_6)^*} = \begin{bmatrix} -d_2 \\ d_1 \\ -d_4 \\ d_3 \end{bmatrix}, \\ \mathbf{x}_m^{(f_7)^*} = \begin{bmatrix} -d_3 \\ d_4 \\ d_1 \\ -d_2 \end{bmatrix}, \mathbf{x}_m^{(f_8)^*} = \begin{bmatrix} -d_4 \\ -d_3 \\ d_2 \\ d_1 \end{bmatrix} \tag{26}$$

For the Tarokh codes, the MBS precoder is applied jointly for $F = 8$ consecutive subcarriers as

$$\mathbf{V}_m{}^T = \left[\begin{array}{cccccccccccccccccccccccccccccccc}
1 & 0 & 0 & 0 & 0 & 1 & 0 & 0 & 0 & 0 & 1 & 0 & 0 & 0 & 0 & 1 & 1 & 0 & 0 & 0 & 0 & 1 & 0 & 0 & 0 & 0 & 1 & 0 & 0 & 0 & 0 & 1 \\
0 & 1 & 0 & 0 & -1 & 0 & 0 & 0 & 0 & 0 & 0 & -1 & 0 & 0 & 1 & 0 & 0 & 1 & 0 & 0 & -1 & 0 & 0 & 0 & 0 & 0 & 0 & -1 & 0 & 0 & 1 & 0 \\
0 & 0 & 1 & 0 & 0 & 0 & 0 & 1 & -1 & 0 & 0 & 0 & 0 & -1 & 0 & 0 & 0 & 0 & 1 & 0 & 0 & 0 & 0 & 1 & -1 & 0 & 0 & 0 & 0 & -1 & 0 & 0 \\
0 & 0 & 0 & 1 & 0 & 0 & -1 & 0 & 0 & 1 & 0 & 0 & -1 & 0 & 0 & 0 & 0 & 0 & 0 & 1 & 0 & 0 & -1 & 0 & 0 & 1 & 0 & 0 & -1 & 0 & 0 & 0 \\
j & 0 & 0 & 0 & 0 & j & 0 & 0 & 0 & 0 & j & 0 & 0 & 0 & 0 & j & -j & 0 & 0 & 0 & 0 & -j & 0 & 0 & 0 & 0 & -j & 0 & 0 & 0 & 0 & -j \\
0 & j & 0 & 0 & -j & 0 & 0 & 0 & 0 & 0 & 0 & -j & 0 & 0 & -j & 0 & 0 & -j & 0 & 0 & j & 0 & 0 & 0 & 0 & 0 & 0 & j & 0 & 0 & -j & 0 \\
0 & 0 & j & 0 & 0 & 0 & 0 & j & -j & 0 & 0 & 0 & 0 & -j & 0 & 0 & 0 & 0 & -j & 0 & 0 & 0 & 0 & -j & j & 0 & 0 & 0 & 0 & j & 0 & 0 \\
0 & 0 & 0 & j & 0 & 0 & -j & 0 & 0 & j & 0 & 0 & -j & 0 & 0 & 0 & 0 & 0 & 0 & -j & 0 & 0 & j & 0 & 0 & -j & 0 & 0 & j & 0 & 0 & 0
\end{array}\right] \tag{27}$$

As seen for the quasi-orthogonal codes, the precoder is also constant and not dependent on the macro-cell channel as verified in Eq. (27), where this condition enables the design of the IA filter at SUEs without any information exchange between the two systems.

## 4. Performance versus information exchange comparison

As discussed in Section 3.1, the system achieves the best performance when full coordination is allowed between the two systems, i.e. the case with the full-coordinated scheme, where it requires the highest amount of information exchange, since the macro-cell system must share $2M_mN_m$ real numbers with small-cell terminals on every TTI. Considering an OFDM-based system, $2M_mN_mN_c$ real number increases the feedback constraints. No information exchange is required for the uncoordinated-static method but this scheme results in worst performance for the macro-cell system. To overcome the limitations of full-coordinated and uncoordinated-static schemes and to achieve a good balance between performance and information exchange, we designed a coordinated $2n$-bit approach [23] that results in reduced information exchange requirements and achieves quite close to the optimal performance. Furthermore, the proposed joint IA and SFBC scheme [25] that has the same information exchange requirement as uncoordinated-static scheme provides much better performance as compared to the uncoordinated-static method. **Table 1** summarizes the information exchange requirements and performance of the proposed methods.

| Methods | Information-exchange requirements | Performance |
|---|---|---|
| Full-coordinated | $2M_mN_mN_c$ Real number | Optimal performance |
| Uncoordinated-static | 0 | Worst performance |
| Coordinated $2n$-bit | $2nM_mN_mN_c$ bits | Close to optimal |
| Joint IA and SFBC scheme | 0 | Much better than uncoordinated-static method |

**Table 1.** Comparison of inter-system information exchange and performance.

## 5. Numerical results and discussion

This section provides the performance assessment of all the methods presented in this chapter. We compare the joint IA and SFBC methods to the full-coordinated, uncoordinated-static and coordinated $2n$-bit schemes with the help of numerical simulations. Furthermore, for the coordinated $2n$-bit scheme, we just consider $n = 1$ to compare the results for macro- and small-cell systems. As it will be seen from the numerical results, the coordinated 2-bit scheme almost provides close to the optimal performance for both the macro-cell and the small-cell systems, which means that by using $n > 1$ the additional performance improvement will be marginal. To perform our simulations, we consider two small-cells (i.e. $K = 2$) sharing the spectrum with macro-cell, since we can completely mitigate the interference irrespective the number of small cells, adding more small cells will not impact the performance of the macro-cell system. Furthermore, the SBSs are able to cooperate through a backhaul network to a CU to perform joint processing of signals. We consider two scenarios:

- Scenario 1: The number of antennas at the MBS, SBSs and SUEs is 2 and single antenna at the MUE, i.e. $M_m = M_s = N_s = 2$, $N_m = 1$.
- Scenario 2: The number of antennas at the MBS, SBSs and SUEs is 4 and 1 at the MUE, i.e. $M_m = M_s = N_s = 4$, $N_m = 1$.

We consider the ITU pedestrian channel model B, with modified tap delays according to the sampling frequency specified in LTE standards. The SNR at the cell edge is defined as $(P_t/\sigma^2)$, where $P_t$ is the transmit power. For the macro-cell, the transmit power is equal to $P_m = 1$ and for the small cells it is equal to $P_s = 1$. We used the following OFDM parameters used for simulating both the macro-cell and small-cell systems: FFT size = 1024 (where only 128 subcarriers are used for both the systems); sampling frequency $f_s = 15.36$MHz; cyclic prefix length $c_p = 5.21\mu$s and subcarrier separation is 15 kHz [23]. We present results for full-coordinated, coordinated 2-bit, uncoordinated-static and three joint IA and SFBCs: IA with a standard Alamouti code [30], IA with a quasi-orthogonal code [31] and IA with a half-rate orthogonal Tarokh code [32]. In order to allow an appropriate comparison, all the considered methods are evaluated for the same spectral efficiency. Therefore, we used QPSK modulation for joint IA and Alamouti code, joint IA and quasi-orthogonal code, coordinated 2-bit, full-coordinated and uncoordinated-static schemes and 16-QAM for the joint IA and Tarokh codes.

Let us start by considering the first scenario, where IA is jointly used with Alamouti code. For this case, we compare the performance of full-coordinated (for both the case of macro-cell/

small-cell coexistence and the case where small-cell system is switched off), coordinated 2-bit, uncoordinated-static and joint IA and Alamouti code schemes. As it can be seen from **Figure 4**, the performance of the coordinated 2-bit approach is quite close to the optimal performance. The BER performance of the joint IA and Alamouti code approach has a gap of around 3 dB as compared to the full-coordinated case, since the SFBC scheme can provide an array gain of 1 [23]. On the other hand, the joint IA and Alamouti scheme provides much better performance (a gap of around 10 dB for a target BER of $10^{-3}$) as compared to the uncoordinated-static method while the information-exchange requirements for both schemes are identical.

In **Figure 5**, we present the BER curve of the first scenario for the small-cell system. In **Figure 5**, we just consider the curves for the full-coordinated (as the performance of full-coordinated, coordinated 2-bit and uncoordinated-static methods is identical) and the joint IA and Alamouti

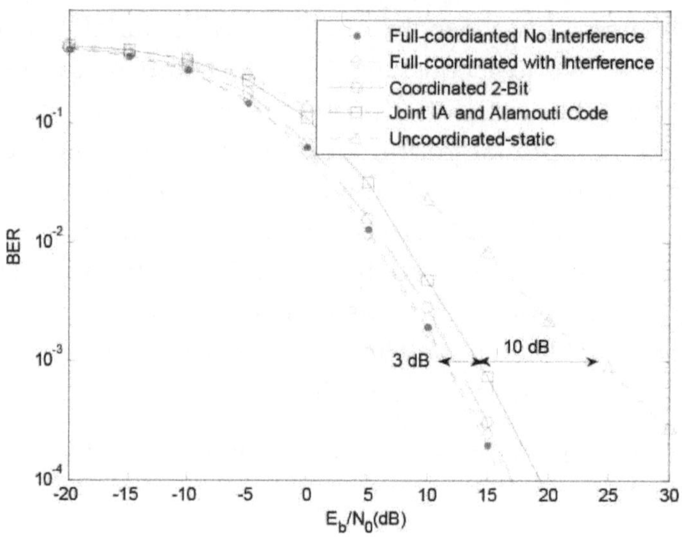


**Figure 4.** BER performance for the macro-cell system (scenario 1).

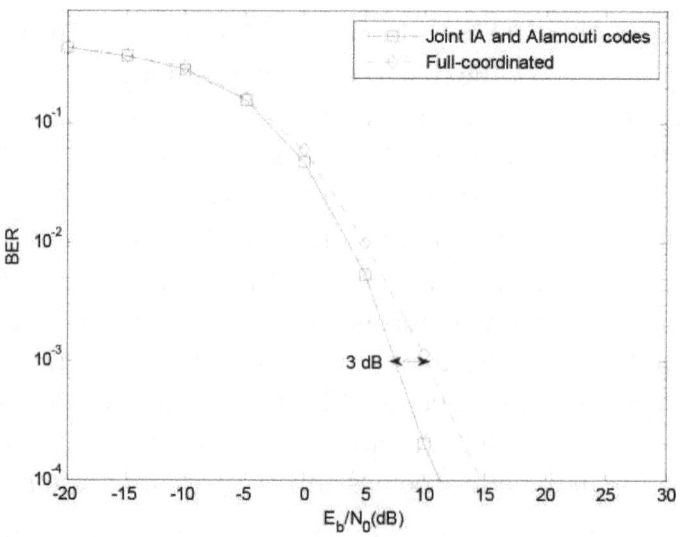


**Figure 5.** BER performance for the small-cell system (scenario 1).

code scheme. This is true, since the design of filter matrix is not dependent on the small-cell channels $[\mathbf{H}_k^f]_{1 \le k \le K}$. Therefore, the equivalent channel preserves the original channel distribution. As seen from **Figure 5**, the joint IA and Alamouti code provides 3 dB which is a better performance as compared to the full-coordinated approach. This is due to the fact that for the SFBC scheme every symbol is transmitted over two subcarriers, contrarily to the full-coordinated method where each symbol only spans one subcarrier [23].

Let us now consider the second scenario where IA is combined with the quasi-orthogonal and Tarokh codes. For this case, we compare the performance of the full-coordinated (for both the case of macro-cell/small-cell coexistence and the case where small-cell system is switched off), coordinated 2-bit, uncoordinated-static, joint IA and quasi-orthogonal code and joint IA and Tarokh code methods. Figures 6 and 7 present the BER performance for the macro-cell and small-cell system, respectively (using QPSK modulation for full-coordinated, coordinated 2-bit uncoordinated-static and joint IA and quasi-orthogonal code curves and 16-QAM modulation for the joint IA and Tarokh code curve). As seen in **Figure 6**, we can notice that the coordinated 2-bit approach provides close to optimal performance. On the other hand, the performance of joint IA and quasi-orthogonal code, joint IA and Tarokh code methods has a gap of around 5 and 3 dB, respectively, as compared to the full-coordinated method and achieves much better performance (a gap of around 14 and 18 dB for a target BER of $10^{-3}$) as compared to the uncoordinated-static scheme, even if the information-exchange requirements of these schemes are identical.

In **Figure 7**, we compare the BER performance of the proposed joint IA and quasi-orthogonal code and joint IA and Tarokh code with the full-coordinated method for the small-cell system. The proposed joint IA and quasi-orthogonal code scheme provides around 3 dB better performance as compared to the case where full coordination is allowed between the two tiers. The performance of the proposed joint IA and Tarokh code scheme is around 1 dB which is better as compared to the full-coordinated case.

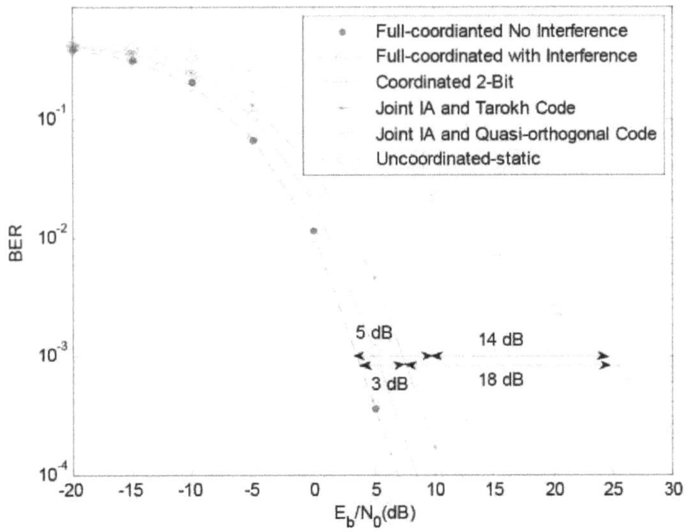


**Figure 6.** BER performance for the macro-cell system (scenario 2).

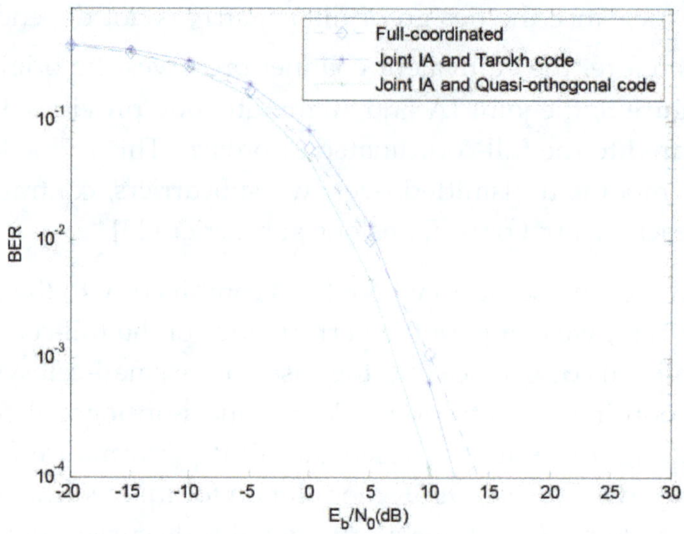


**Figure 7.** BER performance for the small-cell system (scenario 2).

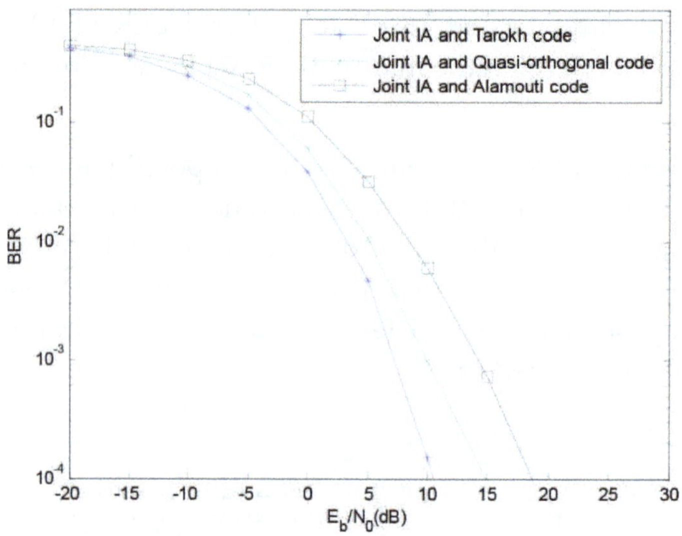


**Figure 8.** BER performance at the macro-cell system for joint IA and Alamouti code/joint IA and quasi-orthogonal code/joint IA and Tarokh code.

In **Figures 8** and **9**, we compare the performance of SFBC schemes at the macro-cell and small-cell systems, respectively. As it can be seen from Figure 8, the joint IA and Tarokh code provides the best performance as compared to the joint IA and Alamouti code/quasi-orthogonal code (i.e. a gap of around 3 and 6dB, respectively). At the small-cell system, the performance of joint IA and Alamouti code/joint IA and quasi-orthogonal code is identical and the performance of joint IA and Tarokh code is around 2 dB which is worse as compared to the other two schemes, as shown in Figure 9. This is due to the fact that the high order modulation (16-QAM) is used for the joint IA and Tarokh code and therefore it is more prone to errors than the other two SFBC schemes that use QPSK modulation.

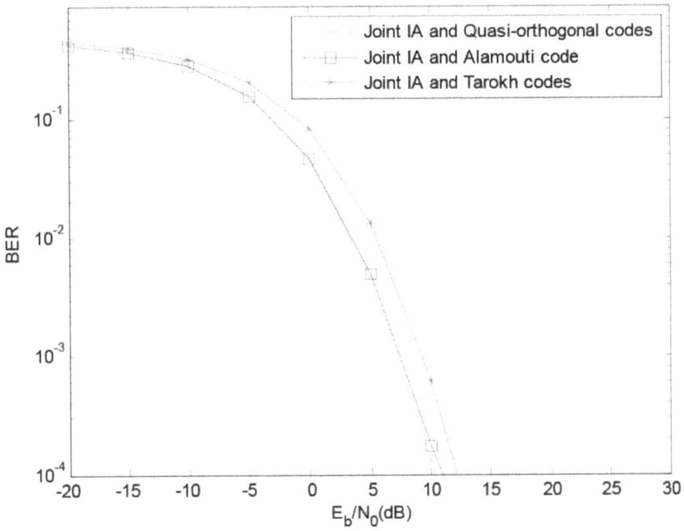


**Figure 9.** BER performance at small-cell system for joint IA and Alamouti code/joint IA and quasi-orthogonal code/joint IA and Tarokh code.

## 6. Conclusions

In this chapter, we presented a general framework of our previously proposed methods for the downlink of heterogeneous-based systems. The system achieves the best performance with full-coordinated scheme, but with very high feedback requirements. For the uncoordinated-static approach, it requires no information exchange between the two systems, but the performance of the macro-cell system is degraded. To overcome the limitations of full-coordinated and the uncoordinated-static methods, we designed the coordinated $2n$-bit scheme and the joint IA and SFBC method that can be applied to any SFBC.

The proposed joint IA and SFBC scheme allows the small-cell system to opportunistically access the free space resources of the macro-cell system without any performance degradation. The proposed joint IA and SFBC method also provides much improved performance with comparable information-exchange requirements to the uncoordinated-static approach. We can say that the proposed method allows the network to achieve the benefits of full-coordinated and uncoordinated-static methods without their main drawbacks. As one of the requirements of 5G is to increase spectral efficiency by a factor about 10, the proposed method will contribute to this goal and thus it can be very useful for the future 5G-based networks.

## Acknowledgements

This work was supported by the Portuguese Fundação para a Ciência e Tecnologia (FCT) under PURE-5GNET project UID/EEA/50008/2013 and FCT grant for the first and second (SFRH/BD/94548/2013) (SFRH/BPD/95375/2013) authors, respectively.

## Author details

Syed Saqlain Ali*, Daniel Castanheira, Adão Silva and Atílio Gameiro

*Address all correspondence to: syedsaqlain@av.it.pt

Instituto de Telecomunicações, University Campus, Aveiro, Portugal

## References

[1] Panzner B, Zirwas W, Dierks S: Deployment and implementation strategies for massive MIMO in 5G. In: Proceedings of the IEEE Global Communications Conference Workshop (GLOBECOM′14); March. IEEE; 2014.

[2] Damnjanovic A: A survey on 3GPP heterogeneous networks. IEEE Transaction on Wireless Communications. 2011;**18**:10–21. DOI: 10.1109/MWC.2011.5876496

[3] Clara-Li Q, Niu H, Tolis-Papathanssiou A, Wu G: 5G network capacity-key elements and technologies. IEEE Vehicular Technology Magazine. 2014; **9**:71–78. DOI: 10.1109/MVT.2013.2295070

[4] Anqi-He: Uplink interference management in massive MIMO enabled heterogeneous cellular networks. IEEE Wireless Communications Letters. 2016;**5**:560-563. DOI: 10.1109/LWC.2016.2601102

[5] Olabisi E, Anthony-Chan H: Radio resource management in heterogeneous cellular networks. Cellular Networks — Positioning, Performance Analysis, Reliability. Dr. Agassi Melikov (Ed.), InTech. 2011. DOI: 10.5772/15428.

[6] Hoadley J, Maveddat P: Enabling small cell deployment with hetnet. IEEE Wireless Communication Magazine. 2012;**19**:4–5. DOI: 10.1109/MWC.2012.6189405

[7] Li H, Guangxi Z, Xiaojiang D: Cognitive femtocell networks: An opportunistic spectrum access for future indoor wireless coverage. IEEE Transaction on Wireless Communications. 2013;**20**:44–51. DOI: 10.1109/MWC.2013.6507393

[8] Zahir T, Arshad K, Nakata A, Moessner K: Interference management in femtocells. IEEE Communications Surveys and Tutorials. 2013;**15**:293–311. DOI: 10.1109/SURV.2012.020212.00101

[9] Cheng S, Lien S, Chu FS, Cheng-chen K: On exploiting cognitive radio to mitigate interference in macro/femto heterogeneous networks. IEEE Wireless Communication Magazine. 2011;**18**:40–47. DOI: 10.1109/MWC.2011.5876499

[10] Srinivasa S, Jafar S: Cognitive radios for dynamic spectrum access – The throughput potential of cognitive radio: A theoretical perspective. IEEE Communication Magazine. 2007;**45**:73–79. DOI: 10.1109/MCOM.2007.358852

[11] Goldsmith A, Jafar S, Maric I, Srinivasa S: Breaking spectrum gridlock with cognitive radios: An information theoretic perspective. Proceedings of the IEEE. 2009;**97:**894–914. DOI: 10.1109/JPROC.2009.2015717

[12] Bharucha Z, Haas H, Auer. G, Cosovic I: Femto-cell resource partitioning. In: Proceedings of the IEEE Global Communications Conference Workshop (GLOBECOM ′09); November 2009; Honolulu, HI: IEEE; 2009. pp. 894–914.

[13] Gra J, Pedersen KI, Szufarska A, Strzyz S: Cell-specific uplink power control for heterogeneous networks in lte. In: Proceedings of the IEEE Vehicular Technology Conference (VTC2010-Fall ′10); 6–9 September 2010; Ottawa, Ontario: IEEE; 2010. pp. 1–5.

[14] Cadambe V, Jafar S: Interference alignment and degrees of freedom of the k-user interference channel. IEEE Transaction on Information Theory. 2008;**54**:3425–3441. DOI: 10.1109/ICC.2008.190

[15] Suh C, Ho M, Tse D: Downlink interference alignment. IEEE Transaction on Communications. 2011;**59**:2616–2626. DOI: 10.1109/TCOMM.2011.070511.100313

[16] Maso M, Debbah M, Vangelista L: A distributed approach to interference alignment in OFDM-based two-tiered networks. IEEE Transactions on Vehicular Technology. 2013;**62**:1935–1949. DOI: 10.1109/TVT.2013.2245516

[17] Castanheira D, Silva A, Gameiro A: Limited inter-System information exchange method for heterogeneous networks. IEEE Communications Letters. 2015;**19**:1656–1659.

[18] Castanheira D, Silva A, Dinis R, Gameiro A: Efficient transmitter and receiver designs for SC-FDMA based heterogeneous networks. IEEE Transaction on Communications. 2015;**63:**2500–2510. DOI: 10.1109/TCOMM.2015.2434383

[19] Castanheira D, Silva A, Gameiro A: Set optimization for efficient interference alignment in heterogeneous networks. IEEE Transaction on Wireless Communications. 2014;**13**:5648–5660. DOI: 10.1109/TWC.2014.2322855

[20] Sharma S, Chatzinotas S, Ottersten B: Interference alignment for spectral coexistence of heterogeneous networks. EURASIP Journal on Wireless Communications and Networking. 2013;**46**. DOI: 10.1186/1687-1499-2013-4"6

[21] Castanheira D, Silva A, Gameiro A: Null-space cognitive precoding for heterogeneous networks. IET Communications. 2014;**8**:555–563. DOI: 10.1049/iet-com.2013.0434

[22] Karakayali M, Foschini J, Valenzuela R: Network coordination for spectrally efficient communications in cellular systems. IEEE Transaction on Wireless Communications. 2006;**13:**56–61. DOI: 10.1109/MWC.2006.1678166

[23] Ali S, Castanheira D, Silva A, Gameiro A: Transmission cooperative strategies for MIMO-OFDM heterogeneous networks. Radioengineering Journal. 2015;**25**:431–441. DOI: 10.13164/re.2015.0431

[24] Ali S, Castanheira D, Silva A, Gameiro A: Downlink cognitive interference alignment for heterogeneous networks. In: Proceedings of the IEEE 21st International Conference on Telecommunications (ICT '14); 4-7 May 2014; Lisbon, Portuga: IEEE; 2014. pp. 236–240.

[25] Ali S, Castanheira D, Silva A, Gameiro A: Joint IA and SFBC macro-cells and small-Cells coexistence under minor information exchange. Mobile Information Systems. 2016;Article ID30478591:1-10. DOI: 10.1155/2016/3047859

[26] Matthé M, Mendes L, Michailow N: Widely Linear Estimation for Space-Time-Coded GFDM in Low-Latency Applications. IEEE Transactions on Communications. 2015;**63**:4501–4509. DOI: 10.1109/TCOMM.2015.2468228

[27] Alamouti SM: A simple transmit diversity technique for wireless communications. IEEE Journal on Selected Areas in Communications. 1998;**16**:1451–1458. DOI: 10.1109/49.730453

[28] Heath R, Paulraj A: Linear dispersion codes for MIMO systems based on frame theory. IEEE Transactions on Signal Processing. 2002;**50**:2429–2441. DOI: 10.1109/TSP.2002.803325

[29] Hassibi B, Hochwald B: High-rate codes that are linear in space and time. IEEE Transactions on Information Theory. 2002;**48**:1804–1824. DOI: 10.1109/TIT.2002.1013127

[30] Tirkkonen O, Boariu A, Hottinen A: Minimal non-orthogonality rate 1 space-time block code for 3+ Tx antennas. In: Proceeding of the IEEE 6th International Symposium on Spread Spectrum Techniques and Applications; September; Parsippany, NJ: IEEE; 2000. pp. 429–432.

[31] Tarokh V, Seshadri N, Calderbank A: Space-time codes for high data rate wireless communication: Performance criterion and code construction. IEEE Transactions on Information Theory. 1988;**44**:744–765. DOI: 10.1109/18.661517

[32] Ayach O, Peters S, Heath R: The practical challenges of interference alignment. IEEE Transactions on Wireless Communications. 2013;**20**:35–42. DOI: 10.1109/MWC.2013.6472197.

9

# Spectral Efficiency Analysis of Filter Bank Multi-Carrier (FBMC)-Based 5G Networks with Estimated Channel State Information (CSI)

Haijian Zhang, Hengwei Lv and Pandong Li

**Abstract**

Filter bank multi-carrier (FBMC) modulation, as a potential candidate for physical data communication in the fifth generation (5G) wireless networks, has been widely investigated. This chapter focuses on the spectral efficiency analysis of FBMC-based cognitive radio (CR) systems, and spectral efficiency comparison is conducted with another three types of multi-carrier modulations: orthogonal frequency division multiplexing (OFDM), generalized frequency division multiplexing (GFDM), and universal-filtered multi-carrier (UFMC). In order to well evaluate and compare the spectral efficiency, we propose two resource allocation (RA) algorithms for single-cell and two-cell CR systems, respectively. In the single-cell system, the RA algorithm is divided into two sequential steps, which incorporate subcarrier assignment and power allocation. In the two-cell system, a noncooperative game is formulated and the multiple access channel (MAC) technique assists to solve the RA problem. The channel state information (CSI) between CR users and licensed users cannot be precisely known in practice, and thus, an estimated CSI is considered by defining a prescribed outage probability of licensed systems. Numerical results show that FBMC can achieve the highest channel capacity compared with another three waveforms.

**Keywords:** filter bank multi-carrier, spectral efficiency, resource allocation, cognitive radio, 5G networks

## 1. Introduction

With the increasing demand of communication quality, the fifth generation (5G) communication networks have shown development needs of high speed, low latency, high spectrum efficiency, etc. [1]. As a result, people anticipate that the final outcome for 5G waveforms may include an adaptive solution, which means using the optimum waveform for any given situation. Nowadays, one key element of the cellular communication system is the multiple access technology that is used. Thus, the multi-carrier modulation (MCM) has been a research hotspot of the communication field due to its ability of suppressing the inter-symbol interference (ISI) and inter-channel interference (ICI). The orthogonal frequency division multiplexing (OFDM) is a typical style of MCM, which has been used in the fourth generation (4G) communication systems [2]. Although OFDM has many advantages, it still cannot satisfy the requirements of 5G networks [1]. With the higher level of processing that will be available, new 5G waveforms are being considered and evaluated for using with the new system. There have been some other MCM waveforms studied by the scholars around the world, including the well-known modulation schemes: filter bank multi-carrier (FBMC), generalized frequency division multiplexing (GFDM), and universal-filtered multi-carrier (UFMC) [3]. Each of them has its own advantages and disadvantages [4]. The modulation schemes used for the future 5G networks should have a significant impact on the whole performance and will play a major role in determining the performance and complexity of communication systems; however, single technique is not likely to meet all the requirements. In order to drive 5G standardization, academia is engaging in various collaborative projects such as METIS [5] and 5GNOW [6]. The purpose is to guarantee that the 5G can achieve commercialization in 2020.

In the future 5G networks, there needs more frequency resource for better communication. This requirement becomes particularly important because we have been facing the problem of frequency scarcity. However, in traditional spectrum management policy, there are a large amount of frequency bands which are not sufficiently utilized in most of the time. This results in a serious conflict between the target for better communication and the fact for spectrum scarcity. Cognitive radio (CR) [7–9] and FBMC [10–16] techniques, which are capable of efficiently exploiting the spectrum hole, can be considered to apply in 5G networks. CR technology is considered to be one of the most important technologies to improve the spectral efficiency. It can utilize the flexible and complex algorithms to control the interference to primary users (PUs). By adopting adaptive software, the CR devices are able to reconfigure their communication functions to the requirements of secondary users (SUs), while FBMC has a negligible frequency spectrum leakage, which has high robustness to the interference resulting from frequency offset. Therefore, it does not need to set the guard band in frequency domain, which greatly improves the spectral efficiency. In addition, FBMC can flexibly control the interference between adjacent subcarriers, which are unsynchronized. These advantages make FBMC more and more popular in the academic field. In recent years, the scholars have studied the FBMC system in terms of the spectral efficiency analysis [17], system complexity analysis [18], prototype filter design [19–21], frequency offset estimation [22], multiple-input multiple-output (MIMO) [23], and so on.

This chapter mainly analyses the spectral efficiency of FBMC in the context of CR systems. The results of other MCM waveforms give a better characterization of performance comparison to FBMC. In order to clarify the desirable property of FBMC, two different network scenarios including single cell and two cells are taken into consideration. Specifically, for single-cell systems, we solve the uplink resource allocation (RA) problem by two sequential steps: subcarrier assignment solved by the average capacity metric (AC-metric) combined with Hungarian algorithm and power allocation, which equals to a nonlinear programming solved by the gradient projection method (GPM). As for two-cell CR systems, we establish a noncooperative game, which performs uplink subcarrier assignment and power allocation among noncooperative CR cells with multiple CR users per cell. Since the optimization formulation for rate maximization of multiple users in each CR cell is an integer optimization problem, the multiple access channel (MAC) technique is applied to transform the integer optimization problem into a concave optimization problem. In practice, the channel state information (CSI) between CR users and licensed users cannot be perfectly known, and thus, an estimated CSI is considered by defining a prescribed outage probability of licensed systems.

The remainder of this chapter is organized as follows. Section 2 provides a systematic introduction of FBMC technique and makes a brief comparison with OFDM, GFDM, and UFMC. In Sections 3 and 4, two RA algorithms for single-cell and two-cell CR systems are presented to well evaluate the spectral efficiency of different multi-carrier modulations, respectively. Finally, conclusions are made in Section 5.

## 2. Multi-carrier modulation (MCM)

MCM is an efficient tool to overcome communication channel challenges by dividing the frequency spectrum into multiple subcarriers [4]. Compared with single carrier modulation (SCM), it is easier to tackle the frequency-selective multipath effect in future communication networks. In this section, the introductions of FBMC and other three MCM waveforms are given, in which the description of FBMC is the main concentration. At the end of this section, the properties of these four waveforms are discussed and some generalizations are summarized for a clear understanding of these waveforms.

### 2.1. Filter bank multi-carrier (FBMC)

The basic concept of FBMC modulation technology was first proposed by Chang and Saltherg in the middle of 1960s [13], but it was not paid much attention by scholars because of its complexity. In the 1990s of the last century, we are familiar with the discrete multi-tone (DMT) modulation and discrete wavelet multi-tone (DWMT) modulation, both of which are the special cases of FBMC modulation. In recent years, along with the increasing demands for high reliability and high-rate communication, while signal processing and electronic equipment have made significant progress, the realization of the principle structure of FBMC is relatively easy. As a result, it has aroused the interest of researchers once again.

Generally, FBMC mainly has three kinds of modulation modes: cosine modulated multi-tone (CMT), filtered multi-tone (FMT), and offset quadrature amplitude modulation-based OFDM (OQAM-OFDM). CMT uses the cosine modulated filter bank, which is the early FBMC modulation technology in the field of digital subscriber line (DSL). It has been applied in the field of wireless applications recently. CMT not only has a high bandwidth efficiency but also has a blind detection capability [14]. Due to the reconstruction performance of CMT, the overlapping adjacent bands can be completely separated when the multiple neighbor frequency bands are transmitted at the same time. FMT is another form of FBMC modulation. Compared to CMT, the subcarriers of the FMT are not overlapping between the adjacent frequency bands. In order to avoid the overlapping of subcarriers, the guard band should be added between the subcarriers. Due to the use of the guard interval, the FMT system will waste some bandwidth. Therefore, the main difference between CMT and FMT lies in the use of special frequency bands. Recent FBMC technique is referred to as OQAM-OFDM. Compared to CMT and FMT, OQAM-OFDM has the highest stop-band attenuation for a fixed filter length and number of subcarriers [15].

According to the characteristics of OQAM, the transmission symbols of the OQAM-OFDM communication system are the real and imaginary parts of the complex quadrature amplitude modulation symbols [16], and the transmission time interval is half of the symbol period between the real and imaginary symbols. In addition, the reasonable design of the prototype filter can ensure that the frequency response of each subcarrier has a better roll-off characteristic, for reducing the spectrum leakage of subcarriers. Many scholars have been designing suitable filters for FBMC. The filter using the frequency sampling technique in Ref. [24] has been considered as the reference prototype filter of the European project PHYDYAS. Le Floch [25] gives an overview of the main features concerning isotropic orthogonal transform algorithm (IOTA). The authors in Ref. [20] formulate a direct optimization problem of the filter impulse response coefficients for the FBMC systems to minimize the stop-band energy and constrain the ISI/ICI. In Ref. [26], it is attempted to design the prototype filter by performing time-frequency analysis on the ambiguity function of isotropic Hermite pulses.

Besides the research of prototype filter, people have made plenty of contributions to improve the performance of FBMC structure. In Ref. [27], a novel architecture for MIMO transmission and reception of FBMC modulated signals under strong frequency selectivity channel is presented. An improved partial transmit sequence (PTS) scheme by employing multi-block joint optimization (MBJO) for the PAPR reduction of FBMC signals is proposed in Ref. [28]. In Ref. [29], a novel scattered pilot method for channel estimation in FBMC is proposed. In Ref. [30], the authors propose a low complexity frequency offset compensation method for FBMC in a context of frequency division multiple access (FDMA). In Ref. [22], a data-aided joint maximum likelihood (ML) estimator of carrier frequency offset (CFO) and channel impulse response for oversampled perfect reconstruction filter banks transceivers are proposed. And the spectral efficiency of FBMC-based CR networks is studied in Ref. [17]. In short, FBMC has made some achievements in various aspects.

To conclude, the above three FBMC techniques could all theoretically offer a significant bandwidth efficiency advantage over OFDM due to their special filter bank based structure and the elimination of cyclic prefix (CP). On the other hand, among different FBMC techniques, OQAM-OFDM is preferred to be a suitable choice for CR applications since FMT and CMT are originally introduced for DSL applications and will be impractical and hard to meet the CR system requirements. In this chapter, unless otherwise stated, the FBMC refers to OQAM-OFDM.

## 2.2. Other multi-carrier waveforms

In order to reflect the spectral efficiency performance of FBMC in CR systems, we compare it with OFDM, GFDM, and UFMC. In the following, we first introduce the other three modulation waveforms, and then, the differences in the four MCM waveforms are summarized. Contrast to FBMC, OFDM has a lower computational complexity [31]. It also can be combined with other technologies easily, such as wavelet orthogonal frequency division multiplexing (WOFDM) and MIMO. However, it has serious out-of-band leakage and high peak-to-average power ratio (PAPR) [32]. Until now, the ways to reduce PAPR are still being researched.

In GFDM system, the use of root-raised cosine (RRC) pulse-shaping filter can greatly reduce the impact of radiation and enhance the system flexibility. In addition, GFDM uses less CP, which improves the spectral efficiency [33, 34]. Similar to FBMC, GFDM can well integrate the spectrum. According to the requirements of the different types of services and applications, GFDM can choose different pulse-shaping filters and insert different types of CP. The subcarriers of GFDM pass through the effective prototype filter to filter and circularly shift both in time and in frequency domain, which reduces the band leakage. However, to meet the requirements of the quality of wireless communication transmission, GFDM technology sacrifices the bit error rate (BER) and the ICI at the cost of eliminating the band radiation [35]. In recent years, the focus of the research on GFDM technology lies in how to improve the BER performance and reduce the computational complexity.

UFMC has the advantages of the FBMC system, and it can also support different types of business [36]. Compared to the prototype filters of FBMC, UFMC uses a shorter filter length, which can support the short burst asynchronous communication [37]. Furthermore, UFMC system has a low requirement about time-frequency calibration and non-orthogonality. However, similar to OFDM system, UFMC suffers the influence of both the Doppler effect and the crystal oscillators of transmitter and receiver, which can result in the CFO. A small CFO will also lead to a sharp decrease in UFMC system performance. Therefore, in order to effectively reduce the interference in UFMC system so that it can improve the transmission reliability and ensure the effectiveness of the signal, interference cancelation has become a hot spot in this field.

In conclusion, according to the previous introductions, we have listed the features about OFDM, FBMC, GFDM, and UFMC in **Table 1** [1–4, 37, 38], including the PAPR, the out of band, and the spectral efficiency we are concerned about. According to these characteristics, we can make a rough comparison of these four kinds of waveforms. And the superiority and inferiority of each waveform are also clearly presented. We can select different waveforms based on various application scenarios.

| | OFDM | FBMC | GFDM | UFMC |
|---|---|---|---|---|
| PAPR | High | High | Low | Medium |
| Out of band | High | Low | Low | Low |
| Spectral efficiency | Medium | High | Medium | High |
| CP | Yes | No | Yes | No |
| Orthogonality | Yes | Yes | No | Yes |
| Synchronization requirement | High | Low | Medium | Low |
| Ease of integration with MIMO | Yes | No | Yes | Yes |
| Latency | Short | Long | Short | Short |
| Effect of frequency offset | Medium | Medium | Medium | Medium |

**Table 1.** Comparison of the features among OFDM, FBMC, GFDM, and UFMC.

It is seen that these four modulation waveforms have their own drawbacks and superiorities. In addition, we have simulated the BER of these four waveforms, which is an important factor to measure the modulation waveforms [39]. The BER performance in different signal noise ratios (SNRs) is shown in **Figure 1**, where the parameters of these modulation waveforms are as follows: the number of subcarriers of FBMC is 128, and the prototype filter of FBMC is the PHYDYAS filter [40]; the number of subcarriers of OFDM is 128, and the prototype filter of

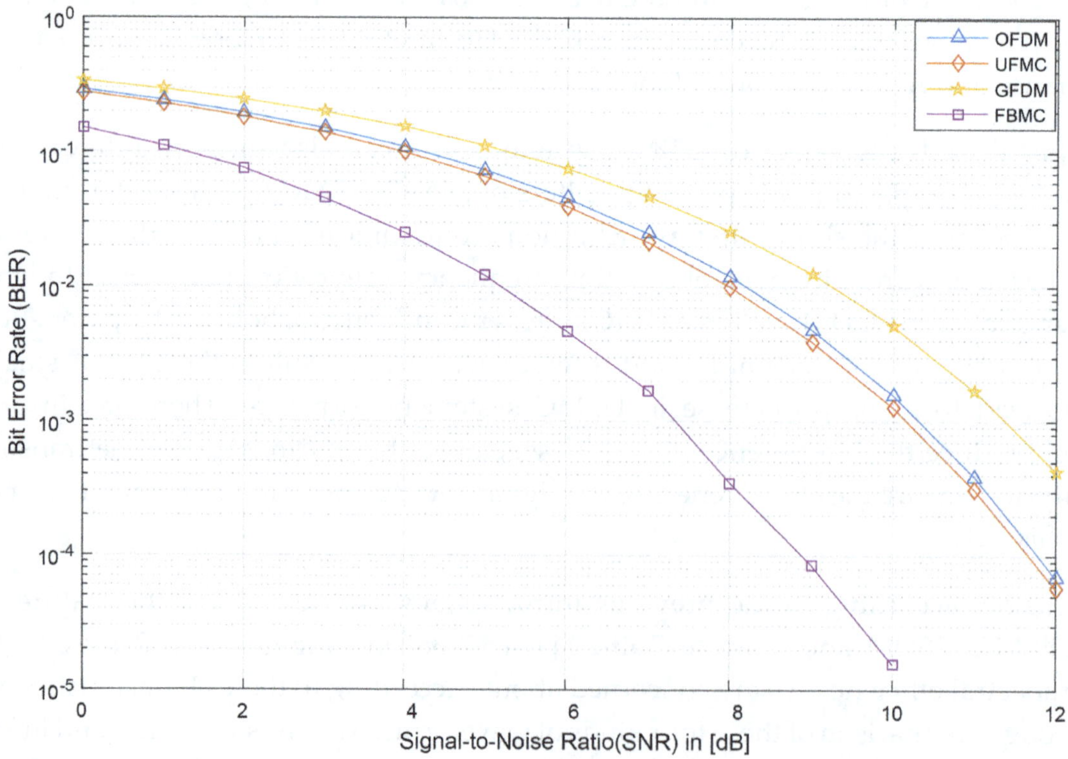


**Figure 1.** BER vs. SNR levels for FBMC-, OFDM-, GFDM-, and UFMC-based systems.

OFDM is the rectangle filter; the number of subcarriers of GFDM is 128, the number of sub-symbols of GFDM is 9 in each subcarrier, and the prototype filter of GFDM is the RRC filter with roll-off coefficient $\alpha = 0.5$; the number of sub-bands of UFMC is 10, the number of subcarriers of UFMC is 12 in each sub-band, and the prototype filter of UFMC is the Dolph-Chebyshev filter [36].

The prototype filter is a key element in the MCM schemes because all synthesis and analysis filters are frequency-shifted versions of the corresponding low-pass prototype filter frequency response. The principle how we select the prototype filter is that the most commonly used prototype filter is chosen in the research of different waveforms. For FBMC, we adopt the prototype filter used in PHYDYAS project [27–30], which can reduce the side-lobe of FBMC effectively. For OFDM, the rectangular filter is chosen as the prototype filter, which is one of the most popular prototype filters in the OFDM theory model. For GFDM, we use RRC filter which has a lower spectrum leakage in the frequency domain if the roll-off coefficient is larger. Normally, when the GFDM system is studied, the RRC filter [34, 35] is widely used as the prototype filter. For UFMC, we adopt Dolph-Chebyshev filter used in Ref. [36], which proposes the method for designing UFMC. Another important reason for the selection of these prototype filters is that they play their respective advantages in different modulation structures. For example, the use of RRC filter makes the GFDM flexible, which might be difficult to realize by other prototype filters.

It is noted from **Figure 1** that the BER of FBMC is the lowest than those of other waveforms in different SNRs, which means FBMC has the best BER performance than other three modulation waveforms. The BER performance of GFDM is the worst, while the BER performance of UFMC is better than that of OFDM.

In this chapter, the interferences of side-lobe radiation in these MCM modulations are the focus of consideration. **Figure 2** shows the frequency responses of prototype filters for FBMC, OFDM, GFDM, and UFMC. Although the energy is mainly located in the main lobe, it is intuitively clear that the four modulations have different side-lobe radiations. FBMC has the minimum out-of-band leakage, and the out-of-band leakage of OFDM is the largest, while the out-of-band leakage of GFDM is larger than that of UFMC, that is, the interference that depends on the out-of-band leakage among subcarriers of different modulations is not the same. The reason why they have different spectrum leakages, to a large extent, depends on the prototype filters they use. Hence, if we want to establish an interference model, it can be based on the side-lobe radiation, which is determined by the power spectral density (PSD) model of multi-carrier signals. According to the PSD-based approach in Ref. [41], the interference values of each modulation scheme can constitute an interference vector. This is an important measure to distinguish different waveforms in the following sections. The interference vectors of FBMC and OFDM are referred in Refs. [41, 42]. Assuming that a single complex symbol with power equals to "1," the element of vectors is the power of out-of-band radiation. The interference vectors of UFMC and GFDM are calculated with the same method in Refs. [41, 42], wherein interference less than $10^{-3}$ is ignored. Thus, the interference vectors are derived as

$$\begin{cases} V^{fbmc} = [6.38e^{-2},0,0,0,0,0,0] \\ V^{ofdm} = [6.31e^{-2},1.07e^{-2},4.42e^{-3},2.52e^{-3},1.73e^{-3},1.31e^{-3},1.02e^{-3}] \\ V^{ufmc} = [12.27e^{-2},0,0,0,0,0,0] \\ V^{gfdm} = [4.80e^{-2},4.18e^{-2},1.40e^{-3},0,0,0,0] \end{cases} \quad (1)$$

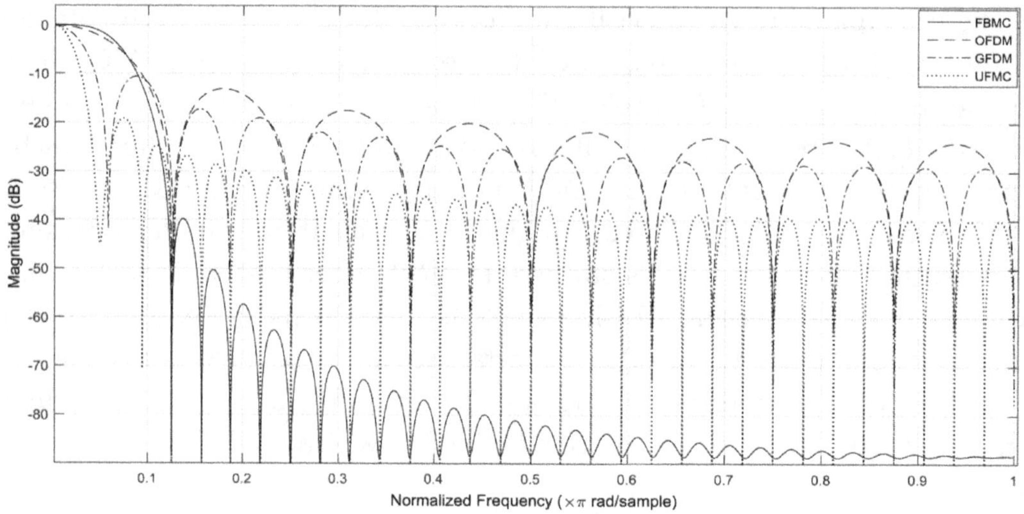


**Figure 2.** Frequency responses of prototype filters for FBMC, OFDM, GFDM, and UFMC.

This section compares the characteristics between FBMC and other three modulation schemes. The interference vectors are also given to quantify the out-of-band radiation, and they will be applied for the comparison of spectral efficiency among these four modulation waveforms in Sections 3 and 4.

## 3. Spectral efficiency comparison in single-cell systems

In this section, the RA of single CR cell with multiple CR users is designed to evaluate the spectral efficiency of FBMC and other three waveforms-based CR networks. The spectral efficiency is measured by the average capacity of available frequency bands, which is mainly determined by the MCM scheme and the RA strategy. Considering the low complexity, the proposed RA algorithm in the context of single CR cell is split into two sequential steps: subcarrier assignment and power allocation. In the following, the detailed statements including the system model and the RA algorithm are presented.

### 3.1. System model

In the context of CR systems, a group of SUs randomly distributed with an accessing point called secondary base station (SBS) constitutes a CR cell. As depicted in **Figure 3**, the uplink scenario of CR systems incorporates a primary cell and a secondary cell with multiple PUs and SUs. Generally, due to the spectral leakage (indicated in **Figure 4**) and imperfect synchronization between SU and PU, the out-of-band radiation of a subcarrier will be regarded as interference. If the spectrum holes adjacent to the PUs are occupied by the SUs, the PUs may

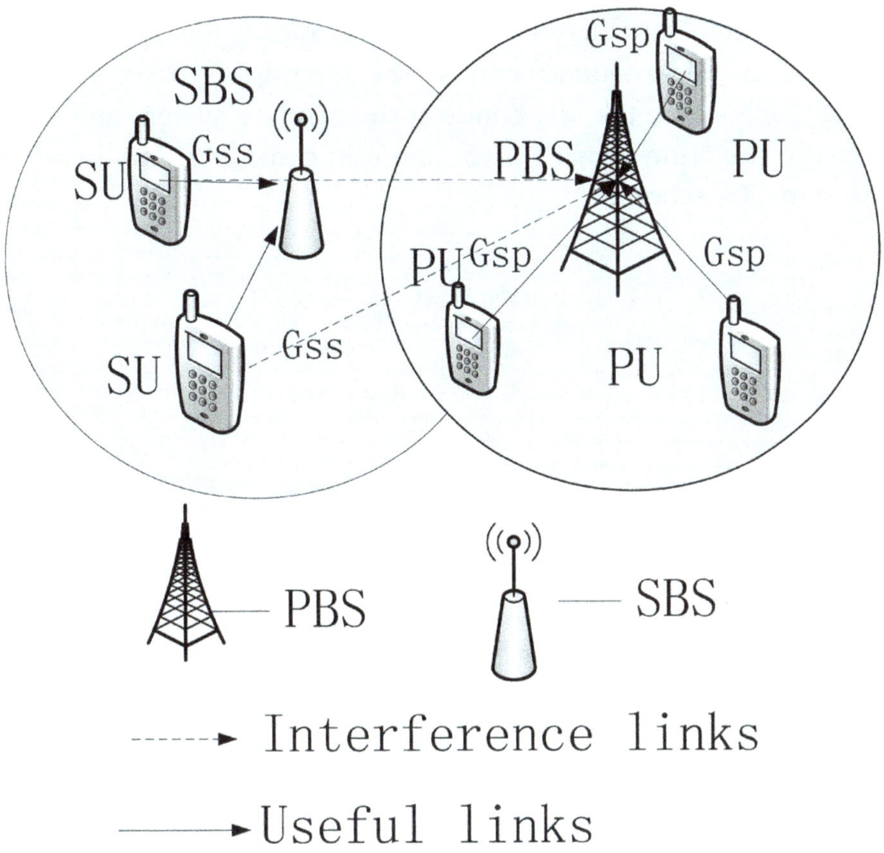


**Figure 3.** The system model of single CR cell including multiple users, the solid lines with arrow stand for the links producing the capacity, the dash lines with arrow stand for the interference links.

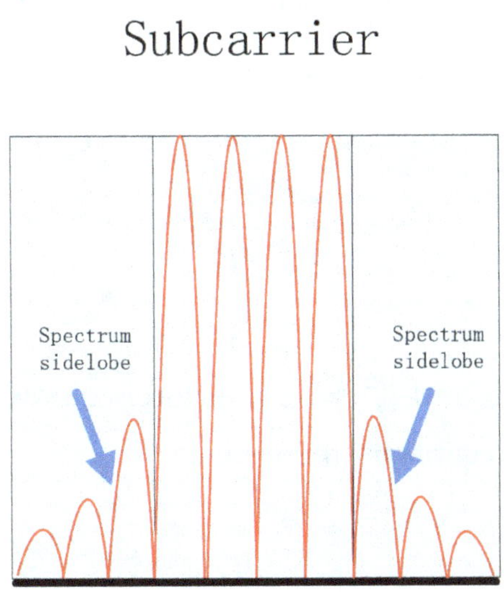


**Figure 4.** The neighbor frequency interference resulting from spectral leakage.

suffer from the intercell interference. For the guarantee of the quality of service (QoS) of PUs, the interference constraint must be considered to limit the interference from SUs. The distributions of PUs and the spectral holes are depicted in **Figure 5**. Assuming that the whole bandwidth is divided into 48 sub-bands, each sub-band includes $L = 18$ subcarriers. A spectrum hole may incorporate many sub-bands. As shown in **Figure 5**, the busy and idle sub-bands are represented by "1" and "0," respectively: "1" means the occupied frequency bands by PUs and "0" means the idle frequency bands to be dynamically accessed by SUs. Assuming that the SBS can perfectly sense the idle bands of the primary system and SUs in the CR cell are synchronized, the spectrum sensing error is not in consideration; therefore, the concentration is located in the RA scheme.

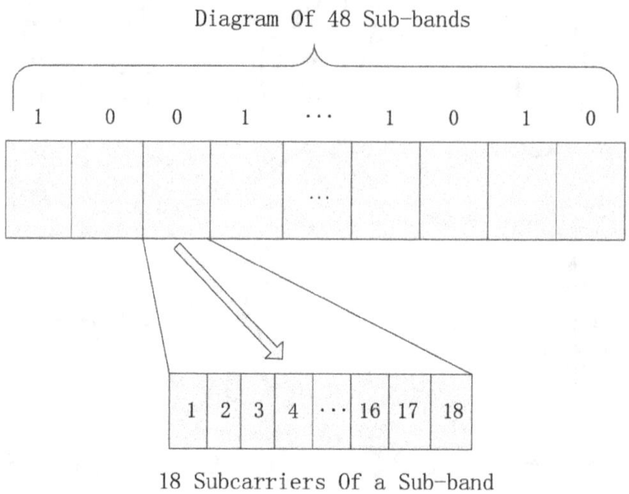


**Figure 5.** The diagram of idle and occupied frequency bands (a sub-band incorporates 18 subcarriers).

According to the above analysis, the CR cell wants to maximize its sum data rate by allocating power into the detected spectrum holes for its own users. Considering the information rate of user $m$ on the $f_{th}$ subcarrier of the $k_{th}$ spectrum hole, the signal-to-interference-plus-noise ratio (SINR) with transmission power $p_m^{kf}$ and transmission gain $G_{ss}^{mkf}$ can be written as

$$\text{SINR} = \frac{p_m^{kf} G_{ss}^{mkf}}{\sigma^2 + I_f^k} \tag{2}$$

where $\sigma^2$ is the power of noise, and $I_f^k$ is the interference power. Therefore, the information rate is obtained by the Shannon capacity theorem as

$$Rate = log_2 \left[ 1 + \frac{p_m^{kf} G_{ss}^{mkf}}{\sigma^2 + I_f^k} \right]. \tag{3}$$

Noticed that whether the subcarrier is assigned to the user $m$ or not, this can be represented through the subcarrier allocation indicator $\theta_m^{kf}$. Assumed that there are $M$ SUs, the number of spectrum holes is $K$, the number of subcarriers in the $k_{th}$ spectrum hole is $F_k$, the problem of maximizing the total information rate can be formulated as

$$Problem1: \max_{p}: C(p) = \sum_{m=1}^{M}\sum_{k=1}^{K}\sum_{f=1}^{F_k}\theta_m^{kf}\log_2\left[1+\frac{p_m^{kf}G_{ss}^{mkf}}{\sigma^2+I_f^k}\right]$$

$$s.t.\begin{cases} \sum_{k=1}^{K}\sum_{f=1}^{F_k}\theta_m^{kf}p_m^{kf} \le P_{th}, & \forall m; (st1) \\ 0 \le p_m^{kf} \le P_{sub}, & \forall m,k,f; (st2) \\ \sum_{k=1}^{K}\sum_{h=1}^{H}\theta_m^{k_{l(r)}h}p_m^{k_{l(r)}h}G_{sp}^{mk_{l(r)}}\sum_{i=1}^{H-h+1}V_{H-i+1} \le I_{th}, & \forall k; (st3) \end{cases} \quad (4)$$

| | |
|---|---|
| $\theta_m^{kf}$ | Binary variable $\theta_m^{kf} \in \{0,1\}$, $\theta_m^{kf} = 1$ implies that the user $m$ uses the $f_{th}$ subcarrier in the $k_{th}$ hole and $\theta_m^{kf} = 0$ means the subcarrier is not accessed by the user |
| $p_m^{kf}$ | User power on the $f_{th}$ subcarrier in the $k_{th}$ hole |
| $G_{ss}^{mkf}$ | The channel gain from user $m$ to SBS |
| $\sigma^2$ | The noise power of each subcarrier |
| $I_f^k$ | Interference from PU on the $f_{th}$ subcarrier in the $k_{th}$ hole |
| $P_{th}$ | The user's sum power limitation |
| $P_{sub}$ | Each subcarrier power limitation |
| $p_m^{k_{l(r)}n}$ | The user power on the left (right) $n_{th}$ subcarrier in the $k_{th}$ hole |
| $G_{sp}^{mk_{l(r)}}$ | The channel gain from SU to primary base station (PBS) on the left (right) primary subcarrier adjacent to the $k_{th}$ hole |
| $V_i$ | The $i_{th}$ element of the interference vectors |
| $V$ | Interference vector |
| $H$ | The length of interference vector |
| $L$ | The number of subcarriers in each sub-band |
| $I_{th}$ | Interference threshold protecting the QoS of PUs |

**Table 2.** Parameter definitions of CR system model.

The parameter definitions are shown in **Table 2**. While the first constraint $st1$ is to limit the sum maximum power of the SUs, the second constraint $st2$ specifies the range of each subcarrier power. The third constraint $st3$ indicates that the interference to PU should not exceed to the interference threshold $I_{th}$. In Eq. (4), the intercell interference from PU to SU $I_f^k$ can be calculated as follows:

$$I_f^k = \begin{cases} \sum_{h=f}^{H} P_p^{k_l} G_{ps}^{k_l f} V_h, & f = 1,2,\ldots H \\ \sum_{h=F_{k-f+1}}^{H} P_p^{k_r} G_{ps}^{k_r f} V_h, & f = F_{k-f+1},\ldots F_k \\ 0, & \text{otherwise} \end{cases} \tag{5}$$

where $P_p^{k_{l(r)}}$ is the transmission power of PU located in the left (right) of the $k_{th}$ spectrum hole, and $G_{ps}^{k_{l(r)}f}$ is the channel magnitude from PU located in the left (right) of the $k_{th}$ spectrum hole to SBS on the $f_{th}$ subcarrier of the $k_{th}$ spectrum hole. The $I_f^k$ can be measured during spectrum sensing by SBS without need to know this information.

The standards of CR systems are still being studied; to the best of our knowledge, the interference threshold $I_{th}$ does not have a common definition in academic field. In order to make a trade-off between the QoS of PUs and the capacity of SUs, an appropriate interference threshold is needed. In this chapter, the interference threshold $I_{th}$ is prescribed by the primary system through the capacity loss coefficient $\lambda$ of PU. If there is no interference from SU, the capacity of PU in a sub-band is computed as follows: (Generally, the SNR in wireless communication systems is $-5 \sim 30$dB; here, we select the SNR = 10 in simulation test.)

$$C = log_2\left(1 + \frac{P_p G_{pp}}{L\sigma^2}\right) \tag{6}$$

$$\text{SNR} = \frac{P_p G_{pp}}{L\sigma^2} = 2^C - 1 \tag{7}$$

where $P_p$ and $G_{pp}$ are the power of PU and the channel gain from PU to primary base station (PBS), respectively. Considering the interference threshold $I_{th}$, the minimal capacity and the SINR of PU are

$$C_2 = log_2\left(1+\frac{P_p G_{pp}}{L\sigma^2 + I_{th}}\right) \tag{8}$$

$$\text{SINR} = \frac{P_p G_{pp}}{L\sigma^2 + I_{th}} = 2^{C_2} - 1. \tag{9}$$

Comparing Eqs. (7) and (9), we obtain

$$\frac{I_{th}}{L\sigma^2} = \frac{2^C - 1}{2^{C_2} - 1} - 1. \tag{10}$$

Defining the tolerable capacity loss coefficient $\lambda$ of primary system, then we have

$$C_2 = (1-\lambda)C. \tag{11}$$

Substituting Eq. (11) into Eq. (10), $I_{th}$ is fully determined by the value of $C$ and the capacity loss coefficient $\lambda$

$$I_{th} = \left(\frac{2^C - 1}{2^{(1-\lambda)C} - 1} - 1\right) L\sigma^2. \tag{12}$$

Defining different values of $\lambda$, there are corresponding different levels of interference threshold. The larger the value of $\lambda$ is, the more interference the primary system can tolerate.

Besides the interference threshold $I_{th}$, another considered parameter is the channel gain $G_{sp}$. In fact, perfect CSI in RA problem [43–45] cannot be obtained because of channel delays and hardware limitation in channel estimation. In Section 3.2.3, we will describe the channel estimation of $G_{sp}^{mk_{l(r)}}$ in detail.

## 3.2. Resource allocation

The RA problem in communication systems generally needs to simultaneously solve two kinds of tasks: the channel assignment and the power allocation. Instead of pursuing an optimal solution, RA algorithms in many existing works search for a suboptimal solution which decomposes the RA problem into two steps: first assigning the subcarriers and then allocating the power [46, 47]. Generally, the solution of the suboptimal algorithm, which has low computational complexity, can be close to that of the optimal one. Therefore, the suboptimal idea is inherited, and an efficient suboptimal algorithm solving the optimization problem in Eq. (2) is presented as follows.

### *3.2.1. Subcarrier assignment*

The first task of subcarrier assignment is the bandwidth allocation. From the view of fairness, the SU which exhibits the minimum average capacity always increases the number of its subcarriers until the total number of sub-bands assigned to SUs equals to the number of free sub-bands. This mechanism helps to promise that each SU can achieve the fairness. Assuming that $F$ is the number of free sub-bands, the number of SUs is $M$ $(F > M)$, and $N^i$ stands for the number of sub-bands assigned to the $i_{th}$ user. Then, the number of sub-bands can be determined according to the steps below:

① First, suppose that each SU has the equal number of sub-bands, given as: $N^i = F/M,\ \forall\ i$ where $x$ denotes the largest integer not exceeding $x$.

② Second, calculate the average capacity of each SU $C^i = N^i log_2\left(1 + \frac{\frac{P_{th}}{G_i N^i}}{\sigma^2}\right)$. Find the SU with minimal capacity: $i' = \mathrm{arg}min_i(C^i)$. And then add the sub-band number of SU $i'$, i.e., $N^{i'} = N^{i'} + 1$.

③ If all available sub-bands are allocated (which means $\sum_{i=1}^{M} N^i = F$), terminate. Else, repeat the step ②.

Next, the relevant subcarrier assignment is completed. In traditional multi-carrier systems, a good channel quality depends on its high SNR. The maximum SNR-metric is widely applied to assign the subcarriers to the user by the value of SNR "$\bar{P}G_{ss}/\sigma^2$" ($\bar{P}$ is the averaged power by dividing the total power budget on the number of the subcarriers), which only considers the channel gain factor. Therefore, the SNR-metric is not accurate enough to assess the average capacity in CR systems.

In order to obtain the average capacity precisely, the average capacity metric (AC-metric) is applied to take more factors into account. The AC-metric is decided by the channel gain $G_{ss}$, the interference threshold $I_{th}$, the user power limitation $P_{th}$, and the channel gain $G_{sp}$. AC-metric makes a balance between all these influence factors. **Figure 6** shows four different styles of sub-bands in idle spectrum holes, and the average capacity of each style can be calculated by Eq. (12), where C1, C2, C3, and C4 stand for the average capacities of the four different sub-bands, respectively. $H$ is the length of interference vectors, $SINR_h^{l(r)}$ is the SINR on the left (right) $h_{th}$ subcarrier of one spectrum hole, $p_h^{l(r)}$ represents the power on the left (right) $h_{th}$ subcarrier of one spectrum hole, $G_{ss}^{l(r)h}$ stands for the channel magnitude of SU to SBS on the left (right) $h_{th}$ subcarrier.

Assuming that there are $K^u$ SUs and $K^c$ idle sub-bands, a $K^c \times K^c$ AC-matrix can be obtained using the bandwidth allocation method. Our task is how to optimally assign these $K^c$ sub-bands to the $K^u$ SUs, which equals to the search of the optimal matching of a bipartite graph, and the Hungarian algorithm [48] can efficiently solve this assignment problem. Therefore, the subcarrier assignment is implemented by means of AC-metric and the Hungarian algorithm.

$$\begin{cases} C1 = \dfrac{\sum_{h=1}^{H} log_2\left(1+SINR_h^l\right)+\sum_{h=1}^{H} log_2\left(1+SINR_h^r\right)+\left(18-2N\right)log_2\left(1+\dfrac{P_{th}-P_l-P_r}{\left(18-2H\right)\sigma^2}\right)}{18} \\ C2 = \dfrac{\sum_{h=1}^{H} log_2\left(1+SINR_h^l\right)+\left(18-H\right)log_2\left(1+\dfrac{P_{th}-P_l}{\left(18-H\right)\sigma^2}\right)}{18} \\ C3 = \dfrac{\sum_{h=1}^{H} log_2\left(1+SINR_h^r\right)+\left(18-H\right)log_2\left(1+\dfrac{P_{th}-P_r}{\left(18-H\right)\sigma^2}\right)}{18} \\ C4 = log_2\left(1+\dfrac{\overline{p}G_{ss}}{\sigma^2}\right) \\ SINR_h^l = \dfrac{p_h^l G_{ss}^{lh}}{\sigma^2+I_n^l} \quad SINR_h^r = \dfrac{p_h^r G_{ss}^{rh}}{\sigma^2+I_l^r} \\ p_h^l = \min\left\{\overline{P}, \dfrac{I_{th}}{LV_h G_{sp}^l}\right\}, p_h^r = \min\left\{\overline{P}, \dfrac{I_{th}}{LV_h G_{sp}^r}\right\} \end{cases} \tag{13}$$

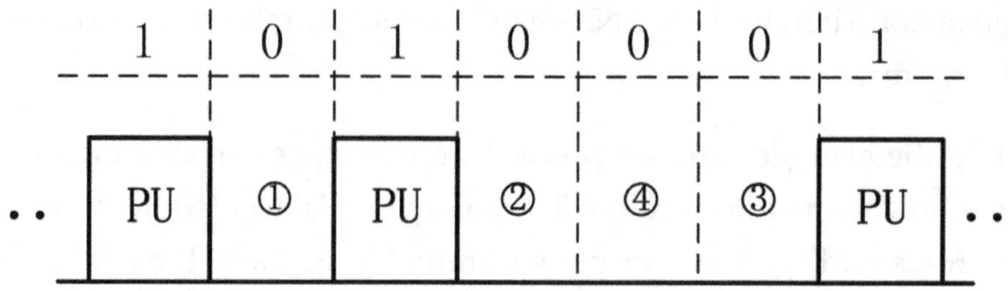


**Figure 6.** Four different types of sub-bands in available spectrum holes.

### 3.2.2. *Power allocation*

The subcarrier assignment has been discussed in Section 3.2.1. In this subsection, the focus is on the problem of power allocation. At the premise of knowing the result of the subcarrier assignment, the power allocation of multiuser system can be virtually regarded as a single-user system. Hence, the task becomes a nonlinear programming, which can be efficiently solved by some algorithms, such as the Lagrangian multiplier and the gradient projection method (GPM) [49]. Considering that the nonlinear programming has the expression as Eq. (14), the Lagrangian multiplier is computational complex if there are extensive multipliers. Instead, the GPM can be applied to obtain the optimal power allocation solution in Eq. (14) with a low computational complexity. The steps of GPM are summarized in **Table 3**.

$$\begin{gathered}\textit{Objective function}: \max f(x) \\ s.t. \begin{cases} Ax < b ; \\ Ex = e ; \end{cases}\end{gathered} \tag{14}$$

| **Steps** | |
|---|---|
| Step 1: | Find the projection matrix $P = I - E^T(EE^T)^{-1}E$. $E$ is the coefficient matrix of active constraints |
| Step 2: | Calculate the next iteration direction $d_{k+1} = P \times \nabla f(x_k)$,$\nabla f(x_k)$ is the gradient of current iteration point |
| Step 3: | If $\|\|d_k\|\| \le \epsilon$ or iteration times equal to the predetermined threshold, quit the iteration. ($\epsilon$ is the threshold of norm) |
| Step 4: | Calculate the range of step size $\alpha_{max}$. $c = b - Ax$, $f = Ad_k \alpha_i = \frac{c_i}{f_i}$ if $f_i > 0$, $\alpha_i = \infty$ if $f_i \le 0$ |
| Step 5: | Find the step size $\alpha$ by line search |
| Step 6: | Compute the next iteration point. $x_{k+1} = x_k + \alpha d_k$ and go to step 2 |
| Step 7: | Quit the iteration |

**Table 3.** Steps of the GPM algorithm to solve nonlinear programming with linear constraints.

### 3.2.3. Estimated channel state information (CSI)

Generally, it is not practical to assume that the perfect CSI in RA problem is available. Due to the channel delays and the inaccuracy of channel gain estimation, there is always an estimation error between estimated CSI and ideal CSI. Thus, the estimated CSI has a more practical significance in communication research than the ideal CSI.

Notice that in $Problem1$, the channel gains include the gain from SU to SBS $G_{ss}$, the gain from SU to PBS $G_{sp}$, and the gain from PU to SBS $G_{ps}$. Although there are multiple types of channel links, there is no need to estimate all kinds of links for channel estimation load. We concentrate on the capacity of secondary cell and control its interference to primary cell. It is reasonable to assume that the channel gain $G_{sp}$ from SU to PBS is not obtained and needs to be estimated, while the other types of links are known by SBS. The interference constraint cannot be tackled without the necessary information of the channel gain $G_{sp}$ from SU to PBS. Although we do not know the channel gain $G_{sp}$, the path loss gain $G_{pl}$ of the link from PBS to SU on the subcarriers used by the primary system can be computed, and through interpolation, the channel gain $G_{sp}$ on free subcarriers can be acquired. If frequency division duplex is applied, the downlink channel gain is not equal to the uplink channel gain. In this case, the downlink channel gain can be used as a rough estimate of the uplink channel gain. In addition, to guarantee the QoS of primary systems, a channel gain margin $G_m$ is added on the pathloss gain $G_{pl}$. Thus, the estimated channel gain $\overline{G_{sp}}$ can be computed by

$$\overline{G_{sp}} = (1 + G_m) G_{pl}. \tag{15}$$

The $G_m$ is associated with the prescribed outage probability $P_{out}$ tolerated by the primary sytem. Based on the implicit hypothesis that there is no difference between the downlink and the uplink path loss, the evaluation of $\overline{G_{sp}}$ only depends on the Rayleigh fading. When the actual channel gain $G_{sp}$ is larger than the estimated channel gain $\overline{G_{sp}}$, we define this case as the outage of primary system. The outage probability $P_{out}$ can be represented as

$$P_{out} = P\left(G_{sp} > \overline{G_{sp}}\right) = P\left(H_{sp}^2 G_{pl} > (1 + G_m) G_{pl}\right) = P\left(H_{sp}^2 > (1 + G_m)\right) \tag{16}$$

where $G_{sp} = H_{sp}^2 G_{pl}$, $H_{sp}$ is the Rayleigh fading frequency response. Since $H_{sp}$ ~ Rayleigh (μ), the $H_{sp}^2$ has a Gamma distribution with shape parameter $\alpha = 1$ and scale parameter $\beta = 2\mu^2$. The cumulative distribution function of $H_{sp}^2$ is the regularized Gamma function. Therefore, Eq. (16) can be further described as

$$1-P_{out}=\frac{\gamma\left(\alpha,\frac{1+G_m}{\beta}\right)}{\Gamma(\alpha)} \tag{17}$$

where $\gamma$ is the lower incomplete gamma function. Then, given a tolerant outage probability $P_{out}$ of primary system, the channel gain margin $G_m$ will be determined by Eq. (17)

$$G_m=2\mu^2 log_e\left(\frac{1}{P_{out}}\right)-1\,. \tag{18}$$

According to the path loss gain $G_{pl}$ and the outage probability $P_{out}$ in Eq. (15), the estimated channel gain can be obtained.

### 3.3. Numerical results

The spectral efficiency of FBMC and the other three modulation waveforms are evaluated by using the abovementioned RA algorithm. We analyze the channel capacity of these waveforms in single CR cell systems from four aspects: the distance $D$ between SBS and PBS, the maximal power of SUs $P_{th}$, the capacity loss coefficient $\lambda$ of PU, and the outage probability $P_{out}$ of PU. The simulation parameters are shown in **Table 4**, and the simulation results are illustrated in **Figures 7–10**.

| Parameters | Value | Unit |
|---|---|---|
| Total bandwidth | 10 | MHz |
| Center frequency | 2.5 | GHz |
| Number of subcarriers | 1024 | – |
| Number of subcarriers in each sub-band | 18 | – |
| Power limitation of each subcarrier | 5 | mW |
| Noise power of each subcarrier | –134.1 | dBm |
| Channel delays | $10^{-9}[0, 110, 190, 410]$ | s |
| Channel powers | $[0, -9.7, -19.2, -22.8]$ | dB |

**Table 4.** Simulation parameters of single CR cell systems.

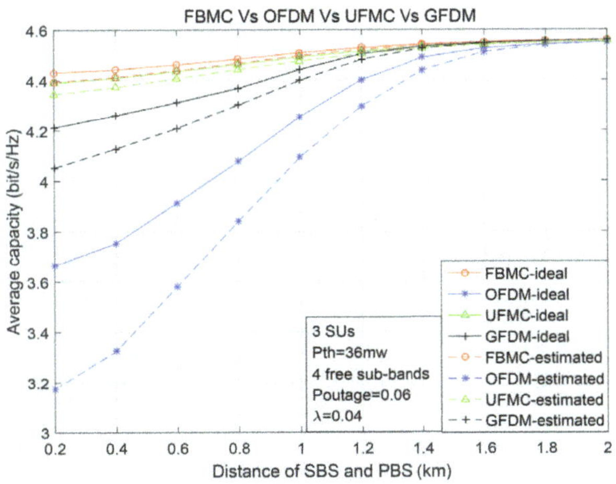


**Figure 7.** The relationship of capacity and distance between PBS and SBS.

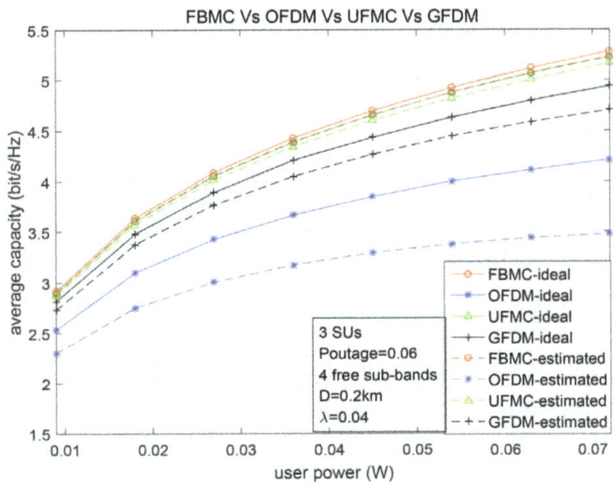


**Figure 8.** The relationship of capacity and the maximal power of SUs.

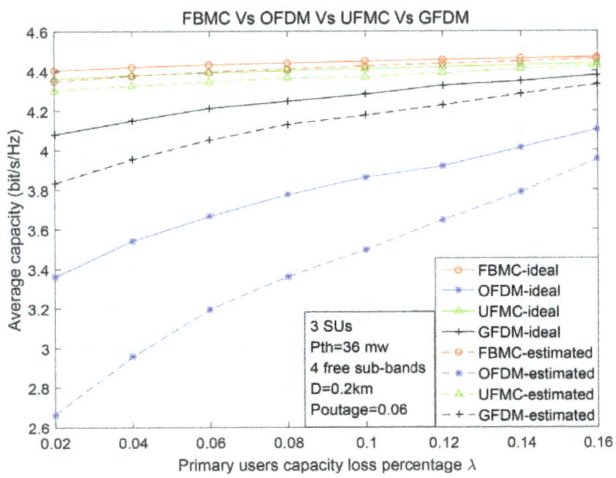


**Figure 9.** The relationship of capacity and interference threshold.

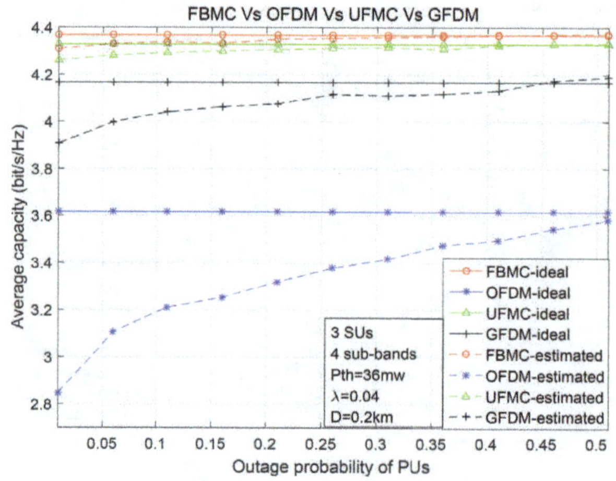


**Figure 10.** The relationship of capacity and the outage probability of PU.

The impact of the distance $D$ between SBS and PBS on spectral efficiency is investigated in **Figure 7**. It can be seen that accompanied by the increase in the distance, the capacities of all waveforms arise and the curves tend to merge asymptotically. The increase in $D$ reduces the mutual interference between the PU and the SU, which is the reason why the average capacity increases. The effect of the maximal power of SU $P_{th}$ is assessed in **Figure 8**. We can obtain that the spectral efficiency of all waveforms increases with the augmentation of $P_{th}$. At the premise of satisfying the constraints, the larger power of SUs means that the more power is allocated to the spectrum holes, which results in the expansion of channel capacity.

**Figures 9** and **10** evaluate the spectral efficiency from the perspective of PUs. **Figure 9** depicts the inherent interaction of average capacity and the capacity loss coefficient $\lambda$ of PU. When less capacity loss is prescribed by PUs, which means a lower interference threshold and better protection for primary system, the achievable capacity degrades due to the more strict access control. **Figure 10** presents the relationship of spectral efficiency and the outage probability $P_{out}$ of PU. Note that the average capacity of OFDM-based CR system with estimated CSI collapses when a low outage probability is prescribed, while other MCM-based CR systems are much less vulnerable to different outage probabilities.

### 3.4. Discussion of spectral efficiency in single-cell systems

From the above simulation results, some distinct conclusions can be drawn:

1. First, the results of **Figures 7–10** exhibit that the average capacity of FBMC outperforms those of other three waveforms. No matter what factor is considered, FBMC always has the highest spectral efficiency on the basis of capacity due to its slightest spectral leakage, UFMC comes second, GFDM takes the third place, and OFDM is the last. It implies that the less spectral leakage leads to the higher spectral efficiency in single CR cell systems.

2. Second, we can see that there is a channel capacity gap between the case of ideal CSI and the case of estimated CSI for the four MCM-based systems. It is easily found that the

spectral leakage property also plays an important role when the estimated CSI is considered. For the OFDM-based CR system, there is a large channel capacity gap between the case with ideal CSI and the case with an estimated CSI, while there is a slight capacity difference by applying the GFDM, UFMC, and especially the FBMC-based CR systems, which could be explained by the fact that when a low outage probability is required, more subcarriers adjacent to PU should be deactivated or underutilized for OFDM due to its significant spectral leakage, which accordingly decreases the channel capacity.

# 4. Spectral efficiency comparison in two-cell CR systems

In Section 3, the comparison of spectral efficiency in single CR cell is implemented easily by a two-step RA algorithm. However, in case of two CR cells, the situation becomes more complicated with a higher dimension of variables. Besides, different cells will compete for the common resource (assuming that the different CR cells will sense the same results of spectral holes). If the two CR cells use the same frequency bands to communicate, the co-channel interference will arise, which makes the RA problem difficult to tackle. In this section, a two-cell RA algorithm is proposed to evaluate the spectral efficiency of different MCM modulations. In the following, the system model is first introduced, and then, the RA optimization algorithm is elaborated. At last, simulation results will be given.

## 4.1. System model

In the scenario of two CR cells, as depicted in **Figure 11**, where the two CR cells with multiple users per cell are symmetrically distributed with the primary system, each CR cell is responsible for the allocation strategy of its users, and it introduces interference to primary system and another CR cell. Assuming that $N$ denotes the number of cells, the number of users per cell is $M$. The aim is still to achieve the sum capacity of available frequency resource. Similar to the formulation of single-cell case, the expression of system model can be presented as

$$Problem2: \max_{p}: C(p)=\sum_{n=1}^{N}\sum_{m=1}^{M}\sum_{k=1}^{K}\sum_{f=1}^{F_k}\theta_{nm}^{kf}\log_2\left[1+\frac{p_{nm}^{kf}G_{SS}^{nmkf}}{\sigma^2+I_{PS}^{nkf}+I_{SS}^{nkf}}\right]$$

$$s.t.\begin{cases}\sum_{k=1}^{K}\sum_{f=1}^{F_k}\theta_{nm}^{kf}p_{nm}^{kf}\le P_{th}, & \forall n,m;\ (st4)\\ 0\le p_{nm}^{kf}\le P_{sub}, & \forall n,m,k,f;\ (st5)\\ \sum_{k=1}^{K}\sum_{h=1}^{H}\theta_{nm}^{k_{l(r)}}p_{nm}^{k_{l(r)}h}G_{sp}^{nmk_{l(r)}}\sum_{i=1}^{H-h+1}V_{H-i+1}\le I_{th}, & \forall n,k;\ (st6)\end{cases} \tag{19}$$

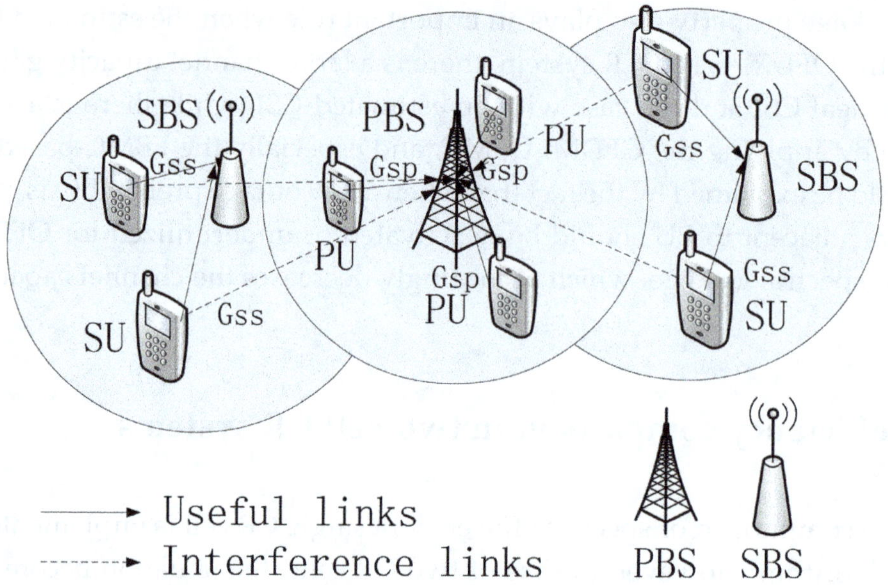


**Figure 11.** System model of two CR cells with multiple CR users per cell.

where the parameter definitions are the same with *Problem*1 in Eq. (4) and $n$ stands for the $n_{th}$ CR cell.

In Eq. (19), the mutual interference and the co-channel interference are computed as

$$I_{PS}^{nkf} = \begin{cases} \sum_{h=f}^{H} P_p^{nk_l} G_{ps}^{nk_l f} V_h, \ f = 1,2,\ldots H \\ \sum_{h=F_{k-f+1}}^{H} P_p^{nk_r} G_{ps}^{nk_r f} V_h, \ f = F_{k-f+1},\ldots F_k \\ 0, otherwise \end{cases} \tag{20}$$

$$I_{SS}^{nkf} = \sum_{n' \neq n}^{N} \sum_{m'=1}^{M} G_{SS}^{n'mkfn} p_{n'm}^{kf}. \tag{21}$$

To solve *Problem* 2 by centralized constrained optimization algorithms, all the channel gain information must be known. This causes large computational complexity and a huge amount of channel estimation overheads. Thus, a distributed RA algorithm is more appropriate than centralized optimization algorithms. Next, we will show our proposed algorithm for solving *Problem* 2 in a distributed manner by establishing a noncooperative game, where the convergence is desired.

### 4.2. Proposed algorithm for solving *Problem* 2

Distributed RA through a noncooperative game [50–53] is preferred where CR users in a single cell can make their own decision based on local information. It can significantly reduce the complexity and show an easier way in solving competition problem. Before the formulation of a noncooperative game, some mathematic preliminaries are given.

**The structure of a noncooperative game** A noncooperative game incorporates three elements: the players, the strategy space, and the utility function. A noncooperative game can be denoted by

$$g = \left\{ N, \{p_n\}_{n \in N}, \{u_n\}_{n \in N} \right\} \tag{22}$$

$N$ is the set of players in a game, $N = \{1, 2, 3, ..., N\}$. $p_n$ is the strategy space of the $n_{th}$ player. $u_n$ is the utilization function of the $n_{th}$ player. The competitive result of a noncooperative game is to obtain the Nash equilibrium (NE).

**Definition of NE**

A strategy profile $p^*$ is NE if no unilateral deviation in strategy by any single player is profitable for that player, that is

$$u_n\left(p_n^*, p_{-n}^*\right) \geq u_n\left(p_n, p_{-n}^*\right) \ \forall n; \tag{23}$$

where $p_n^*$ is the strategy of the $n_{th}$ player on the NE point and $p^*_{-n}$ is the strategy profile except for the $n_{th}$ player on the NE point.

**Existence of NE**

*Theorem*: For a utility function $u_n(p_n, p_{-n})$ with a support domain which is a nonempty convex set, and $u_n$ is continuous and quasiconvex or quasiconcave, at least a pure strategy NE point exists [50].

#### 4.2.1. Formulation of the noncooperative game

Notice that the formulation in Eq. (19) is a mixed integer optimization problem, and the existence of the channel indicator $\theta_{nm}^{kf}$ does not satisfy the condition of converging to the NE. Therefore, our interest is casted on how to transform the foregoing problem Eq. (19) into a concave optimization problem. In Refs. [52, 54], the MAC technique [55–56] is advocated for the formulation of a nonlinear programming, which gives an idea of formulating a concave optimization problem.

Simple schemes like time division multiplexing access (TDMA) and FDMA are generally used in many practical situations. The MAC technique allowing more users to access the same channel assists to remove the indicator $\theta_{nm}^{kf}$. In an MAC-based system, a channel via which two (or more) users send information to a common receiver, larger capacity region can be obtained than that achieved by TDMA or FDMA by using a common decoder for all the users of this system. Assuming that there are $m$ senders denoted by $\{x_1, x_2, x_3, ..., x_m\}$ sending to a common single receiver with the power $\{p_1, p_2, p_3, ..., p_m\}$, $\{G_1, G_2, G_3, ..., G_m\}$ stands for the channel gains, and $N_0$ is the power of noise. MAC can get a large capacity region for these senders, and the capacity region can be calculated as

$$\sum_{i=1}^{m} R_i \le log_2\left(1+\frac{p_1G_1+p_2G_2+p_3G_3+\cdots+p_mG_m}{N_0}\right) \tag{24}$$

MAC can realize the channel assignment and eliminate the non-concave property which results from the channel indicator $\theta_{nm}^{kf}$. Therefore, the task turns into being the power control on each subcarrier of users. With the help of MAC, the noncooperative game is formulated as

$$Problem3:\ max:\ u_n\left(p_n, p_{-n}\right)=\sum_{k=1}^{K}\sum_{f=1}^{F_k} log_2\left[1+\frac{\sum_{m=1}^{M} p_{nm}^{kf} G_{ss}^{nmkf}}{\sigma^2+I_{PS}^{nkf}+I_{SS}^{nkf}}\right]$$

$$s.t.\begin{cases} \sum_{k=1}^{K}\sum_{f=1}^{F_k} p_{nm}^{kf} \le P_{th}, & \forall n,m;(st7) \\ 0 \le p_{nm}^{kf} \le P_{sub}, & \forall n,m,k,f;(st8) \\ \sum_{k=1}^{K}\sum_{h=1}^{H} p_{nm}^{k_{l(r)}h} G_{sp}^{nmk_{l(r)}} \sum_{i=1}^{H-h+1} V_{H-i+1} \le I_{th}, & \forall n,k;(st9) \end{cases}. \tag{25}$$

Notice that the objective function in *Problem* 3 is the summation of logarithmic functions, and the logarithmic function has the following style

$$f(x)=log_2\left(1+a_1x_1+a_2x_2+a_3x_3+\cdots+a_mx_m\right) \tag{26}$$

with parameters $[x_1, x_2, x_3, ..., x_m] \ge 0$. The summation of concave functions is still concave; therefore, if Eq. (26) is proved to be concave, then the objective function in Eq. (25) is also concave.

**Proof:** The Hessian matrix of $f(x)$ at point $x$ is

$$\nabla^2 f(x) = -\frac{(1+a_1x_1+a_2x_2+a_3x_3+\cdots+a_mx_m)^{-2}}{log2} \times \begin{bmatrix} a_1^2 & a_1a_2 \cdots & a_1a_m \\ a_1a_2 & a_2^2 \cdots & a_2a_m \\ \cdots & \cdots \ddots & \cdots \\ a_1a_m & \cdots\cdots & a_m^2 \end{bmatrix} \tag{27}$$

$$= -\frac{(1+a_1x_1+a_2x_2+a_3x_3+\ldots+a_mx_m)^{-2}}{log2} \boldsymbol{a}\boldsymbol{a}^T$$

where $\boldsymbol{a} = [\boldsymbol{a}_1\boldsymbol{a}_2\cdots\boldsymbol{a}_m]^T$. For arbitrary row vector $\mathbf{P}$ with $m$ elements, there are

$$\boldsymbol{P}\nabla^2 f(x)\boldsymbol{P}^T = -\frac{(1+a_1x_1+a_2x_2+a_3x_3+\ldots+a_mx_m)^{-2}}{log2} \times \| \mathbf{Pa} \|^2 \leq 0 \tag{28}$$

Therefore, $f(x)$ in Eq. (26) is concave which also indicates that the objective function in *Problem* 3 is concave, which satisfies the existence condition of NE point, and thus, the convergence of *Problem* 3 is promised.

### *4.2.2. Determination of each cell's strategy*

After the formulation of the noncooperative game, the next work is to determine the specific power allocation scheme. In the game theory-based algorithm, the power allocation scheme of each player is determined sequentially. It is observed that *Problem* 3 is a nonlinear programming with the same constraints as *Problem* 2. It has been shown that GPM is a useful tool to solve the nonlinear programming in the scenario of single CR cell. In order to solve *Problem* 3, GPM is still applied in the scenario of two CR cells. The steps of GPM have been presented in **Table 3**; for the sake of saving space, it is not restated. Readers are encouraged to review **Table 3** again if not familiar with GPM.

### *4.2.3. Estimated CSI of two CR cells*

Based on the same assumption of single CR cell, each SBS has the perfect knowledge of its cell but does not have the CSI knowledge to PBS. Thus, the CSI in the link from SU to PBS is estimated. By means of estimating the channel gain $G_{ps}$ in the inverse link from PBS to SU, the estimated CSI can be obtained with $G_{ps}$ and $G_m$. It should be noticed that both of the two CR cells should conduct the CSI estimation. Since the specific process of channel state estimation has been stated in Section 3.2.3, there is no need for overmuch repeat.

## 4.3. Numerical results of two CR cells

With the same simulation parameters, the comparison of spectral efficiency between FBMC and other modulation waveforms in two CR cells is still assessed from the four aspects. According to the proposed RA algorithm of two CR cells, the simulation results are shown in **Figures 12–15**.

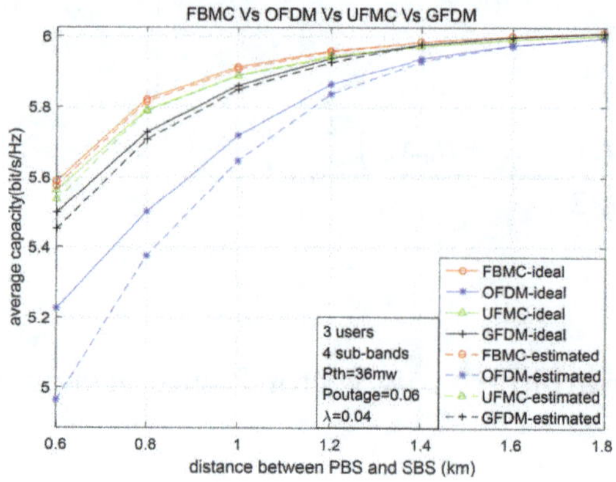


**Figure 12.** The relationship between distance and average capacity.

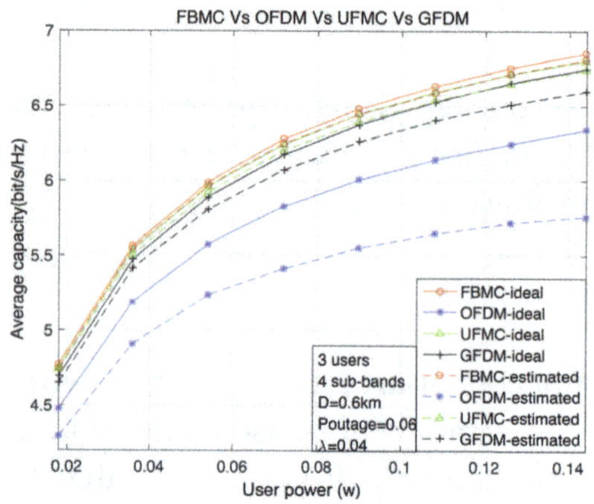


**Figure 13.** The relationship between interference threshold and average capacity.

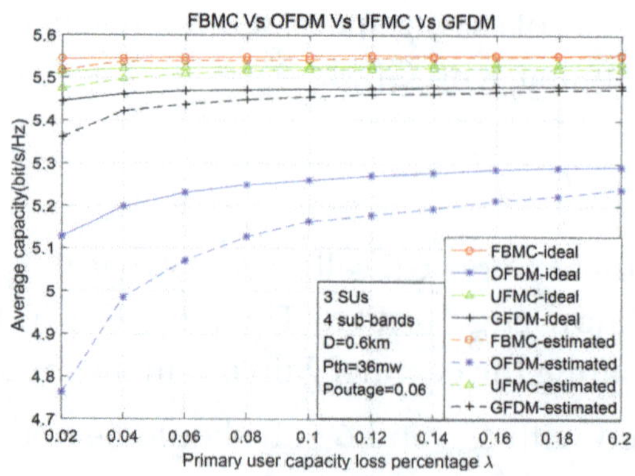


**Figure 14.** The relationship between user power and average capacity.

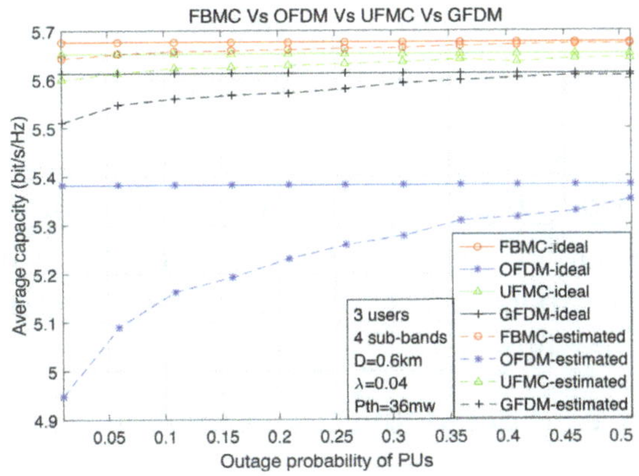


**Figure 15.** The relationship between outage probability and average capacity.

**Figure 12** gives the impact of distance $D$ between SBS and PBS in the context of two CR cells. We can find that the average capacity enlarges as the increase in distance similar to the case of single CR cell. However, compared to single CR cell, there is a clear difference that the span between the highest channel capacity and the lowest one of two CR cells is larger than that of single CR cell. This results from the co-channel interference; when the distance is small, there is an intense interference between the two CR cells in the common channel, which contributes to the dropping of capacity. When the distance becomes large gradually, both of the mutual interference and the co-channel interference wane with $D$, which explains why the curves merge. **Figure 13** assesses the spectral efficiency of two CR cells in terms of maximal user power. Although in the two CR cells, the user with larger power always can access by allocating more power to subcarriers and achieve a higher capacity, this explains the variation tendency of the capacity curves.

The relationship between capacity and the capacity loss coefficient $\lambda$ of two CR cells is presented in **Figure 14**. Similar to the case of single CR cell, there is a slight capacity difference as the $\lambda$ decreases for FBMC. If the primary system needs a strict protection for QoS, which means a low capacity loss coefficient $\lambda$, there is no doubt that FBMC is more able to meet the requirement. **Figure 15** shows the influence of average capacity and the outage probability of PU in the scenario of two CR cells. It is seen that FBMC has the best capacity performance with the slightest capacity difference between ideal and estimated CSIs if the same outage probability is considered. Although the performance curves of FBMC, UFMC, and GFDM are closer to each other than that in the case of single CR cell, the three waveforms show the dramatic difference from OFDM.

### 4.4. Discussions of spectral efficiency in two-cell systems

Based on the simulation results of single CR cell systems and two CR cells systems, the following discussions are presented.

1. Considering the case of two CR cells, we can conclude the same result as in single CR cell that FBMC shows the best spectral efficiency performance from any of the four aspects: the distance $D$ between SBS and PBS, the interference threshold $I_{th}$, the maximal power

of SU $P_{th}$, and the outage probability $P_{out}$ of PU. Moreover, compared to other waveforms, FBMC exhibits the best advantage when estimated CSI is considered.

2. The gaps of waveforms in two-cell CR system are smaller than those in the case of single cell. This can be explained as the existence of co-channel interference, which reduces the relative difference in total interference that a subcarrier can suffer. Compared to the single cell, the maximal average capacity of two cells is larger, which results from the application of the MAC technique that allows a large capacity region.
3. If more CR cells (>2) are considered, the co-channel interference will become larger and larger, and further narrow the difference in total interference. Therefore, it can be deduced that the spectral efficiency curves will be closer to each other.

Based on the above discussions, FBMC not only can achieve the largest channel capacity in the same constraints but also has the slightest capacity difference gap between perfect CSI and estimated CSI compared to other three MCM waveforms. As a consequence, FBMC technology providing the best system performance has been recommended in the future 5G communication networks.

## 5. Conclusion

In this chapter, the spectral efficiency comparison is conducted by analyzing the achievable channel capacity among four different multi-carrier modulations. Two RA algorithms with the practical consideration of estimated CSI are designed for evaluating and comparing the capacity performance. Simulation results show that in our scenarios, FBMC can offer the highest channel capacity and can achieve much more performance gain if rough estimated channel state information is considered. As a result, we conclude that the little spectral leakage of FBMC plays an essential role in achieving high spectral efficiency, and further verify that FBMC is a competitive candidate for 5G physical layer data communication.

## Acknowledgements

This work was supported by the National Natural Science Foundation of China under Grant 61501335.

## Author details

Haijian Zhang*, Hengwei Lv and Pandong Li

*Address all correspondence to: haijian.zhang@whu.edu.cn

School of Electronic Information, Wuhan University, Wuhan, China

## References


[1] Banelli P, Buzzi S, Colavolpe G, Modenini A, Rusek F, Ugolini A. Modulation formats and waveforms for 5G networks: who will be the heir of OFDM? An overview of alternative modulation schemes for improved spectral efficiency. IEEE Signal Processing Magazine. 2014;31:80–93. doi:10.1109/MSP.2014.2337391

[2] Liu Q P, Yang Y N, Li W X. Application of OFDM technology in 4G mobile network. Applied Mechanics & Materials. 2014;631:851–855. doi:10.4028/www.scientific.net/AMM.631-632.851

[3] Tao Y, Liu L, Liu S, Zhang Z. A survey: several technologies of non-orthogonal transmission for 5G. China Communications. 2015;12:1–15. doi:10.1109/CC.2015.7315054

[4] Sahin A, Guvenc I, Arslan H. A survey on multicarrier communications: prototype filters, lattice structures, and implementation aspects. IEEE Communications Surveys & Tutorials. 2014;16(3):1312–1338. doi:10.1109/SURV.2013.121213.00263

[5] FP7 European Project 317669 METIS (Mobile and Wireless Communications Enablers for the Twenty-Twenty Information Society). 2012. Available from: https://www.metis2020.com/

[6] FP7 European Project 318555 5G NOW (5th Generation Non-Orthogonal Waveforms for Asynchronous Signalling). 2012. Available from: http://www.5gnow.eu/

[7] Lien S Y, Chen K C, Liang Y C, Lin Y. Cognitive radio resource management for future cellular networks. IEEE Wireless Communications. 2014;21:70–79. doi:10.1109/MWC.2014.6757899

[8] Sharma S K, Bogale T E, Chatzinotas S, Ottersten B, Le L B, Wang X. Cognitive radio techniques under practical imperfections: a survey. IEEE Communications Surveys & Tutorials. 2015;17:1858–1884. doi:10.1109/COMST.2015.2452414

[9] Zhang H J, Le Ruyet D, Roviras D, Sun H. Polyphase filter bank based multi-band spectrum sensing in cognitive radio systems. International Journal of Communication Systems. 2014;29:1844–1862. doi:10.1002/dac.2798

[10] Zhang L, Xiao P, Zafar A, Quddus A. ul Tafazolli R. FBMC System: an insight into doubly dispersive channel impact. IEEE Transactions on Vehicular Technology. doi:10.1109/TVT.2016.2602096

[11] Gregoratti D, Mestre X. Uplink FBMC/OQAM-based multiple access channel: distortion analysis under strong frequency selectivity. IEEE Transactions on Signal Processing. 2016;64:4260–4272. doi:10.1109/TSP.2016. 2566601

[12] Savaux V, Bader F, Palicot J. OFDM/OQAM blind equalization using CNA approach. IEEE Transactions on Signal Processing. 2016;64:2324–2333. doi:10.1109/TSP.2016.2519000

[13] Vaidyanathan P P. Multirate digital filters, filter banks, polyphase networks, and applications: a tutorial. Proceedings of the IEEE. 1990;78:56–93. doi:10.1109/5.52200

[14] Farhang-Boroujeny B. Multicarrier modulation with blind detection capability using cosine modulated filter banks. IEEE Transactions on Communications. 2003;51:2057–2070. doi:10.1109/TCOMM.2003.820753

[15] Amini P, Kempter R, Djoko-Kouam M. A comparison of alternative filterbank multicarrier methods for cognitive radio systems. Proceeding of the SDR Technical Conference and Product Exposition. 2006:1–6.

[16] Aminjavaheri A, Farhang A, Rezazadehreyhani A, Farhangboroujeny B. Filter bank multicarrier modulation: a waveform candidate for 5G and beyond. Advances in Electrical Engineering. 2014;2014:1–26. doi:10.1155/2014/482805

[17] Zhang H, Le Ruyet D, Terre M. Spectral efficiency comparison between OFDM/OQAM- and OFDM-based CR networks. Wireless Communications & Mobile Computing. 2009;9:1487–1501. doi:10.1002/wcm.704

[18] Kang A S, Vig R. Computational complexity analysis of FBMC-OQAM under different strategic conditions. Recent Advances Engineering and Computational Sciences (RAECS). 2014:1–6.

[19] Renfors M, Yli-Kaakinen J, Harris F J. Analysis and design of efficient and flexible fast-convolution based multirate filter banks. IEEE Transactions on Signal Processing. 2014;62:3768–3783. doi:10.1109/TSP.2014.2330331

[20] Chen D, Qu D, Jiang T, He Y. Prototype filter optimization to minimize stopband energy with NPR constraint for filter bank multicarrier modulation systems. IEEE Transactions on Signal Processing. 2013;61:159–169. doi:10.1109/TSP.2012.2222397

[21] Wu Y, Chen D, Jiang T. Efficient branch and bound algorithms for prototype filter optimization in OQAM-OFDM systems. International Journal of Communication Systems. 2015. doi:10.1002/dac.3031

[22] Rahimi S, Champagne B. Joint channel and frequency offset estimation for oversampled perfect reconstruction filter bank transceivers. IEEE Transactions on Communications. 2014;62:2009–2021. doi:10.1109/TCOMM.2014.2318717

[23] Caus M, Pérez-Neira A I. Multi-stream transmission for highly frequency selective channels in MIMO-FBMC/OQAM systems. IEEE Transactions on Signal Processing. 2014;62:786–796. doi:10.1109/TSP.2013.2293973

[24] Bellanger M G. Specification and design of a prototype filter for filter bank based multicarrier transmission. In Proceedings ICASSP, IEEE International Conference on Acoustics, Speech and Signal Processing; Salt Lake City, UT. 2001, pp. 2417–2420.

[25] Floch B L, Alard M, Berrou C. Coded orthogonal frequency division multiplex. Proceedings of the IEEE. 1995;83:982–996. doi:10.1109/5.387096

[26] Prakash J A, Reddy G R. Efficient prototype filter design for filter bank multicarrier (FBMC) system based on ambiguity function analysis of hermite polynomials. In International Multi-Conference on Automation, Computing, Communication, Control and Compressed Sensing, IEEE; Kottayam. 2013, pp. 580–585.

[27] Mestre X, Gregoratti D. Parallelized structures for MIMO FBMC under strong channel frequency selectivity. IEEE Transactions on Signal Processing. 2016;64:1200–1215. doi:10.1109/TSP.2015.2493988

[28] Qu D, Lu S, Jiang T. Multi-block joint optimization for the peak-to-average power ratio reduction of FBMC-OQAM signals. IEEE Transactions on Signal Processing. 2013;61:1605–1613. doi:10.1109/TSP.2013.2239991

[29] Cui W, Qu D, Jiang T, Farhang-Boroujeny B. Coded auxiliary pilots for channel estimation in FBMC-OQAM systems. IEEE Transactions on Vehicular Technology. 2016;65:2936–2946. doi:10.1109/TVT.2015.2448659

[30] Doré J B, Cassiau N, Ktenas D. Low complexity frequency domain carrier frequency offset compensation for uplink multiuser FBMC receiver. In European Conference on Networks and Communications; Bologna. 2014, pp. 1–5.

[31] Agnès N, Dooguy K A, Mohamed F S, Samuel O. Performance of a broadband multiple access system by spreading of OFDM symbols in a multipath channel. In 2015 World Congress on Information Technology and Computer Applications Congress (WCIT-CA); Hammamet. 2015, pp. 1–8.

[32] Hung Y C, Tsai S H L. PAPR analysis and mitigation algorithms for beamforming MIMO OFDM systems. IEEE Transactions on Wireless Communications. 2014;13:2588–2600. doi:10.1109/TWC.2014.031914.130347

[33] Fettweis G, Krondorf M, Bittner S. GFDM – generalized frequency division multiplexing. In IEEE 69th Vehicular Technology Conference (VTC Spring); Barcelona. 2009, pp. 1–4.

[34] Michailow N, et al. Generalized frequency division multiplexing for 5th generation cellular networks. IEEE Transactions on Communications. 2014;62:3045–3061. doi: 10.1109/TCOMM.2014.2345566

[35] Sendrei L, Marchevský S, Michailow N, Fettweis G. Iterative receiver for clipped GFDM signals. In Radioelektronika (RADIOELEKTRONIKA), 2014 24th International Conference; Bratislava. 2014, pp. 1–4.

[36] Vakilian V, Wild T, Schaich F, ten Brink S, Frigon J F. Universal-filtered multi-carrier technique for wireless systems beyond LTE. In 2013 IEEE Globecom Workshops (GC Wkshps); Atlanta, GA. 2013, pp. 223–228.

[37] Schaich F, Wild T, Chen Y. Waveform contenders for 5G – suitability for short packet and low latency transmissions. In 2014 IEEE 79th Vehicular Technology Conference (VTC Spring); Seoul. 2014, pp. 1–5.

[38] Long F. Signal processing techniques for 5G: an overview. ZTE Communications. 2015;2015:20–27. doi:10.3969/j.issn.1673-5188.2015.01.003

[39] Medjahdi Y, Terre M, Le Ruyet D, Roviras D, Dziri A. Performance analysis in the downlink of asynchronous OFDM/FBMC based multi-cellular networks. IEEE Transactions on Wireless Communications. 2011;10:2630–2639. doi:10.1109/TWC.2011.061311.101112

[40] FP7 European Project 211887 PHYDYAS, 2010 (PHYsical layer for DYnamic AccesS and cognitive radio). Available from: http://www.ict-phydyas.org

[41] Medjahdi Y, Terré M, Le Ruyet D, Roviras D. On the accuracy of PSD-based interference modeling of asynchronous OFDM/FBMC in spectrum coexistence context. In 2014 11th International Symposium on Wireless Communications Systems (ISWCS); Barcelona. 2014, pp. 638–642.

[42] Medjahdi Y, Terré M, Le Ruyet D, Roviras D. Interference tables: a useful model for interference analysis in asynchronous multicarrier transmission. EURASIP Journal on Advances in Signal Processing. 2014;2014:1–17. doi:10.1186/1687-6180-2014-54

[43] Wang X, Zhu P, Zheng F C, Meng C, You X. Energy-efficient resource allocation in multi-cell OFDMA systems with imperfect CSI. In 2015 IEEE 82nd Vehicular Technology Conference (VTC Fall); Boston, MA. 2015, pp. 1–5.

[44] Zhang Y Y, Wang Z W, Liu L H. Adaptive multi-user uplink resource allocation with imperfect CSI. In 11th International Conference on Wireless Communications, Networking and Mobile Computing (WiCOM 2015); Shanghai. 2015, pp. 1–5.

[45] Van Hecke J, Del Fiorentino P, Giannetti F, Lottici V, Vandendorpe L, Moeneclaey M. Resource allocation for multicarrier cooperative cognitive radio networks with imperfect channel state information. In 2014 IEEE 25th Annual International Symposium on Personal, Indoor, and Mobile Radio Communication (PIMRC); Washington, DC. 2014, pp. 653–658.

[46] Zhang H J, Le Ruyet D, Roviras D, Sun H. Spectral efficiency comparison of OFDM/FBMC for uplink cognitive radio networks. Eurasip Journal on Applied Signal Processing. 2010;2010:1–14. doi:10.1155/2010/621808

[47] Zhang H J, Le Ruyet D, Roviras D, Sun H. Capacity analysis of OFDM / FBMC based cognitive radio networks with estimated CSI. In 2010 Proceedings of the Fifth Interna-

tional Conference on Cognitive Radio Oriented Wireless Networks and Communications; Cannes. 2010, pp. 1–5.

[48] Kuhn H W. The Hungarian method for the assignment problem. Naval Research Logistics. 2005;52:7–21. doi:10.1002/nav.20053

[49] Barzaraa M S. Nonlinear Programming: Theory and Algorithms, 2nd ed. Hoboken, NJ: Wiley. 1993.

[50] Fudenberg D, Tirole J. Game Theory. Cambridge, MA: MIT Press. 1991.

[51] Wang B, Wu Y, Liu K J R. Game theory for cognitive radio networks: an overview. Computer Networks. 2010;54:2537–2561. doi:10.1016/j.comnet.2010.04.004

[52] Zhang H J, Le Ruyet D, Roviras D, Sun H. Noncooperative multicell resource allocation of FBMC-based cognitive radio systems. IEEE Transactions on Vehicular Technology. 2012;61:799–781. doi:10.1109/TVT.201 1.2180743

[53] Huang J, Sun Y, Chen Q. GALLERY: a game-theoretic resource allocation scheme for multicell device-to-device communications underlaying cellular networks. IEEE Internet of Things Journal. 2015;2:504–514. doi:10.1109/JIOT.2015.2419632

[54] Zhang H J. Filter bank based multi-carrier (FBMC) modulation for cognitive radio. Doctor Thesis, Conservatoire National des Arts et Metiers and Wuhan University. 2011.

[55] El Gamal A, Kim Y H. Network Information Theory. Cambridge, UK: Cambridge University Press. 2012.

[56] Mamandipoor B, Moshksar K, Khandani A K. Capacity-achieving distributions in Gaussian multiple access channel with peak power constraints. IEEE Transactions on Information Theory. 2014;60:6080–6092. doi:10.1109/TIT.2014.2342218

10

# Non-Orthogonal Multiple Access (NOMA) for 5G Networks

Refik Caglar Kizilirmak

**Abstract**

In this chapter, we explore the concept of non-orthogonal multiple access (NOMA) scheme for the future radio access for 5G. We first provide the fundamentals of the technique for both downlink and uplink channels and then discuss optimizing the network capacity under fairness constraints. We further discuss the impacts of imperfect receivers on the performance of NOMA networks. Finally, we discuss the spectral efficiency (SE) of the networks that employ NOMA with its relations with energy efficiency (EE). We demonstrate that the networks with NOMA outperform other multiple access schemes in terms of sum capacity, EE and SE.

**Keywords:** non-orthogonal multiple access (NOMA), energy efficiency, power efficiency

## 1. Introduction

In this chapter, we explore the concept of non-orthogonal multiple access (NOMA) method for the upcoming 5G networks. All of the current cellular networks implement orthogonal multiple access (OMA) techniques such as time division multiple access (TDMA), frequency division multiple access (FDMA) or code division multiple access (CDMA) together. However, none of these techniques can meet the high demands of future radio access systems.

The characteristics of the OMA schemes can be summarized as follows. In TDMA, the information for each user is sent in non-overlapping time slots [1], so that TDMA-based networks require accurate timing synchronization, which can be challenging, particularly in the uplink. In FDMA implementations, such as orthogonal frequency division multiple access (OFDMA), information for each user is assigned to a subset of subcarriers [1]. CDMA utilizes codes in order to separate the users over the same channel [1]. NOMA is fundamentally different than these multiple access schemes which provide orthogonal access to the users either in time, frequency, code or space. In NOMA, each user operates in the same band and at the same time where they are distinguished by their power levels. NOMA uses superposition coding at the transmitter such that the successive interference cancellation (SIC) receiver can separate the users both in the uplink and in the downlink channels.

NOMA was proposed as a candidate radio access technology for 5G cellular systems [2, 3]. Practical implementation of NOMA in cellular networks requires high computational power to implement real-time power allocation and successive interference cancellation algorithms. By 2020, the time that 5G networks are targeted to be deployed, the computational capacity of both handsets and access points is expected to high enough to run NOMA algorithms.

In this chapter, we present the fundamentals and capacity limits of NOMA as a future radio access technology. The imperfectness in the SIC receiver and its impact on the overall capacity is also presented. We further contribute to the literature by demonstrating the improved energy and spectral efficiencies with NOMA over-conventional OFDMA.

## 2. Non-orthogonal multiple access (NOMA)

We consider orthogonal frequency division multiplexing (OFDM) as the modulation scheme and NOMA as the multiple access scheme. In conventional 4G networks, as natural extension of OFDM, orthogonal frequency division multiple access (OFDMA) is used where information for each user is assigned to a subset of subcarriers. In NOMA, on the other hand, all of the subcarriers can be used by each user. **Figure 1** illustrates the spectrum sharing for OFDMA and NOMA for two users. The concept applies both uplink and downlink transmission.

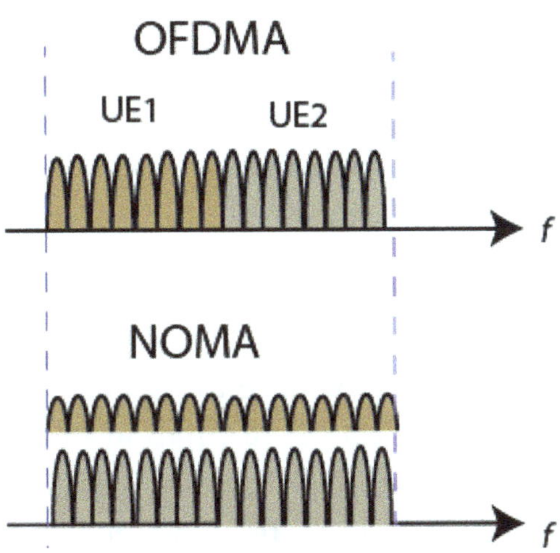


**Figure 1.** Spectrum sharing for OFDMA and NOMA for two users.

Superposition coding at the transmitter and successive interference cancellation (SIC) at the receiver makes it possible to utilize the same spectrum for all users. At the transmitter site, all the individual information signals are superimposed into a single waveform, while at the receiver, SIC decodes the signals one by one until it finds the desired signal. **Figure 2** illustrates the concept. In the illustration, the three information signals indicated with different colors are superimposed at the transmitter. The received signal at the SIC receiver includes all these three signals. The first signal that SIC decodes is the strongest one while others as interference. The first decoded signal is then subtracted from the received signal and if the decoding is perfect, the waveform with the rest of the signals is accurately obtained. SIC iterates the process until it finds the desired signal.

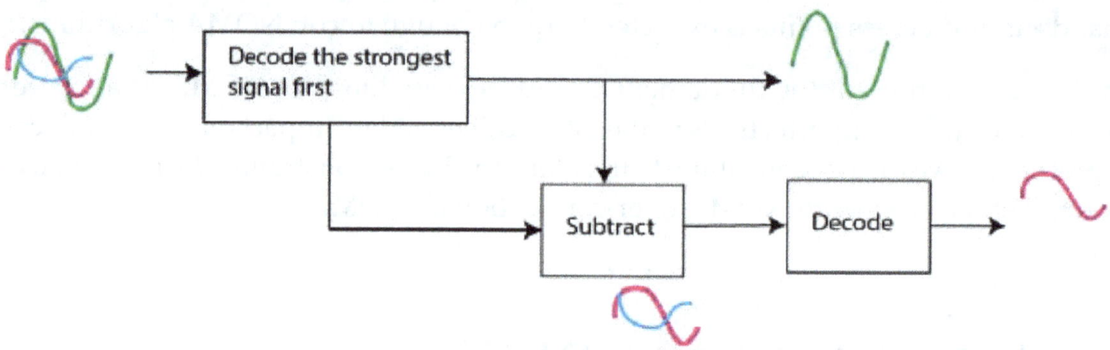


**Figure 2.** Successive interference cancellation.

The success of SIC depends on the perfect cancellation of the signals in the iteration steps. The transmitter should accurately split the power between the user information waveforms and superimpose them. The methodology for power split differs for uplink and downlink channels.

### 2.1. NOMA for downlink

In NOMA downlink, the base station superimposes the information waveforms for its serviced users. Each user equipment (UE) employs SIC to detect their own signals. **Figure 3** shows a BS and K number of UEs with SIC receivers. In the network, it is assumed that the $UE_1$ is the closest to the base station (BS), and $UE_K$ is the farthest.

The challenge for BS is to decide how to allocate the power among the individual information waveforms, which is critical for SIC. In NOMA downlink, more power is allocated to UE located farther from the BS and the least power to the UE closest to the BS. In the network, all UEs receive the same signal that contains the information for all users. Each UE decodes the strongest signal first, and then subtracts the decoded signal from the received signal. SIC receiver iterates the subtraction until it finds its own signal. UE located close to the BS can cancel the signals of the farther UEs. Since the signal of the farthest UE contributes the most to the received signal, it will decode its own signal first.

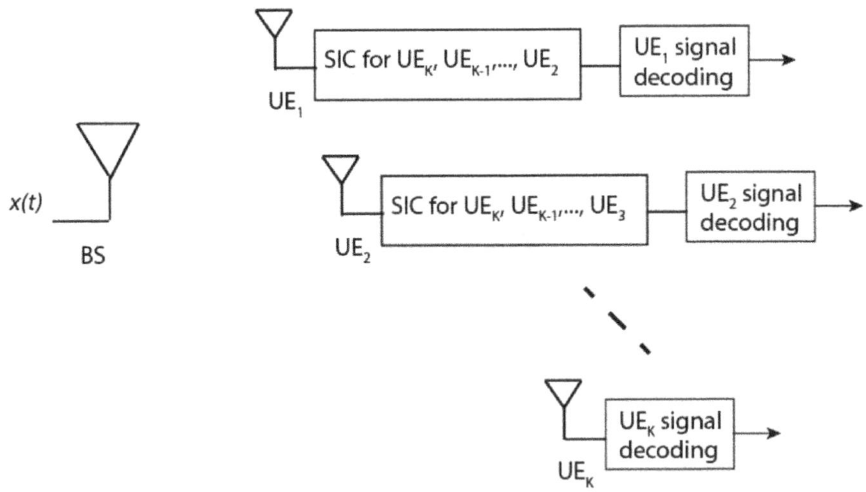


**Figure 3.** Downlink NOMA for K users.

The transmitted signal by the BS can be written as

$$x(t)=\sum_{k=1}^{K}\sqrt{\alpha_k P_T}\,x_k(t) \tag{1}$$

where $x_k(t)$ is the individual information conveying OFDM waveform, $\alpha_k$ is the power allocation coefficient for the $UE_k$, and $P_T$ is the total available power at the BS. The power allocated to each $UE_k$ then becomes $P_k = \alpha_k P_T$. The power is allocated according to the distance of UEs to the BS: $UE_1$ is the closest to the BS, so it is allocated the least power, whereas $UE_K$ is the farthest one, therefore it has the highest power.

The received signal at the $UE_k$ is

$$y_k(t)=x(t)g_k+w_k(t) \tag{2}$$

where $g_k$ is the channel attenuation factor for the link between the BS and the $UE_k$, and $w_k(t)$ is the additive white Gaussian noise at the $UE_k$ with mean zero and density $N_0$ (W/Hz).

Let us consider the farthest user first. The signal it decodes first will be its own signal since it is allocated the most power as compared the others. The signals for other users will be seen as interference. Therefore, the signal-to-noise ratio (SNR) for $UE_K$ can be written as [1]

$$SNR_K=\frac{P_K g_K^2}{N_0 W+\sum_{i=1}^{K-1} P_i g_K^2} \tag{3}$$

where $W$ is the transmission bandwidth.

For the closest $UE_1$, the last signal it decodes will be its signal. Assuming perfect cancellation, the SNR for $UE_1$ becomes

$$SNR_1 = \frac{P_1 g_1^2}{N_0 W}. \tag{4}$$

In general, for the $UE_k$, the SNR becomes

$$SNR_k = \frac{P_k g_k^2}{N_0 W + \sum_{i=1}^{k-1} P_i g_k^2}. \tag{5}$$

When NOMA is used, the throughput (bps) for each UE can be written as

$$R_k = W log_2 \left( 1 + \frac{P_k g_k^2}{N + \sum_{i=1}^{k-1} P_i g_k^2} \right). \tag{6}$$

In OFDMA, on the other hand, UEs are assigned to a group of subcarriers in order to receive their information. When the total bandwidth and power are shared among the UEs equally, the throughput for each UE for OFDMA becomes

$$R_k = W_k log_2 \left( 1 + \frac{P_k g_k^2}{N_k} \right) \tag{7}$$

where $W_k = \frac{W}{K}$ and $N_k = N_0 W_k$.

The sum capacity for both OFDMA and NOMA can be written as

$$R_T = \sum_{k=1}^{K} R_k. \tag{8}$$

We further define fairness index as [4]

$$F = \frac{\left(\sum R_k\right)^2}{K \sum R_k^2} \tag{9}$$

which indicates how fair the system capacity is shared among the UEs, that is, when $F$ gets close to 1, the capacity for each UE gets close to each other.

We can set the objective of the power allocation mechanism as to maximize the sum capacity $R_T$ under a fairness constraint for NOMA systems. The optimization problem is then formulated as

$$\underset{\alpha_k}{\text{maximize}}\, W log_2\left(1+\frac{P_k g_k^2}{N+\sum_{i=1}^{k-1} P_i g_k^2}\right) \quad \text{subject to: } \begin{array}{l}\sum_{k=1}^{K} \mathrm{P_k} \le P_T \\ \mathrm{P_k} \ge 0, \forall k \\ \mathrm{F} = \mathrm{F'}\end{array} \tag{10}$$

where $F'$ is the target fairness index in the network. The power allocation coefficients $\alpha_k$ for each $\mathrm{UE_k}$ can be obtained with exhaustive search.

## 2.2. NOMA for uplink

Uplink implementation of NOMA is slightly different than the downlink. **Figure 4** depicts a network that multiplexes $K$ UEs in the uplink using NOMA. This time, BS employs SIC in order to distinguish the user signals.

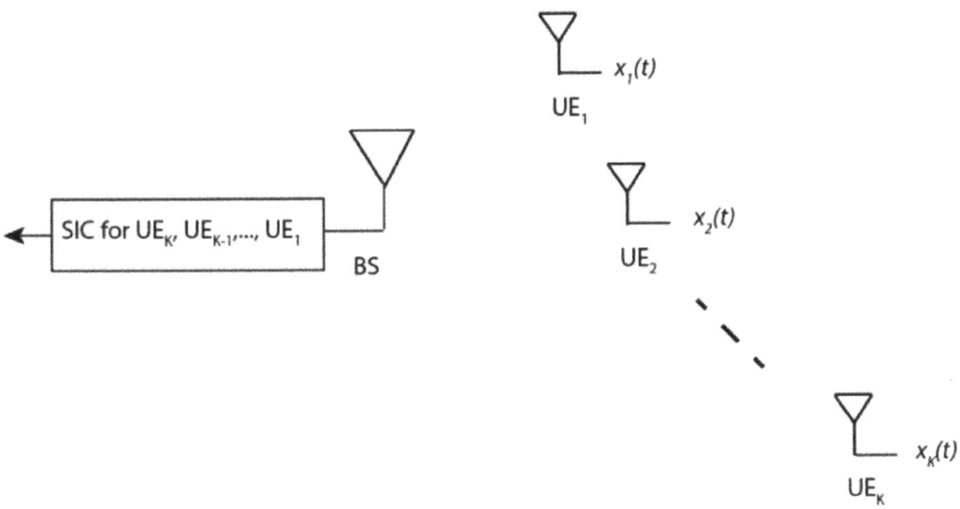


**Figure 4.** Uplink NOMA for K users.

In the uplink, the received signal by the BS that includes all the user signals is written as

$$y(t) = \sum_{k=1}^{K} x_k(t) g_k + w(t) \tag{11}$$

where $g_k$ is the channel attenuation gain for the link between the BS and the $UE_k$, $x_k(t)$ is the information waveform for the $k$th UE, and $w(t)$ is the additive white Gaussian noise at the BS with mean zero and density $N_0$ (W/Hz). In the uplink, the UEs may again optimize their transmit powers according to their locations as in the downlink. However, here we assume that the users are well distributed in the cell coverage, and the received power levels from different users are already well separated. This assumption is more natural from practical point of view, since power optimization requires connection between all the UEs which may be difficult to implement.

At the receiver, the BS implements SIC. The first signal it decodes will be the signal from the nearest user. The SNR for the signal for the $UE_1$ can be written as, including others as interference,

$$R_1 = \frac{Pg_1^2}{N + \sum_{i=2}^{K} Pg_i^2} \tag{12}$$

where $P$ is the transmission power of UEs and $N = N_0 W$.

The last signal that the BS decodes is the signal for the farthest user $UE_K$. Assuming perfect cancellation, the SNR for $UE_K$ can be written as

$$SNR_K = \frac{Pg_K^2}{N}. \tag{13}$$

Generally, for the $k$th UE, the SNR becomes,

$$SNR_k = 1 + \frac{Pg_k^2}{N + \sum_{i=k+1}^{K} Pg_i^2}. \tag{14}$$

The throughput (bps) for each UE can be written as

$$R_k = W\log_2\left(1 + \frac{Pg_k^2}{N + \sum_{i=k+1}^{K} Pg_i^2}\right). \tag{15}$$

In OFDMA, on the other hand, UEs are allocated orthogonal carriers in order to receive their information. When the total bandwidth and power are shared among the UEs equally, the throughput for each UE for OFDMA becomes

$$R_k = W_k log_2\left(1+\frac{P_k g_k^2}{N_k}\right) \tag{16}$$

where $W_k = \frac{W}{K}$ and $N_k = N_0 W_k$.

The sum capacity for both OFDMA and NOMA can be written as

$$R_T = \sum_{k=1}^{K} R_k. \tag{17}$$

## 3. Imperfectness in NOMA

Our discussions so far in the previous sections assume perfect cancellation in the SIC receiver. In actual SIC, it is quite difficult to subtract the decoded signal from the received signal without any error. In this section, we revisit the NOMA concept with cancellation error in the SIC receiver.

Here, we consider the downlink only; however, the discussions can easily be extended for the uplink. Recall that SIC receiver decodes the information signals one by one iteratively to obtain the desired signal. In SIC, after decoding the signal, one should regenerate the original individual waveform in order to subtract it from the received signal. Although it is theoretically possible to complete this process without any error, in practice, it is expected to experience some cancellation error.

In downlink, the SNR for the $k$th user with cancellation error is written as [5]

$$SNR_k = \frac{P_k g_k^2}{N_0 W + \sum_{i=1}^{k-1} P_i g_k^2 + \epsilon \sum_{i=k+1}^{K} P_i g_k^2}. \tag{18}$$

where $\epsilon$ is cancellation error term that represents the remaining portion of the cancelled message signal. In the previous section, the third term in the denominator is not included since perfect cancellation is assumed there.

## 4. Spectral efficiency and energy efficiency

Most analysis so far included the throughput performance of the network. In addition to spectral efficiency (SE) of NOMA, in this section, we analyze the energy efficiency (EE) of

NOMA systems. In our analysis, we incorporate the static power consumption of the network due to the power amplifiers in addition to the power consumed for the information waveform.

The total power consumption at the transmitter can be represented as the sum of the information signal power and the power consumed by the circuits (mainly by power amplifiers). Considering the downlink, the total power consumed by the BS can then be written as

$$P_{total} = P_T + P_{static} \tag{19}$$

where $P_T$ is the total signal power as mentioned earlier and $P_{static}$ is the power consumed by the circuitry.

Energy efficiency (EE) is defined as the sum rate over the total consumed power of the base-station [6]

$$EE = \frac{R_T}{P_{total}} = SE\frac{W}{P_{total}} \text{ (bits/joule)} \tag{20}$$

where SE is the spectral efficiency ($R_T/W$) in terms of bps/Hz.

The energy efficiency and spectral efficiency relationship (EE-SE) in Shannon theory does not consider the power consumption of the circuit and consequently is monotonic where a higher SE always results in a lower EE. When the circuit power is considered, the EE increases in the low SE region and decreases in the high SE region. The peak of the curve (or the corresponding derivative of the EE-SE relationship) is where the system has the maximum energy efficiency. This point is called "*green point*" [6–8]. For a fixed $P_{total}$, the EE-SE relationship is linear with a positive slope of $R_T/P_{total}$ where an increase in SE simultaneously results in an increase in EE. As we demonstrate in the next section, NOMA provides higher energy efficiency than OFDMA.

## 5. Results

### 5.1. Rate pairs

We assume that there are two users in the network for the sake of discussion and analyze the boundaries of the achievable rate regions for these two users. We consider a symmetric downlink channel so that the users are at equal distance to the BS. $SNR_1 = SNR_2 = 10dB$. **Figure 5** shows the boundaries of the achievable rate regions $R_1$ and $R_1$ for NOMA and OFDMA. As illustrated in **Figure 5**, NOMA achieves higher rate pairs than the OFDMA except at the corners points (where the rates are equal to the single user capacities). When the fairness is high, both users experience 1.6 bps/Hz throughputs with both NOMA and OFD-

MA. However, when the fairness is lower, both sum capacity and individual throughputs are higher with NOMA. **Figure 6** shows rate pairs when the channel is asymmetric, that is, $SNR_1 = 20dB$ and $SNR_2 = 0dB$. NOMA achieves much higher rate pairs than OFDMA, particularly for the farther user, $UE_2$.

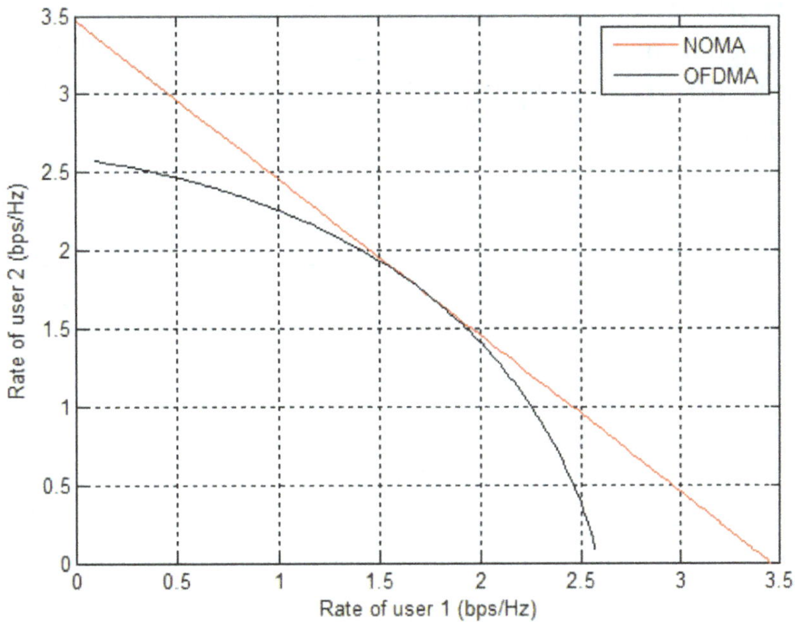


**Figure 5.** Rate pairs with OFDMA and NOMA for downlink NOMA, $SNR_1 = SNR_2 = 10dB$.

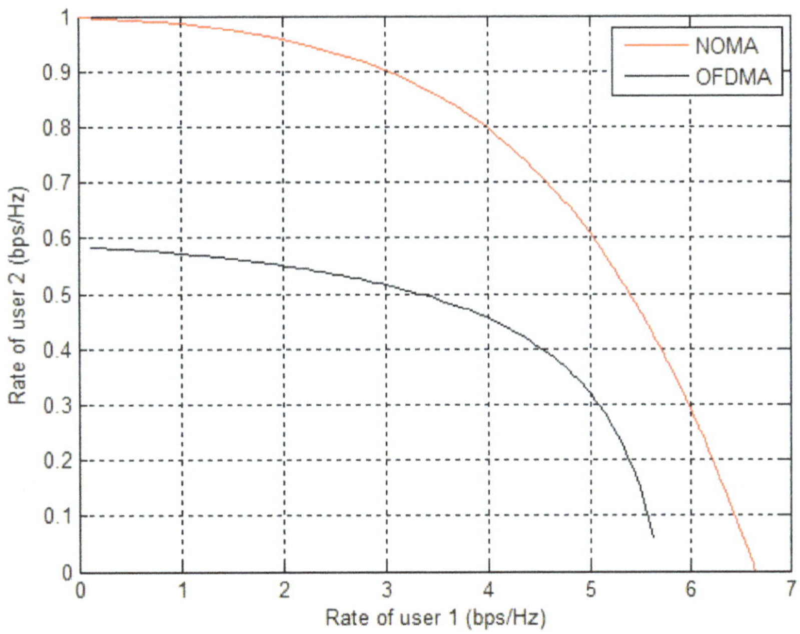


**Figure 6.** Rate pairs with OFDMA and NOMA for downlink NOMA, $SNR_1 = 20dB$ and $SNR_2 = 0dB$.

### 5.2. Impact of imperfect cancellation

In **Figure 7**, we repeat the same conditions for the asymmetric downlink channel in the previous section with imperfectness in SIC. The case for perfect cancellation is given as reference which is the same as the results in **Figure 6**. We then analyze the impact of imperfect cancellation by setting the cancellation error term ($\epsilon$) at 1, 5 and 10%. For instance, when $\epsilon = 1\%$, $UE_1$ cannot perfectly cancel the signal for $UE_2$ in the first iteration, and 1% of the power of the second user's signal still remains as interference. When $\epsilon = 1\%$, the individual rate pairs and accordingly overall capacity slightly reduce. When $\epsilon = 10\%$, on the other hand, the reduction is more distinct.

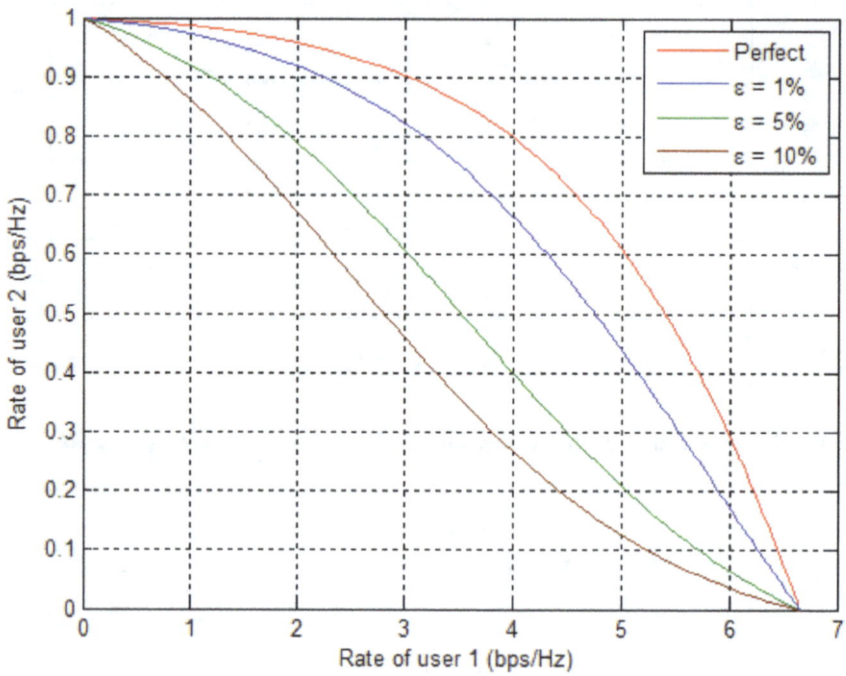


**Figure 7.** Impact of imperfect cancellation in SIC.

### 5.3. SE-EE trade-off with NOMA

Here, we compare the EE and SE of NOMA with OFDMA. We again consider the downlink. The system bandwidth is taken as $W = 5$ MHz. The channel gains for UE1 and UE2 are, respectively, taken as $g_1^2 = -120dB$ and $g_2^2 = -140dB$. Noise density $N_0$ is taken as −150 dBW/Hz. We assume that the static power consumption at the BS is $P_{static} = 100W$. **Figure 8** shows the obtained EE-SE curves for this setup. It is seen that NOMA achieves higher EE and SE than OFDMA system. The green-points occur for NOMA and OFDMA when $P_T$ is at 17 W and 18 W, respectively. At these points, both systems achieve their maximum EE. NOMA clearly outperforms OFDMA at green point and beyond for both EE and SE.

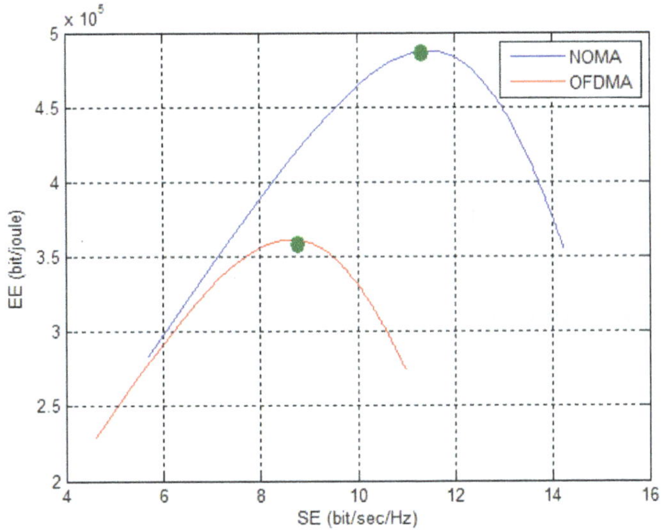


**Figure 8.** EE-SE trade-off curves for NOMA an OFDMA.

## 6. Conclusion

In this chapter, we have presented the fundamentals of NOMA and demonstrated its superior performance over conventional OFDMA in terms of sum capacity, energy efficiency and spectral efficiency. We have further mentioned the impact of imperfectness at the SIC receiver on the system performance. With its distinct features, NOMA stays as the strongest candidate for the future 5G networks. There are, however, still some challenges for successful implementation of NOMA. First of all, it requires high computational power to run SIC algorithms particularly for high number of users at high data rates. Second, power allocation optimization remains as a challenging problem, particularly when the UEs are moving fast in the network. Finally, SIC receiver is sensitive to cancellation errors which can easily occur in fading channels. It can be implemented with some other diversity techniques like multiple-input-multiple-output (MIMO) or with coding schemes in order to increase the reliability and accordingly reduce the decoding errors. There are recent works that implement MIMO for NOMA [9, 10]; the impact of channel state information (CSI) is studied in [11], capacity maximization problem is discussed in [11], and outage probability expressions are derived in [12]. The current state of the art for NOMA, however, is still far from its potential and requires further investigation.

### Appendix

**MATLAB code for Figure 5.**

```
clear all;

clc;
```

```
%%% NOMA parameters
P = 1;
G1 = 10;
G2 = 10;
count = 1;
for alpha = 0:0.01:1 %power splitting factor
P1 = P*alpha;
P2 = P - P1;
R1(count) = log2(1 + P1*G1);
R2(count) = log2(1 + P2*G2/(P1*G2 + 1));
count = count + 1;
end
hold on;
plot (R1,R2,'r');
grid on;
count = 1;
for alpha = 0:0.01:1 %bandwidth splitting factor
P1 = P/2;
P2 = P/2;
R1(count) = alpha*log2(1 + P1*G1/alpha);
R2(count) = (1-alpha)*log2(1 + P2*G2/(1-alpha));
count = count + 1;
end
hold on;
plot(R1,R2,'k');
xlabel('Rate of user 1 (bps/Hz)');
ylabel('Rate of user 2 (bps/Hz)');
grid on;
box on;
legend('NOMA','OFDMA')
```

**MATLAB code for Figure 8.**

```
clear all;
clc;
B = 5*10^6; %bandwidth Hz
N0 = 10^-21; %-150 dBw/Hz
N = N0*B; % dBW
G1 = 10^-12; %-120 dB
G2 = 10^-14; %-140 dB
Pcircuit = 100; %watt
%%%%%%%%%%%%%%%%%%%%%%%%%%%%%%%%%%%%%%%%
%% NOMA
count = 1;
for p = 1:1:100 %W
P1 = p*0.1; %allocate less power to UE1
P2 = p - P1;
R1 = B*log2(1 + P1*G1/N);
R2 = B*log2(1 + P2*G2/(P1*G2 + N));
R = R1 + R2;
SE(count) = R/B; % bit/sec/Hz
EE(count) = (R/(Pcircuit + p)); % bit/watt.sec
count = count + 1;
end
hold on;
plot(SE,EE,'k');
xlabel('SE (bit/sec/Hz)');
ylabel('EE (bit/joule)');
grid on;
% OFDMA
count = 1;
```

```
greenpoint = 0;
maxEE = -1000;
for p = 1:1:100 %Watt
P1 = p/2;
P2 = p/2;
R1 = (B/2)*log2(1 + P1*G1/(N0*B/2));
R2 = (B/2)*log2(1 + P2*G2/(N0*B/2));
R = R1 + R2;
SE_line(count) = R/B; % bit/sec/Hz
EE_line(count) = (R/(Pcircuit + p)); % bit/watt.sec = Mbit/joule
count = count + 1;
end
hold on;
plot(SE_line,EE_line,'g-');
xlabel('SE (bit/sec/Hz)');
ylabel('EE (bit/joule)');
grid on;
```

## Author details

Refik Caglar Kizilirmak

Address all correspondence to: caglar.kizilirmak@gmail.com

Nazarbayev University, Kazakhstan

## References

[1] D. Tse and P. Vishwanathan, Multiuser Capacity and Opportunistic Communication, Fundamentals of Wireless Communication, Cambridge University Press, 2005.

[2] Y. Saito et al, System Level Performance Evaluation of Downlink Non-Orthogonal Multiple Access (NOMA), in Proceedings of IEEE Symposium on Personal, Indoor and Mobile Radio Communications (PIMRC), Sept. 2013.

[3] Y. Saito et al, Non-Orthogonal Multiple Access (NOMA) for Future Radio Access, in Proceedings of IEEE Vehicular Technology Conference (VTC Spring), pp. 1–5, Sept. 2013.

[4] Jain, Raj, Dah-Ming Chiu, and William R. Hawe. A Quantitative Measure of Fairness and Discrimination for Resource Allocation in Shared Computer System. Vol. 38. Hudson, MA: Eastern Research Laboratory, Digital Equipment Corporation, 1984. pp. 20–21.

[5] J.G. Andrews and T.H. Meng, Optimum Power Control for Successive Interference Cancellation with Imperfect Channel Estimation, IEEE Trans. Wireless Comm., vol. 2, no. 2, pp. 375–383, 2003.

[6] C. Xiong, G.Y.Li, S. Zhang, Y.Chen, and S. Xu, 2011. Energy-and Spectral-Efficiency Tradeoff in Downlink OFDMA Networks. IEEE transactions on wireless communications, 10(11), pp.3874–3886.

[7] S. Han, C.-H. I, Z. Xu, Q. Sun, Energy Efficiency and Spectrum Efficiency Co-Design: From NOMA to Network NOMA, IEEE COMSOC MMTC E-Letter, vol. 9, no. 5, pp. 21–25, 2014.

[8] I. Chih-Lin, C. Rowell, S. Han, Z. Xu, G. Li, and Z. Pan, Toward Green and Soft: A 5G Perspective, IEEE Communications Magazine, vol. 52, no. 2, pp. 66–73, 2014.

[9] Z. Ding, F. Adachi, and H. V. Poor, The Application of MIMO to Non-Orthogonal Multiple Access, IEEE Trans. Wireless Commun., vol. 15, no. 1, pp. 537–552, 2016.

[10] Z. Ding and H. V. Poor, "Design of Massive-MIMO-NOMA with Limited Feedback," IEEE Signal Processing Letters, vol. 23, no. 5, pp. 629–633, 2016.

[11] Q. Sun, S. Han, C.-L. I, and Z. Pan, "On the Ergodic Capacity of MIMO NOMA Systems," IEEE Wireless Communications Letters, vol. 4, no. 4, pp. 405–408, 2015.

[12] Z. Yang, Z. Ding, P. Fan, and G.K. Karagiannidis, "On the Performance of Non-Orthogonal Multiple Access Systems with Partial Channel Information," IEEE Trans. Commun., vol. 64, no. 2, pp. 654–667, 2016.

# Permissions

All chapters in this book were first published in T5GWNAPLP, by InTech Open; hereby published with permission under the Creative Commons Attribution License or equivalent. Every chapter published in this book has been scrutinized by our experts. Their significance has been extensively debated. The topics covered herein carry significant findings which will fuel the growth of the discipline. They may even be implemented as practical applications or may be referred to as a beginning point for another development.

The contributors of this book come from diverse backgrounds, making this book a truly international effort. This book will bring forth new frontiers with its revolutionizing research information and detailed analysis of the nascent developments around the world.

We would like to thank all the contributing authors for lending their expertise to make the book truly unique. They have played a crucial role in the development of this book. Without their invaluable contributions this book wouldn't have been possible. They have made vital efforts to compile up to date information on the varied aspects of this subject to make this book a valuable addition to the collection of many professionals and students.

This book was conceptualized with the vision of imparting up-to-date information and advanced data in this field. To ensure the same, a matchless editorial board was set up. Every individual on the board went through rigorous rounds of assessment to prove their worth. After which they invested a large part of their time researching and compiling the most relevant data for our readers.

The editorial board has been involved in producing this book since its inception. They have spent rigorous hours researching and exploring the diverse topics which have resulted in the successful publishing of this book. They have passed on their knowledge of decades through this book. To expedite this challenging task, the publisher supported the team at every step. A small team of assistant editors was also appointed to further simplify the editing procedure and attain best results for the readers.

Apart from the editorial board, the designing team has also invested a significant amount of their time in understanding the subject and creating the most relevant covers. They scrutinized every image to scout for the most suitable representation of the subject and create an appropriate cover for the book.

The publishing team has been an ardent support to the editorial, designing and production team. Their endless efforts to recruit the best for this project, has resulted in the accomplishment of this book. They are a veteran in the field of academics and their pool of knowledge is as vast as their experience in printing. Their expertise and guidance has proved useful at every step. Their uncompromising quality standards have made this book an exceptional effort. Their encouragement from time to time has been an inspiration for everyone.

The publisher and the editorial board hope that this book will prove to be a valuable piece of knowledge for researchers, students, practitioners and scholars across the globe.

# List of Contributors

**Wei Feng, Ning Ge and Jianhua Lu**
Department of Electronic Engineering, Tsinghua National Laboratory for Information Science Technology, Tsinghua University, Beijing, PR China

**Md. Hashem Ali Khan and Moon Ho Lee**
Division of Electronics and Information Engineering, Chonbuk National University, Jeonju, Republic of Korea

**Mohammad-Hossein Golbon-Haghighi**
School of Electrical and Computer Engineering, University of Oklahoma, Norman, OK, USA

**Evren Çatak**
Department of Electronics and Communications Engineering, Yıldız Technical University, Istanbul, Turkey

**Lütfiye Durak-Ata**
Informatics Institute, Istanbul Technical University, Istanbul, Turkey

**Md Sipon Miah and Md Mahbubur Rahman**
Department of Information and Communication Engineering, Islamic University, Kushtia, Bangladesh

**Heejung Yu**
Department of Information and Communication Engineering, Young Man University, Busan, South Korea

**Ayesha Ijaz, Lei Zhang, Pei Xiao and Rahim Tafazolli**
Institute for Communication System (ICS), Home of 5G Innovation Centre (5GIC), University of Surrey, Guildford, UK

**Saheed A. Adegbite**
School of Engineering and Built Environment, Glasgow Caledonian University, Glasgow, UK

**Brian G. Stewart**
Department of Electronics and Electrical Engineering, University of Strathclyde, Glasgow, UK

**Syed Saqlain Ali, Daniel Castanheira, Adão Silva and Atílio Gameiro**
Instituto de Telecomunicações, University Campus, Aveiro, Portugal

**Haijian Zhang, Hengwei Lv and Pandong Li**
School of Electronic Information, Wuhan University, Wuhan, China

**Refik Caglar Kizilirmak**
Nazarbayev University, Kazakhstan

# Index